Advances in Construction Safety

N. A. Siddiqui · Bikarama Prasad Yadav ·
S. M. Tauseef · S. P. Garg · E. R. Devendra Gill
Editors

Advances in Construction Safety

Proceedings of HSFEA 2020

Springer

Editors
N. A. Siddiqui
Centre of Excellence-Occupational Health, Safety, Fire and Environment
GD Goenka University
Gurgaon, India

Bikarama Prasad Yadav
Department of Health Safety, Environment and Civil Engineering
University of Petroleum and Energy Studies
Dehradun, India

S. M. Tauseef
University of Petroleum and Energy Studies
Dehradun, India

S. P. Garg
GAIL (India) Limited
New Delhi, India

E. R. Devendra Gill
Delhi Metro Rail Corporation
New Delhi, India

ISBN 978-981-19-4000-2 ISBN 978-981-19-4001-9 (eBook)
https://doi.org/10.1007/978-981-19-4001-9

This Springer imprint is published by the registered company Springer Nature Singapore Pte Ltd.
The registered company address is: 152 Beach Road, #21-01/04 Gateway East, Singapore 189721, Singapore

Contents

About the Editors

Dr. N. A. Siddiqui currently working as a Professor and Director, of Centre of Excellence-Occupational Health, Safety, Fire and Environment, GD Goenka University, India. Before joining GD Goenka University Prof. Siddiqui has worked as Professor and Program Director of HSE and Civil Engineering Department at UPES. He pursued his postgraduate degree in environmental science and doctorate in environmental biology. In addition, he also has an International Diploma in Occupational Health (IDip. NEBOSH, UK), Diploma in Industrial Safety, and Postgraduate Diploma in Environmental Impact Assessment. His research interests lie in environmental impact assessment. He specializes in the areas of safety management, occupational health and safety, environmental pollution, environmental monitoring, and control techniques. He has more than 170 research papers to his credit and has participated in several national and international conferences. He has authored 14 books and has guided more than 200 M.Tech. theses and 21 Ph.D. theses.

Dr. Bikarama Prasad Yadav is an Senior Associate Professor with the Sustainability Cluster at the University of Petroleum and Energy Studies (UPES), Dehradun, India. He holds good experience in industry and academia with a core HSE educational background such as B.Tech. in Safety and Fire Engineering, MBA in Safety and Environment, M.Tech. in Health Safety and Environmental (HSE) Engineering, and Ph.D. in Safety Engineering. He has also earned a few other certificates and diplomas in HSE and other fields. His research interests are fire and process safety, construction safety, risk analysis and management, behavioural-based safety, and safety psychology. He is a fellow of the Institution of Engineers, India (IEI), Chartered Engineer, TSP (BCSP, USA), and also a member of the Fire and Security Association of India. During his academic career, he has published more than 50 research papers along with 7 patents (02 granted and 05 published) and 2 books. He has guided more than 50 M.Tech. students for their dissertation work, and 05 Ph.D. research scholars (01 awarded). Before joining UPES, he was associated with various construction companies for Delhi Metro Rail Corporations projects.

Dr. S. M. Tauseef is a Professor and Associate Dean (R&D) at the University of Petroleum and Energy Studies (UPES), Dehradun, India. He holds an M.S. in chemical engineering (process design) from Sharif University of Technology, Iran, and a Ph.D. in environmental engineering from Pondicherry University. His research interests include process safety, especially forecasting and consequence assessment of accidents in chemical process industries. He has developed methodologies for the application of computational fluid dynamics (CFD) to dense gas dispersion, vapor cloud explosions, and single and multiple pool fires. He is interested in solid waste management, especially the design of continuously operable vermireactor systems for assorted wastes and high-rate anaerobic digesters. He has published more than 100 research articles, co-authored/edited eleven books, published three patents, and presented papers at 30 conferences.

Dr. S. P. Garg is a Safety Professional and an engineering graduate from Nagpur University. After pursuing his MBA in HR Management, he pursued his doctorate in "Safety Management System" from the University of Petroleum and Energy Studies (UPES), Dehradun, India. He superannuated from GAIL (India) Limited, a Government of India Enterprise, as Executive Director (HSE), in 2020 but continues to be engaged with GAIL, as Advisor (HSE). He has served the oil and gas industry in India for more than 38 yrs. He has chaired and co-chaired my technical sessions at several national/international conferences and presented more than 40 technical papers. He has co-authored a book. He is also the president of the Indian Society of HSE Professionals (i-HSE) and a member of the Board of Governors of the National Safety Council, Mumbai.

Shri. E. R. Devendra Gill, FIE, FIFireE (UK) is a mechanical engineer by qualification having qualified membership examinations of the Institution of Fire Engineers, London. He is an MBA in Disaster Management from Guru Govind Singh IP (GGSIP) University and also did his post-graduation in Environmental Laws Management from Indian Law Institute (ILI), Delhi. He is presently pursuing his doctoral study at the University of Petroleum and Energy Studies (UPES), Dehradun. After serving Defence R&D Organisation (DRDO) for 17 years (1990–2007) where he rose to the position of Deputy Director/Scientist "D", he is working with Delhi Metro Rail Corporation since 2007. Currently, he is working as Executive Director Safety looking after safety in construction projects in Delhi, Patna, and Mumbai. He has more than 31 years of experience in fire science and safety pertaining to defence needs as well as construction safety management in metro projects in major cities of India. He was conferred with the All India Best Safety Engineer Award in the year 2012 by Safety and Quality Forum (SQF), the Institution of Engineers, India (IEI) for his outstanding work related to safety in engineering. During his tenure in Defence R&D Organisation, he participated in search and rescue operations after January 2001 earthquake tragedy in Ahmedabad, Gujarat. He was also assigned a special task to conduct a fire safety study for forest fires of Shivalik Range, Garhwal Hills and Slum Cluster Fires at Yamuna Pushta, Delhi. He has received specialized training in Metro Construction Safety in Tokyo, Japan, and Dubai, UAE.

Bridge Scour Analysis and Protection Methods

Check for updates

Raja and Durga Prasad Panday

1 Introduction

Scouring is the most used term in the field of bridge construction. What importance does this term have while going through the construction of bridges? A satisfying answer to this question would be very simple based on the data. The estimates from the record reveal that around 60% of bridge failures occur due to scour. Therefore, the word scour has a lot of importance in itself. Bridge scour is termed as the process in which the sand and gravel are present at the foot of pier or abutments are removed due to continuous water flow. The water that is moving has a large amount of current in itself which causes scour hole around the pier or abutments. Bridge scour when not monitored properly results in a severe disaster. Statistics show that around 46 out of 86 bridges failed due to scour in the United States between the years 1961–1976.

Scour affects this frontal portion of the pier. All the sand, silt, gravel and aggregates that are present in front of the pier get eroded hence resulting in the formation of scour hole. The erosion takes place due to the loose deposits at the foot of the pier. Most of the time, the extent of scouring is that high that it exposes the foundation of the pier. Therefore, the footing type and depth of the pier should be decided only after monitoring the scour and scouring conditions.

Various factors affect the rate of scouring. One of the major factors is the shape of the pier. It is seen that the most efficient shape out of square, rectangular, and circular was found to be circular as the rate of the scour is very much less when compared to other shaped piers. This is because a circular shape cuts the water current very efficiently hence producing very less impact at the foot of the piers.

The following research paper discusses the research done by different authors for mitigating scour, observations that are made in the scour depth experiments,

Raja (✉) · D. P. Panday
HSE and Civil Engineering Department, UPES, Dehradun, Uttarakhand, India
e-mail: rajayad1632@gmail.com

N. A. Siddiqui et al. (eds.), *Advances in Construction Safety*,
https://doi.org/10.1007/978-981-19-4001-9_1

establishment of equilibrium scour depth with time and lastly, it provides some economical solutions for reducing/mitigating the scour.

Hong-WuTANG performed research and experiment to find out the effectiveness of tetrahedral frames around the bridge piers during 3D flow [1]. It was observed that bridge pier scours with the help of tetrahedral frames could be mitigated as it alters the current of river flow. Flow around the tetrahedral frame can be categorized into three regions, namely the de-acceleration region near the sediment bed, an acceleration region in the middle of the water death region and restoration reason near the water surface. The region, which is directly affecting the scouring volume, is the deceleration region. The velocity magnitudes, turbulent intensities and vortices decreased in the deceleration region. Tetrahedral frames can dissipate the energy associated with the downfall and the horseshoe vortex generated around the bridge pier. The scour depth is reduced up to 50% or more when compared with the normal flow condition.

Alabi concluded from their research on bridge collar that the use of oblong collars could decrease local scouring around bridge pier that occurs because of flow separation and vortex developed around the pier [2]. It was found that collar efficacy increases with the increasing collar width and provides little long-term benefits at reducing the maximum scour depth. Maximum reduction in scours depth is found to be 78% of the scour depth without the collar (Fig. 1).

Kothyari [3] discussed the three-dimensional flow (horseshoe vortex, downward flow of water after striking) that takes place around the piers thus developing scour hole. The author discusses the Lacey–Inglis method for the computation of scour

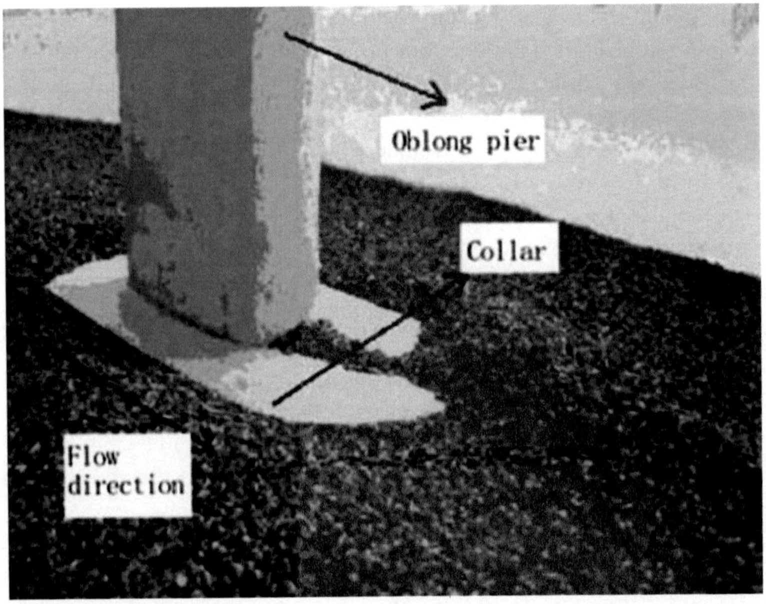

Fig. 1 Bridge collar around bridge piers for scour protection. *Source* Alabi [2]

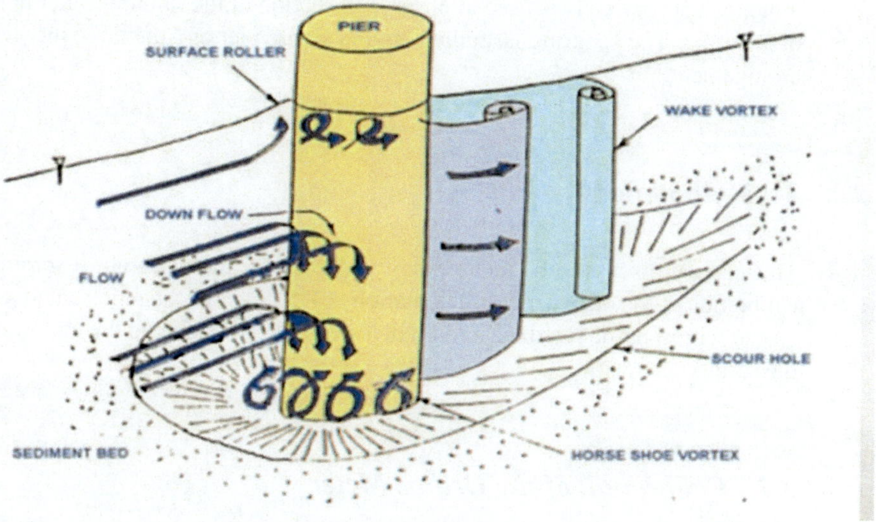

Fig. 2 Flow structure at bridge pier. *Source* Kothyari [3]

depth. This method is recommended by IRC and Indian Railways for design purpose. The author also discusses the effect of flood on the scour depth at different intervals (Fig. 2).

R. J. Garde, U. C. Kothyari and others discussed the phenomenon of scour around the bridge pier and the needs of determining the scour [4–6]. The author enumerates the factors affecting scouring like depth of flow, the shape of pier nose, angle of inclination of the pier, incoming flow characteristics whether it is clear water flow or carries some sediment. The author talks about the relevance of the Lacey–Inglis equation and the conditions where these are used for the determination of scouring (moving bed condition). Studies have also shown that there exists a relationship between sediment size, flow depth, pier dia (max 13 m) and scour depth and shows a variation of scour depth based on these factors. At last, the author talks about the scour control and prevention method like provision of caissons or well below the pier, riprap protection method [7, 8].

2 Experimental Setup for the Research

The experimental setup consists of the hydraulic tilting flume with dimensions (10 m × 0.6 m × 0.75 m as length, width and depth, respectively). The flume is filled with D50 particle size (calculation is shown in the methodology section). The orifice meter is used to measure the discharge and its calibration is also shown in the methodology section. The three piers made of Perspex sheet (diameter: 5 cm and depth: 65 cm) arranged in the triangular pattern is used to study the scour. A high-resolution camera mounted appropriately is used to observe the action of the running water in the form of scouring. The scouring is studied around single pier and the three pier triangular arrangement.

3 Methodology

This section discusses the methodology adopted for the research. It starts with the calibration of the orifice meter, calculation of D50 sizes as the representative particle size to represent the riverbed, and finally, scour observation and the interpretation of the results.

3.1 Calibration of the Orifice Meter

It is the process in which the reading of an instrument is matched to the standard values to improve the accuracy and performance of the instrument. The hydraulic tilting flume is calibrated using the standard and obtained coefficient of discharge was matched with the standard values (Table 1).

The coefficient of discharge comes out to be 0.67 (Range = 0.62–0.68). Therefore, the flume was successfully calibrated.

Table 1 Shows the manometer readings at different time intervals

S. No.	Manometric reading		h, cm of water = (y1 − y2) × 13.6	Qt (cumecs)	A (Area of collecting tank)	Time taken for 10 cm rise	Qa, m^3/s	Cd = Qa/Qt
	y1	y2						
1	58	90	435.2	0.0630	6.605	14.36	0.0459	0.7299
2	42	107	884	0.0898	6.605	11.56	0.0571	0.6362
3	39	110	965.6	0.0938	6.605	11.49	0.0574	0.6124
4	39	110	965.6	0.0938	6.605	10.07	0.065	0.6988
			Average Cd					0.669

3.2 *Calculation of D50 Mean Size of Aggregate Sand at Flume Bed*

Different size aggregates were laid on the bed of the flume to notice the scour pattern around the pier. As per IRC 78-2014; the mean size of the particle (D50) should be considered for an anticipated depth of scour.

Sieve of 4.75, 2.36, 1.18, 0.6, 0.3, 0.15 and 0.075 was taken, and sand was passed through it successively. Retained weight, percentage retained, cumulative weight, cumulative percent and pass percent were calculated and noted down after the sand was passed from all the sieves (Table 2).

A log scale graph plotting is done between pass% and sieve size to find out D50 as shown below.

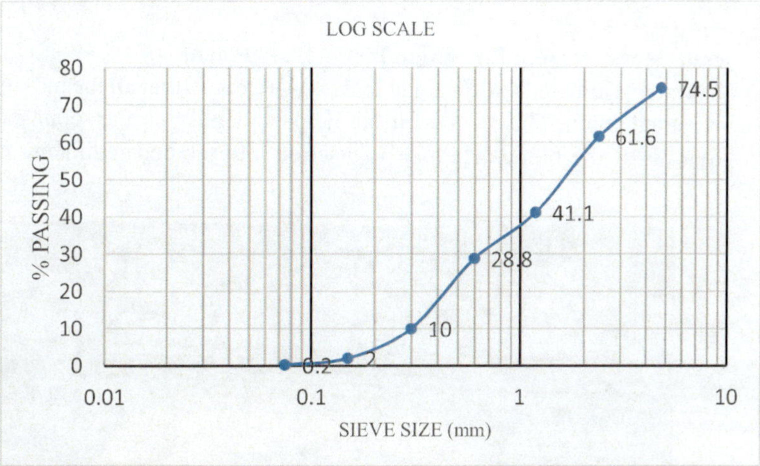

The value of D50 was plotted in the logarithmic graph and it came out to be **1.692** mm.

Table 2 Shows the different pass % of sieved sand aggregate present at the flume bed

Sieve	Retained weight (kg)	% Retained	Cumulative weight	Cumulative %	Pass %
4.75	0.255	25.5	0.255	25.5	74.5
2.36	0.129	12.9	0.384	38.4	61.6
1.18	0.205	20.5	0.589	58.9	41.1
0.6	0.123	12.3	0.712	71.2	28.8
0.3	0.188	18.8	0.9	90	10
0.15	0.080	8	0.98	98	2
0.075	0.018	1.8	0.998	99.8	0.2
PAN	0.002	0.2	1	100	0

3.3 Scour Measurement

To understand the scouring pattern around the bridge piers, experiments were conducted at the flume bed with pier's arranged individually and in triangular pattern. A prototype pier made of plastic material with height (65 cm), diameter (50 mm) and thickness (5 mm) is used along with circular acrylic sheets. Acrylics are weather as well as impact resistant. Therefore, acrylic sheets are used for the footing of the piers. These acrylics are arranged in such a way that the pier footing resembles as stepped footing as shown in Figs. 3 and 4.

A high-resolution camera is used for recording the flow pattern around the bridge pier. Coarse aggregates of size 20 mm are used at the entrance of the flume to suppress the fine aggregate sand so that when the water enters into the flume no extra amount of sand gets eroded due to the water current. Below are the figures depicting the mounting of the camera and coarse aggregate laying (Figs. 5 and 6).

Scour Measurement for Single Pier Arrangement

The water is allowed to flow for at least 20–30 min so that all the moving silt particles get settled down. This would further help in identifying the scour pattern from the sidewalls of the flume. The pier is inserted into the bed sediment, and the surface

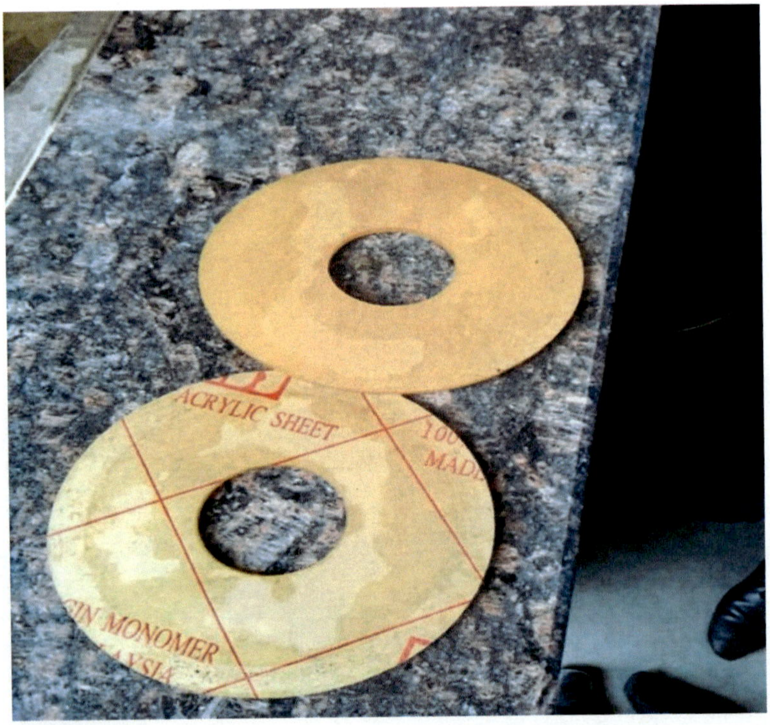

Fig. 3 Circular acrylic sheets with inner radius 50 mm and thickness of 5 mm

Fig. 4 Pier equipped with stepped footing

Fig. 5 Camera mounted above the pier for recording the flow pattern

Fig. 6 Coarse aggregate laid at the water entry point in the flume to prevent erosion

around it is leveled. The camera is mounted just above the pier for video recording. The water is allowed to flow at 3 valve rotation and the scour pattern is noticed for each 3 min interval.

Scour Measurement with Triangular Pier Arrangement

The water is allowed to flow for at least 20–30 min so that all the moving silt particles get settled down. The pier is inserted into the bed sediment in a triangular pattern with a center-to-center distance of 20 cm as shown in Fig. 7 and the surface around it is leveled. The camera is mounted just above the pier for recording the flow pattern. At first, the water is allowed to flow at 4 valve rotations followed by 8 valve rotations. Consequently the scour pattern was noticed for each 5, 10 and 15 minutes interval (Figs. 8, 9, 10, 11 and 12).

4 Result

Results for the scour depth measurements at each pier are presented at different intervals (Fig. 13; Table 3).

Fig. 7 Arrangement of piers in triangular pattern with c/c distance 20 cm (time interval 10 min)

10 min Interval (P1, P2 and P3 stands for Pier 1, Pier 2 and Pier 3)

From the graph, it can be seen that the scour depth has attained an equilibrium depth after a certain time interval. The exposed bed layer after a certain time interval is hard and well compacted, therefore, scouring at these surfaces is slow (Fig. 14; Table 4).

5 min Interval (P1, P2 and P3 stands for Pier 1, Pier 2 and Pier 3)

From the figure, it can be seen that equilibrium scour depth is not attained perfectly even after 10 readings. Therefore, equilibrium depth is much dependent on time. Equilibrium depth is attained after a long time once the flow starts (Fig. 15; Table 5).

15 min interval (P1, P2 and P3 stands for Pier 1, Pier 2 and Pier 3).

Fig. 8 Vortex formed at 4 rotations. The water strikes the pier directly and spreads up in three different directions (three-dimensional flow, time interval 10 min)

Fig. 9 Valve rotation 8. The footing of the pier is exposed as seen (time interval 10 min)

Fig. 10 Due to silting and decrease in the velocity, scour depth gets decreased (10–20 min interval)

Fig. 11 After continuous run of 90 min, footing 1 and 3 were more visible as compared to footing 2 because the width of footing 2 was more than the other two

Fig. 12 At 8 valve rotations, water started eroding the fine aggregates present at the bed

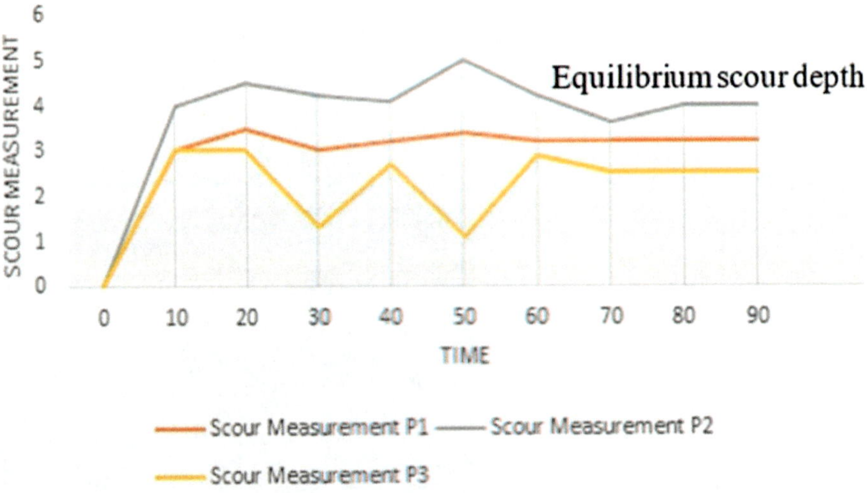

Fig. 13 Shows the graph between scouring depth and time

Table 3 Scour depth measurements at each pier after 10 min interval for 8 valve rotations

Time (min)	Valve rotation	SCOUR measurement		
		P1 (mm)	P2 (mm)	P3 (mm)
0	8	0	0	0
10	8	3	4	3
20	8	3.5	4.5	3
30	8	3	4.2	1.3
40	8	3.2	4.1	2.7
50	8	3.4	5	1.1
60	8	3.2	4.2	2.9
70	8	3.2	3.6	2.5
80	8	3.2	4	2.5
90	8	3.2	4	2.5

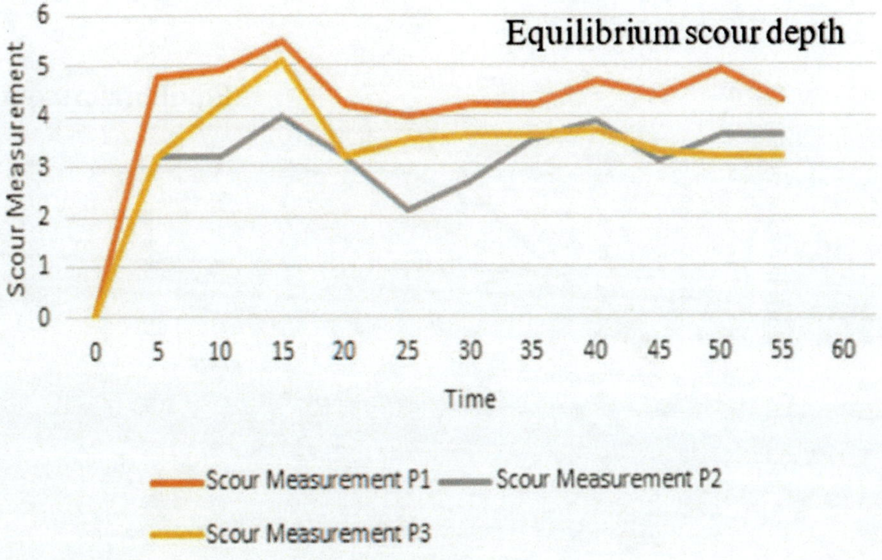

Fig. 14 Shows the graph between scouring depth and time

5 Discussion and the Conclusions from the Experiment

Different observations were made during each experiment. Each observation is listed below based on the experiments accordingly.

Scour Measurement around Single Pier pattern

- When the water is allowed to flow through the pier at a lower speed, the rate of the scour around the bridge pier is observed to be less than the rate at the higher

Table 4 Scour depth measurements at each pier after 5 min interval for 4 valve rotations

Time (min)	Valve rotation	Scour measurement		
		P1 (mm)	P2 (mm)	P3 (mm)
0		0	0	0
5	4	4.8	3.2	3.2
10	4	4.9	3.2	4.2
15	4	5.5	4	5.1
20	4	4.2	3.2	3.2
25	4	4	2.1	3.5
30	4	4.2	2.7	3.6
35	4	4.2	3.5	3.6
40	4	4.7	3.9	3.7
45	8	4.4	3.1	3.3
50	8	4.9	3.6	3.2
55	8	4.3	3.6	3.2

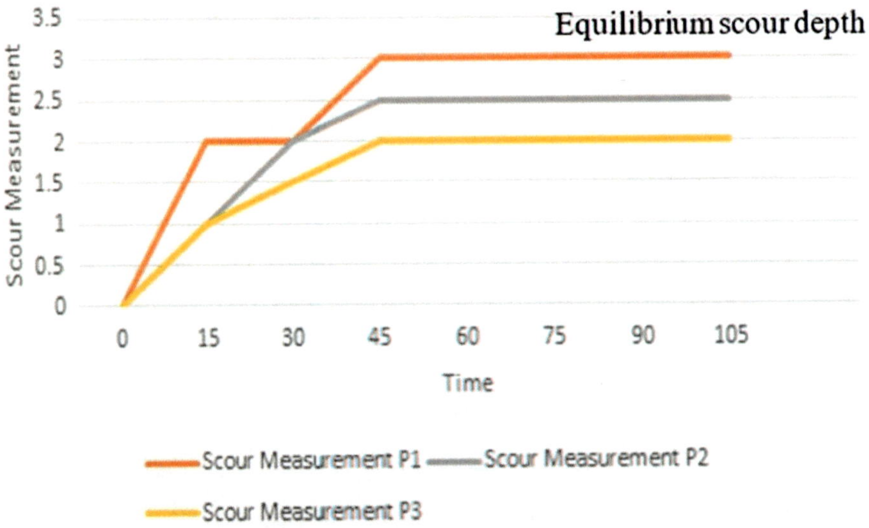

Fig. 15 Shows the graph between scouring depth and time

flow of water. Therefore, the speed of water striking the pier surface plays a major role in the formation of scour depth.

- The exit gate of the flume was closed gradually to increase the water level in the flume for observing the effect of water level increment on the scouring. It is seen that there is no/negligible change in the rate of scour. Therefore, from this, it was concluded that the scour depth remains undisturbed during the time of flood when the water level is at its peak.

Table 5 Scour depth measurements at each pier after 15 min interval for 4 valve rotations

Time (min)	Valve rotation	Scour measurement		
		P1 (mm)	P2 (mm)	P3 (mm)
0	4	0	0	0
15	4	2	1	1
30	4	2	2	1.5
45	4	3	2.5	2
60	4	3	2.5	2
75	4	3	2.5	2
90	4	3	2.5	2
105	4	3	2.5	2

- After a flow of around 30 min, there is barely any increase in the scour depth. This depth is called equilibrium scour depth.
- Scouring of bed materials occurred at the front and the side portion of piers. The rate of scouring at the backside of the pier (opposite to the flow) was observed to be very low.

Scour Measurement around Triangular Pier Pattern

- The rate of the scour increased with time and gradually slowed down after a certain time of flow.
- Radii of scour around the bridge piers increased gradually with the passing time of flow.
- The lighter bed materials easily moved away from the course materials creating scour hole around the foot of piers.
- Therefore, from the previous point, it can be concluded that the rate of scouring also depends upon the type of bed materials present.
- The bed materials eroded from the frontal pier were deposited at the centroid of the triangular pattern thus increasing the level of the bed sediments.
- Variation in the depth of footing (2 piers with deep footing and 1 pier with shallow footing) was done and a large amount of water was allowed to flow suddenly through the flume. The results showed the pier with shallow footing could not resist the sudden shock.
- The above condition can be related to the areas where a large quantity of water flows suddenly through the channel due to cloud burst. Experimentally, it is concluded that in those conditions, the depth of the foundation should be at 1/7th of the height of the pier from the surface of the riverbed.
- In the continuous flow, if the velocity of the river is reduced suddenly, the heavy bed materials flowing along the river settle down thus reducing the depth to a very less extent.

Acknowledgements The authors would like to express their sincere gratitude to all the referred research papers and acknowledge the support provided by UPES through their advanced hydraulic engineering lab facility to carry out the research.

References

1. Hong-Wu, T., Bing, D., Chiew, Y. M., & Shi-Long, F. A. N. G. (2009). Protection of bridge piers against scouring with tetrahedral frames. *International Journal of Sediment Research, 24*(4), 385–399.
2. Alabi, P. D. (2006). *Time development of local scour at a bridge pier fitted with a collar* (Doctoral dissertation).
3. Kothyari, U. C. (2007). Indian practice on estimation of scour around bridge piers—A comment. *Sadhana, 32*(3), 187–197.
4. Garde, R. J., & Kothyari, U. C. (1998). Scour around bridge piers. *Proceedings-Indian National Science Academy Part A, 64*, 569–580.
5. Shen, H. W., Schneider, V. R., & Karaki, S. (1969). Local scour around bridge piers. *Journal of the Hydraulics Division, 95*(6), 1919–1940.
6. Laursen, E. M., & Toch, A. (1956). *Scour around bridge piers and abutments* (Vol. 4). Iowa Highway Research Board.
7. Qadar, A. (1981). The vortex scour mechanism at bridge piers. *Proceedings of the Institution of Civil Engineers, 71*(3), 739–757.
8. Griffith, W. M. (1944). SILT transportation and its relation to regime channel sections. *Journal of the Institution of Civil Engineers, 22*(6), 107–120.

Construction Safety Culture Models and Their Effectiveness in Construction Industry: A Review

Bikarama Prasad Yadav, Agrapu Yerneesh Yeswanth, Albin C. Suresh, N. A. Siddiqui, and Devendra Gill

1 Introduction

We all know that construction sites have more hazards and hazardous jobs, so for improving the performance, we need to follow safety culture. The topmost importance of every construction company is to execute and enhance the safety of the construction site. The construction sector is distinct from other businesses in that building activities are frequently carried out outside, in hazardous situations. Construction companies all across the world are working to improve site safety [1]. Safety culture influences the overall safety of the organization, however, it is not completely clear about how it is effected and eventually influences the efficacy of the overall safety within the construction sector, and it is important to define what really means about safety culture. To ensure the safety of everyone specially the extrinsic and intrinsic elements of the culture that change people's attitudes and behaviors. All activities involving internal and external parts of the organization will be imbued with organizational culture. This information will then be distributed to all members of the organization. Construction sector works, managerial practices and the structure of work vary differently from other sectors and compared to other sectors maintaining safety at construction sector is challenging mainly due to inconsistent and dynamic working environment throughout the organization [2].

Construction industries have its own unique characteristics when compared with other industries due to the worker's attitude towards safety, the construction organizational structure, the managerial attitudes, the way of carrying out the construction

B. P. Yadav (✉) · A. Y. Yeswanth · A. C. Suresh
University of Petroleum and Energy Studies, Dehradun, India
e-mail: bikarama@gmail.com

N. A. Siddiqui
GD Goenka University, Gurgaon, India

D. Gill
Delhi Metro Rail Corporation, New Delhi, Delhi, India

N. A. Siddiqui et al. (eds.), *Advances in Construction Safety*,
https://doi.org/10.1007/978-981-19-4001-9_2

projects as well as the environment in which they are working and developing a model towards the construction safety culture depends on these characteristics, and the effectiveness of the model depends on how these characteristics are valued. In most of the construction sites, various activities will be carried out in parallel sharing of the work spaces simultaneously as a result, the construction works may shift from place to place as the project proceeds and due to these characteristics, the workers may need to take important decisions individually and may result in undesired output. The responsibility of maintaining safety at the construction site is dependent on various factors like procedures, methods and techniques used and the management attitude towards safety. As a result, managing safety at construction sites is rather difficult compared to other industries [2].

"Safety culture" first came from the nuclear report of 1986 Chernobyl disaster, now, it is used commonly throughout industries as a term to determine the industry's atmosphere towards safety. At high-risk industries like construction sites, safety must be a number one priority and the safety culture attitude is taken from the corporate culture including personal mentality towards the safety. Safety culture is not only affected by the corporate culture but also the other non-safety-related operations of the organization. Safety culture has its own importance and impact in a construction industry but understanding the measurement of safety culture as well as managerial practices, the organizational attitude towards safety, the measurement of safety culture are the issues regarding maintaining the safety culture in a construction sector. The reputation of the construction sector is determined by its safety culture and improving the safe work environment will increase the total productivity of the company eventually. In this paper, various models of construction safety culture are studied along with the factors influencing them to improve the overall health and safety performance of the construction site [3, 4].

2 Literature Review

This paper describes what safety culture is, and how it influences behaviour of the worker. According to R. M. Choudhry, S. Mohamed and D. Fang, for the construction industry, the phrase safety culture is a different study but it has major attention because it deals with both the behaviour of management and attitude of the worker. This paper also describes a systematic conceptual framework, firmly rooted in appropriate academic and applied literature, to deal with an absence of verifiable process for analysing the culture of building safety. It investigates the principle of safety culture, as well as other but related concepts including the security environment, behaviour-based safety and the security system. The model in this paper is also compared with currently existing security culture models in order to emphasize its functionality on construction sites. According to Glendon, organizational culture is defined as the difference between management and workers' way of thinking. It also describes that further studies should focus on some new strategy to describe the patterns by which safety culture deviates or be influenced by other safety factors [2, 5].

There are many other models discussed in this paper like "Total safety culture model", "Socio-Technical model", "Safety Maturity Model" and "Reciprocal Safety Culture", each models has their own methodology and all models deals with the same theme like behaviours of management and workers. The total safety culture model consists of a safety triad which includes three factors like environment, person and behaviour. Likewise, different models have different factors. On the other hand, there are definitions like safety climate and safety culture. Organizational culture theory indulges with practices that affect the workers' behaviour. Almost all industries are trying to increase the importance of safety culture in return it helps the organization to minimize their accident potential [6, 7].

3 Safety Culture in Construction Industry

Present new findings in the sector of safety management have given rise to a new perspective on safety. Implementing safety culture within the construction industry is related to many issues regarding the workers as the problems include the skill level of workers, and the unsafe acts that are performed by the workers and the increase of worker turnover within the construction sector. More focus is being placed on ensuring that everyone recognizes the significance of safety, with the difficult challenge of altering attitudes and behaviour. Safety is not solely the duty of the manager; everyone must contribute [8, 9].

As we all know that construction projects conducted are with a high-risk atmosphere, there is a difference between construction safety culture and traditional organizational safety culture. Throughout the formation of safety culture, both external and internal factors of culture will have effect on the organization's culture. This is not to say that the safety system is not important in our present world, but it will only work properly if the firm has built a safety culture. Over time, the law has evolved to place a greater emphasis on workplace safety. The laws and regulations are still being improved today to ensure a safe working environment. Aside from the impact of laws, several safety activism variables, including the active engagement of trade unions, consumerism and the legal battles waged by accident/incident victims, influence modern managers' safety. All of these reasons are prompting management to rethink their safety policies because the safety in the workplace will be improved [1, 10].

4 Definitions

4.1 *Safety Culture*

Safety culture is defined by many authors and has the same theme which goes through how worker's attitude is affected by companies' culture. The accident on 26 April 1986 at Chernobyl was a major one and the meeting conducted after this accident submitted a report in which the phrase safety culture was used first. Simply we can say that safety culture is common practice given to a group of employees and this includes both the behaviour of the management and behaviour of the employee. Mohamed described that organization culture has a sub-branch which is safety culture [2, 11].

4.2 *Safety Climate*

Safety climate is defined as a summation of the insight that workers share about their workplace environment, and safety climate gives an actual idea of how safety culture is carried out in an organization. Safety culture as well as safety climate are not much different from definitions as these two are different approaches towards creating an importance of safety within the organization, and if the safety culture is efficient, then it directly reflects the safety climate of the organization [2, 12].

4.3 *Safety Commitment*

One of the best ways to make a workplace safe is by start practicing all the rules and procedures set by the organization. In order to make that possible, each and every personnel starting from the management to workers, everyone should practise these safety norms in their everyday life when they step into the workplace. For achieving this, a strong commitment from each and every personnel in the organization is necessary. Developing a stringent commitment towards safety from the workforce can reduce the injuries and fatalities by a huge margin. According to a study conducted by OSHA, the organizations performing safety awareness, safety training, etc. tend to have a decline in the injury and fatality rate. So by developing this safety commitment to ourselves and to others, each and every one in the organization starts to feel like a family. Moreover, it will induce a positive culture in the organization. A good and harmonious workplace atmosphere will create a sense of belongings to the workers which will increase the morale of the workers. So after all, the word "Safety Commitment" is the stepping stone towards achieving a safer workplace [13].

5 Measures of Safety Performance

Safety performance measures to provide information on how well the safety is carried out within the organization as well as current performance level. In a construction sector, due to various activities, various hazards will be present and these hazard levels are measured and these inputs will give information regarding the effectiveness of the control measures applied to reduce the hazards. The health and safety management system will evolve from the initial principles as per time and is measured by evaluating the existing policies and the organizational structure regarding the planning and implementation of safety. Different parts of the organization will have various plans and these must be aligned together in order to provide overall aim towards the safety of the organization. The health and safety management systems: measuring health and safety at workplace is a continuous process and these techniques must be carried out in a effective manner. The frequency of these measurements which are to be carried out must be planned [2, 10].

6 Safety System

Safety system of work is a written procedure that is made from evaluating the present and possible hazards and risks associated with in the construction work site. This results in the identification of a safe system of work to be carried out, safe operating procedures (SOP) will establish a fixed method for executing the activities. SOP is a written procedure which describes the step by step instructions on how to perform a task with safety including sequence of jobs to be carried out, evaluating potential hazards and risks, recommended control measures, personal protective equipment (PPE) to be carried with the task and how to perform the task. Safety policy and goals, safety regulations and objectives, scheduling and work organization, execution and regular operating practice, tracking, responses and inspections, remedial action, evaluation and continuous improvements are the vital components of a system that represent safety in a construction sector are shown in Fig. 1. The system encompasses more than documentation areas as it describes the real execution of various activities and practices. A safety system includes all policies, objectives, roles, duties, accountability, codes, regulations, interactions, methods, techniques, instruments, information and records for safely managing site operations [2].

7 Behaviour-Based Safety (BBS)

This approach mainly deals with systematic approach of utilization, studies on psychology and human behaviour. BBS method is described as a bottom up approach along with the support of safety leaders. It is a data-driven or analytic method in which

Fig. 1 Model of a safety management system's basic elements [2]

key behaviours of the workforce are detected and focused to make some impact that could create positive change in the workforce. It is based on principles of engaging, assisting and reinforcing safety behaviours. The changing behaviour of workers is on certain safety-related activities that are routinely carried out by labourers. Behaviour-based safety is more focused on increasing safety behaviours rather than focusing on the time period of injury. Positive feedback is an effective way to reinforce safety behaviour and corrective feedback for risky observed behaviours. Activities carried out by the employee are routinely assessed in order to determine baseline ratings. Goal-setting sessions with the input of staff are planned to set reasonable and realistic performance standards with these scores. To maintain a safe working atmosphere, employees are emboldened. BBS is a continuous process rather than one time process because the safety officer must continuously promote BBS to get desired results and help the workers to perform safely as a product of various safety behaviours and improve the overall safety culture [14, 15].

8 Factors Influencing the Model Development

This model, defined below, aims to make an objective study in safety culture and make it easier to improve safety in the construction sector. According to Cooper, the role of a relation between behaviour of management and worker behaviour is recognized in accident causation models. This model provides an overall framework and a multi-methodology approach for better evaluation and improving the overall safety of the construction site and the definition of safety culture model is formed from various factors including individual as well as organizational factors [3].

(A) ***Personal Factor***

The personal factors that influence safety include communication skills, the personal skills and competency. The overall attitude of the person as well as the personal traditions, the company practices. Each personal attitude will vary but the only attitudes that are common to the work practices might be shared throughout the construction company. Worker enforcement programmes will help in developing personal factors which influence safety [16].

(B) ***Management Factor***

The managerial factors include the management policies as well as the mission and vision of the construction company including the routine supervision activities carried out within the organization. Organizational support is very important to effectively manage the safety system and maintain a healthy relationship between the supervisors and employees. An effective safety management system includes both safety behaviour and safety leadership to maintain safety and these should be included in the safety codes and standards of the company policies [16].

(C) ***Behavioural Factor***

Behavioural safety evaluation will give an overall idea of the workers' mentality related to safety and how they approach the safety measures, and it is done by conducting behavioural checklist inspections for unsafe and safe. The safe and dangerous behaviours will then be measured regularly by trained and experienced observers to determine the baseline safety scores. The behavioural safety results are valid only for a certain time period and it must be evaluated and checked frequently and revised regularly, and these results will help to develop safety awareness and training classes to be provided to the workers to improve the safety culture of the construction company [2].

(D) ***Relationship Factor***

Relationship factors include the overall relation that the company holds between the stakeholders and interpersonal skills that are practised by the organization. Globalization and market competition between the companies have made many changes in terms of technology as well as the importance to maintain a safe environment towards the workers within the construction site. Globalization has improved the working nature, industrial relations and expansion of the company resulted in an increase of overall productivity of the company [16].

Safety Culture Model

Safety culture included in the practices of organization is termed as safety culture model, and there are two theories for determining the person's psychological behaviour management: they are social learning and cognitive theories. Safety culture model interplays with both the behaviours of management and workers. According to Bandura, the factors like environment and individual persons are mainly responsible for workers' behaviour engagement. As we all know that the connection between the organization and safety culture is not up to the mark and there are some issues like organizational transparency, decision-making skill, coordination principles and handled safety culture as a different category. "Socio-technical model" was developed by linking the safety culture and safety socio-technical systems which includes worker, organization and technology for a better safety practice. By following this model, it recovered the gaps between the organization's things and characteristics. By conducting an audit using this model in the other sectors like petrochemical-based plants, the results were convincing and the organizational factors for developing the safety culture for the construction project were found by using this model [6].

Another popular safety model known as "Total safety culture Model" which also includes the concept of safety culture but the factors are different compared with other models, and the factors are environment factor, person factor and behaviour factor and the concept is known as safety triad, this concept includes the informative composition of the individual factors. As discussed above, the Bandura's model, there is other model based on this it is known as "Reciprocal Safety Culture" used to define the concept of safety culture in this model there are three different factors like observable ongoing safety-related behaviours, subjective internal behavioural factors and objective situational features. For measuring the safety culture, there are measurement tools like safety management audits, safety climate survey and safety management inspection [6, 17].

Safety Culture Maturity Model

To identify the safety culture maturity model of an organization, Fleming developed a model based on various elements including management and cooperative involvement based on their capability, productivity of the organization, the management visible safety willingness, effective communication, safety resources, the nature of working environment, the level of job satisfaction, inter-relational qualities and trust, the level of training provided to the workers and the level of maturity is determined by the ratings given to these elements and deciding at which level is the organization currently is appropriate for its evaluation. The improvement of the weaker levels will determine the progress level of the organization. However, this safety culture model proposed by Fleming was developed as a diagnostic tool but needs more research regarding this model [18, 19].

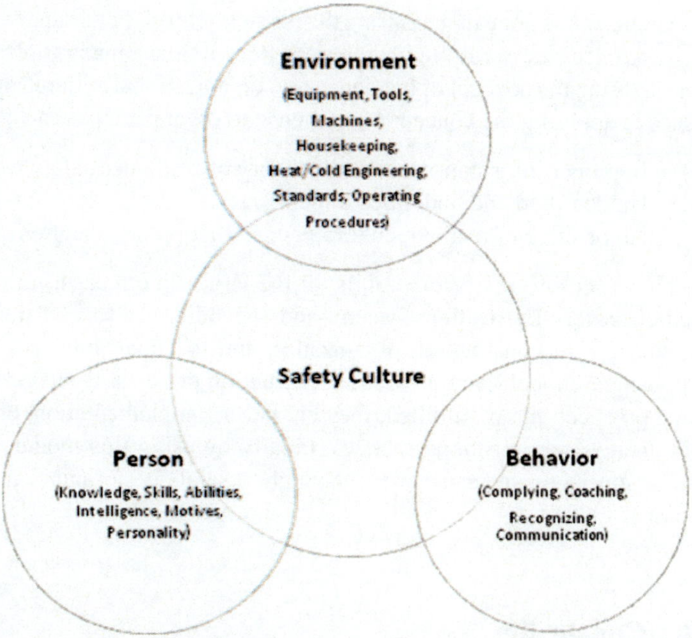

Fig. 2 Total safety culture model [14]

Hudson developed a safety culture maturity model based on pathological problems created by the workers, reactive actions from authorities, calculative management approach with collection of data, proactive workforce involvement towards safety, generative active participation and the framework of this model is applied to countries like Oman and the United Kingdom. When trying to improve the performance of the company, these different stages of safety culture must be taken into consideration. In other point of view, an organization cannot state the fact that it has a safety culture without passing through these elements of the model. Implementing these safety models takes time due to the nature of organization and the development may vary from one sector to another. Many organizations have achieved a good safety performance for their development but not all have achieved the highest stage of safety culture maturity. The importance of this safety culture model is that it gives a reference to the current level of safety culture the organization is following and can take appropriate measures to improve them [18, 20] (Fig. 2).

9 Innovation of the Conceptual Model

In the past, many studies are one on safety culture, and at last, in the end, every model failed to some extent and has some limitations. So now with the increase in technology, all the limitations of the previous approach have been modified. Some research work is done in measuring the resilience level. For a super safe organization, the resilience safety culture is preferred. Recent times, many trials and research are done to bring the concept of resilience into organizational culture, and there are many issues in applying the concept of resilience to organization some of the issues are.

(a) Improper information about the theory of resilience safety culture
(b) Unidentified and undefined dimensions
(c) No practice of resilient culture in construction organization.

The model in this theory fulfils all the issues like measuring the dimension of resilient safety by resilient system, and secondly, it also explains the practice of the concept in construction organization and the final drawback is overcome by proposing a model like this in the construction projects. By using this final model, three practices can be fulfilled; they are hazard implementation, error management and creative organization practices. Finally by using this model, the organization can be super safe and can achieve the highest safety performance in the construction sector [21, 22].

10 Conclusion

In summary, this paper contains the concepts of safety culture, organizational culture and the new model of resilient safety culture. The traditional concept of resilient safety culture has some minor issues regarding the implementation in construction projects but later on, all the drawbacks of traditional resilient safety culture have been overcome and a new model of resilient safety culture is developed. The factors that influence modelling of the construction safety culture are personal factors, management factors, behavioural factors and relationship factors. The model serves as a conceptual foundation for deciding what to analyse and measure in terms of construction safety culture and how to do so. By following several sub goals, it allows you to follow a goal-setting paradigm. Safety climate surveys, system audits/inspections, measures of safety performance and peer observations may all be used to measure employee safety behaviours, attitude and situational or environmental factors eventually determine the overall safety of the construction site. By using the theory of resilient safety culture, the organizational safety performance can be improved. Apart from safety culture and resilient safety culture model, there are other models like "Total safety culture model", "Socio-Technical model", "Safety Maturity Model" and "Reciprocal Safety Culture", each models has their own methodology and all models deals with the same theory which indulges with the management behaviour and workers' attitude. As discussed in the literature review, safety culture is a combination of the workers' attitude and the rules held by the private groups. Evaluation of people's perceptions is done by using safety surveys and interviews.

References

1. Misnan, M. S., & Mohammed, A. H. (2007, January). Development of safety culture in the construction industry: A conceptual framework. In *Association of Researchers in Construction Management ARCOM 2007—Proceedings of the 23rd Annual Conference*, (Vol. 1, pp. 13–22).
2. Choudhry, R. M., Fang, D., & Mohamed, S. (2007). Developing a model of construction safety culture. *Journal of Management in Engineering, 23*(4), 207–212. https://doi.org/10.1061/(asce)0742-597x(2007)23:4(207)
3. Cooper, M. D. (2000). Towards a model of safety culture. *Safety Science, 36*(2), 111–136. https://doi.org/10.1016/S0925-7535(00)00035-7
4. Lee, T., & Harrison, K. (2000). Assessing safety culture in nuclear power stations. *Safety Science, 34*, 61–97. https://doi.org/10.1016/S0925-7535(00)00007-2
5. Heffernan, C. J. (1988). Social foundations of thought and action: A social cognitive theory. *Behaviour Change, 5*(1), 37–38. https://doi.org/10.1017/S0813483900008238
6. Fang, D., & Wu, H. (2013). Development of a safety culture interaction (SCI) model for construction projects. *Safety Science, 57*, 138–149. https://doi.org/10.1016/j.ssci.2013.02.003
7. Maloney, W. F., & Smith, G. R. (2003). Reciprocal determinism model of safety. In *Construction Research Congress: Wind of Change: Integration and Innovation*, 2003, pp. 1–8.
8. Choudhry, R. M., Fang, D., & Mohamed, S. (2007). The nature of safety culture: A survey of the state-of-the-art. *Safety Science, 45*(10), 993–1012. https://doi.org/10.1016/j.ssci.2006.09.003
9. Grote, G., & Künzler, C. (2000). Diagnosis of safety culture in safety management audits. *Safety Science, 34*(1), 131–150. https://doi.org/10.1016/S0925-7535(00)00010-2
10. Cohen, J. M. (2002). Measuring safety performance in construction. *Occupational Hazard, 64*(6), 41–48.
11. Guldenmund, F. W. (2000). The nature of safety culture: A review of theory and research. *Safety Science, 34*(1–3), 215–257.
12. Flin, R., Mearns, K., O'Connor, P., & Bryden, R. (2000). Measuring safety climate: Identifying the common features. *Safety Science, 34*(1–3), 177–192. https://doi.org/10.1016/S0925-7535(00)00012-6
13. Molenaar, K. R., Park, J.-I., & Washington, S. (2009). Framework for measuring corporate safety culture and its impact on construction safety performance. *Journal of Construction Engineering and Management, 135*(6), 488–496. https://doi.org/10.1061/(asce)0733-9364(2009)135:6(488)
14. Geller, E. S. (1996). *The psychology of safety: How to improve behaviors and attitudes on the job.* Chilton Book Company.
15. Choudhry, R. M. (2014). Behavior-based safety on construction sites: A case study. *Accident Analysis and Prevention, 70*, 14–23. https://doi.org/10.1016/j.aap.2014.03.007
16. Ismail, Z., Doostdar, S., & Harun, Z. (2012). Factors influencing the implementation of a safety management system for construction sites. *Safety Science—SAF SCI, 50*. https://doi.org/10.1016/j.ssci.2011.10.001
17. Geller, E. S. (1994). Ten principles for achieving a total safety culture. *Professional Safety, 39*(9), 18.
18. Filho, A. P. G., Andrade, J. C. S., & de O. Marinho, M. M. (2010). A safety culture maturity model for petrochemical companies in Brazil. *Safety Science, 48*(5), 615–624. https://doi.org/10.1016/j.ssci.2010.01.012
19. Laxmi Ramanan, D., Pragadeeswar, N., & Yadav, B. P. (2020). The Perk of safety management system: A meticulous description. 63–72. https://doi.org/10.1007/978-981-15-6852-7_5
20. Goncalves Filho, A. P., & Waterson, P. (2018). Maturity models and safety culture: A critical review. *Safety Science, 105*, 192–211.

21. Trinh, M. T., Feng, Y., & Jin, X. (2018). Conceptual model for developing resilient safety culture in the construction environment. *Journal of Construction Engineering and Management, 144*(7), 06018003. https://doi.org/10.1061/(asce)co.1943-7862.0001522
22. Mohit Reddy, P. K., & Singh, A. K. (2019). E-waste trends and sustainable practices in India. *Journal of Industrial Safety Engineering, 6*(3), 14–20.

Flood Frequency Analysis and the Canal Design for the Barnigad Region in the Yamuna River

Chetan Agrawal, Durga Prasad Panday, and Karan Singh Dhawai

1 Introduction

In the arranging and plan of water assets ventures, designers and organizers are frequently intrigued to decide the extent and recurrence of floods that will happen at the undertaking territories. Flood frequency analysis is the assessment of how regularly a predefined occasion will happen. Before the assessment is done, investigation of the stream information assumes a vital part to acquire a likelihood dispersion of floods. Flood frequency analysis (FFA) is most usually utilized by designers and hydrologists worldwide and fundamentally comprises assessing flood top amounts for a bunch of non-exceedance probabilities [1].

Flood represents more human misfortunes than some other fiascos. To conquer these misfortunes, flood investigation is completed. It is for the most part utilized by engineers worldwide to gauge flood top amounts for a bunch of non-surpassing probabilities. The model boundaries setup is utilized to anticipate the limit occasions of enormous repeat stretch. Accurate information with respect to the recurrence interval of floods and their probability will serve as a potent tool for policymakers to make decisions regarding integrated water resources management.

Mean is the arithmetic average of a bunch of numbers or appropriation. It is the most usually utilized proportion of the focal inclination of a bunch of numbers. Mode is the number that shows up most often in an informational collection. A bunch of numbers might have one mode, more than one mode, or no mode by any means. Median is a better estimator as it is not biased towards the extreme values of a dataset. Variance captures the spread of the dataset. The amount of the squared distances of each term in the conveyance from the mean (μ), partitioned by the quantity of terms in

C. Agrawal · K. S. Dhawai
HSE & Civil Engineering Department, UPES, Dehradun, India

D. P. Panday (✉)
Sustainability Cluster, UPES, Dehradun, Uttarakhand, India
e-mail: dpanday@ddn.upes.ac.in

N. A. Siddiqui et al. (eds.), *Advances in Construction Safety*,
https://doi.org/10.1007/978-981-19-4001-9_3

the dispersion (N), is known as variance ($\sigma 2$). Standard deviation is the square root of the variance, and its symbol is σ. Kurtosis is defined as the measure of "tailedness" of the distribution. Skewness can be quantified as a representation of the extent to which a given distribution varies from a normal distribution. If the curve is shifted to the left or the right, it is said to be skewed.

Methods for estimation of missing data

Missing data is a typical issue looked at by researchers in ecological investigations. Environmental data information, especially precipitation information is profoundly powerless against be missed, which is because of a few reasons, for example, breakdown instrument, off base estimations, and movement of stations. Precipitation information is likewise influenced by the presence of exceptions because of the fleeting and spatial inconstancy of precipitation estimations. These issues may hurt the nature of precipitation information and in this manner, produce incorrectness in the results of the examination. Some of these methods are explained below.

The arithmetic-mean based method to fill the missing precipitation utilizes the concept that if the normal precipitation variation among all the stations is less than 10% of the concerned station, then this method can be chosen. The ordinary proportion strategy shows the typical precipitations fluctuate extensively then missing precipitation (P_x) is assessed by weighting the precipitation at different stations by the proportions of ordinary yearly precipitation. The typical proportion technique is changed to join the impact of distance in the assessment of missing precipitation. This strategy is known as the modified normal ratio method. The inverse distance strategy has been supported to be the most dependable technique as contrast with the other two techniques talked about above.

The average rainfall over a space might be considered as the principal contribution on watershed demonstrating measure, particularly of those which manage surface overflow, mostly because, as a rule, the downpour is the lone climatic variable that can clarify quick expanding stream. In any case, a few examinations show that spatial fluctuation of precipitation over the region and their conveyance design have a significant impact on the runoff reaction created. Here, some methods are shown below.

The arithmetical-mean method calculates areal precipitation using the arithmetic mean of all the point or areal measurements considered in the analysis. Thiessen-mean method calculates station weights based on the relative areas of each measurement station in the Thiessen polygon network. The individual loads are multiplied by the station perception and the qualities are added to acquire the areal normal precipitation. Isohyetal method includes drawing assessed lines of equivalent precipitation over an area based on point estimations. The magnitude and degree of the resultant precipitation spaces of inclusion are then considered versus the territory being referred to appraise the areal precipitation value.

Another very important parameter which is used for the design of hydraulic structures such as irrigation canal in the present case is **return period or recurrence interval**. It is a normal time or an expected normal time between occasions like quakes, floods, avalanches, or waterway release streams to happen. It is a factual estimation normally founded on notable information over a drawn out period and is utilized typically for risk examination. Models incorporate concluding whether an undertaking ought to be permitted to go on in a zone of a particular gamble or planning constructions to endure occasions with a particular bring return period.

$$\text{Recurrence Interval} = \frac{n+1}{m} \tag{1}$$

where n is the number of years on record, and m is the position of observed occurrences when arranged in descending order. The formula is based on the Weibull distribution method. It is an extreme value of probability distribution that is frequently used to model the reliability, endurance, wind speeds, and different information. The main motivation to utilize Weibull distribution is a result of its adaptability. Weibull distribution reliability is estimated with the assistance of boundaries [2]. The main motivation to utilize Weibull distribution is a direct result of its adaptability. Weibull distribution reliability is estimated with the assistance of boundaries. The probability P of an event equalled to is given by Weibull Formula:

$$P = \frac{m}{n+1} \tag{2}$$

where m is rank, and n is no. of years of record.

For the design, we have considered the extreme value as given by the Gumbel distribution method. Gumbel's distribution is a measurable technique regularly utilized for anticipating outrageous hydrological occasions like floods. In this concentrate on it has been applied for flood recurrence investigation because of the space is less coordinated, subsequently, is not fundamentally impacted by repository activities, redirections, or urbanization. Stream information are homogeneous and free along these lines need long stretch examples, top stream information covers a moderately lengthy record (over 10 years) and is of satisfactory quality [3].

The equation for Gumbel's distribution as well as to the procedure with a return period T is given as,

$$X_T = \overline{X} + K.\sigma_x \tag{3}$$

where σ_x = Standard deviation of the sample size.

K = frequency factor, which is expressed as, $K = Yt - Y\overline{\overline{n}}/Snx$.

In which, Y_T = reduced variate, $Y_T = -$ [$Ln\ Ln.\ (T/T - 1)$].

For the design of the irrigation canal, two theories are used. In Kennedy's theory, there can be as many sections for the given discharge. Simply stated that CVR varied according to the silt grade but did not give any method to measure CVR. Adoption of Kutter's N and Chezy's equation to work out mean velocity incorporated the limitations of those relationships [4]. Kutter considered depth as variable and gave critical **velocity formula in terms D**,

$$V_0 = 0.55 \text{ my}^{0.64} \tag{4}$$

Considered channel section trapezoidal. The section is tighter and deeper. Considered all channels in the regime when they neither silt nor scour. Did not consider the importance of bed width depth ratio, wood's B-D rations are not rigid. Applicable to irrigation channels only. In Lacey's theory, only one regime section for given discharge and silt factor. Lacey did not leave silt grade as guesswork but correlated silt factor f to the diameter of silt and rugosity coefficient. Lacey discarded arbitrary factors and gave general regime equation independent of rugosity coefficient [5].

$$V = 10.8\, R^{2/3} S^{1/3} \tag{5}$$

Considered hydraulic mean depth R as relevant variable and derived velocity formula in terms of R:

$$V = 0.639(f * R)^{1/2} \tag{6}$$

Determined regime slope for the given release and silt factor. Silt is kept in suspension by vertical parts of Eddies brought about by grating against the wetted edge. Considered channel segment semi-circular. The part is more extensive and shallower. Thought about that diverts cannot be in the genuine system and arranged them into an underlying and last regime. Given the formula for the distinct wetted perimeter for the given discharge.

$$P = 4.75\, Q^{1/2} \tag{7}$$

2 Study Area and Data Collection

The study area is the Banigrad which is the region in the stretch of the Yamuna river. The absolute length of the Yamuna River from its source up to its intersection with the Ganges at Allahabad is 1376 km. The length of the stream along its navigation from its source up to the proposed redirection weir site is about 69.58 kms. During this cross, the waterway dives from a rise of 6387 to 940.00 m msl—the bed level at the redirection site. The bowl of the stream is invested with thick blended wilderness and the waterway streams in its upper compasses through the rough territory and at lower comes to through genuinely thick blended wilderness mostly comprising pine trees. The catchment region caught at the proposed redirection site, which is roughly 400 m upstream of its intersection with Barnigad stream on the New Delhi-Yamunotri National Highway No. 123 at almost two km before Kuwa and almost 27 km before Barkot, is 1183.17 km^2 out of which 36.23 km^2 is under permanent snow. The climate of the Yamuna river basin at the project site is hilly monsoon type. The climate comprises a mild summer, severe cold in winter and moderate to heavy rainfall during the monsoon (rainy) season. The cold season starts at about the start of October and continues up to the second week of April [6]. The summer begins from the second week of April and extends up to the end of June. The south-west monsoon covers the period from July through September and receives most of the rainfall due to tropical storms and depressions originating in the Bay of Bengal (Fig. 1).

The data for the present study is taken from the Uttarakhand Jal Vidyut Nigam Limited (UJVNL) [3]. The time period of the data is from 1981–2008 and the scale of the data is monthly.

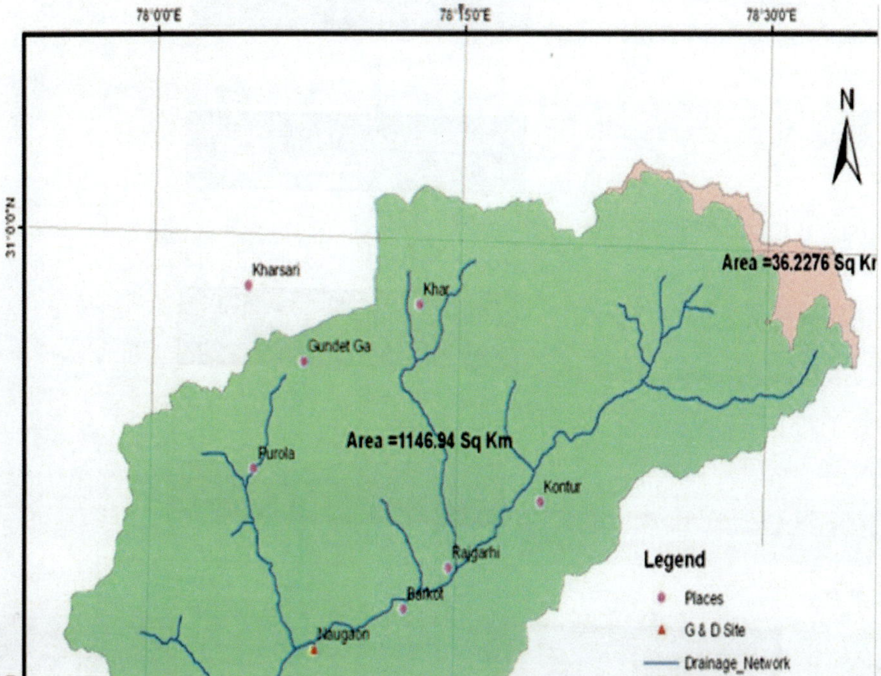

Fig. 1 Catchment area plan. *Source* I-WRIS: India Water Resources Information System [7]

3 Methodology

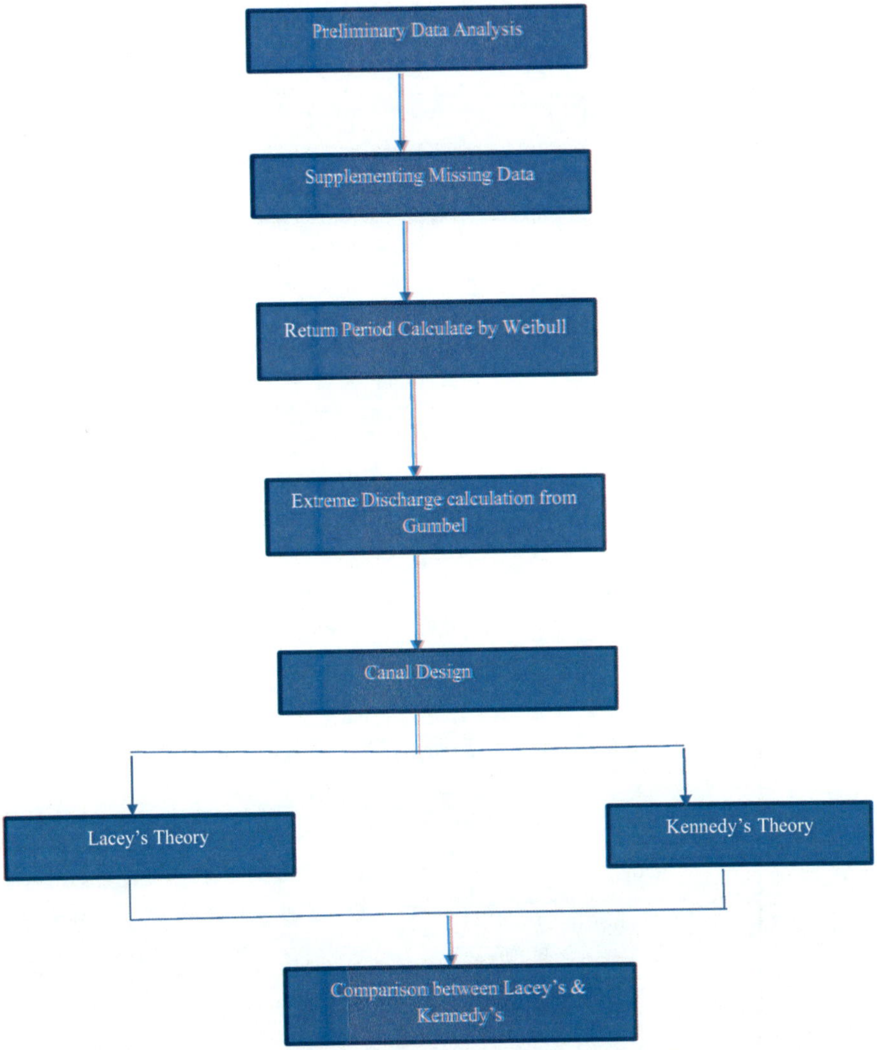

The calculations are shown below that are required for designing the channel section.

(i) **Using Kennedy's Theory**

$$\text{Discharge}(Q) = 276.7333,\ S = 1/6300, n = 0.023, m = 1$$

Now using critical velocity formula that is

$$\mathrm{V_0 = 0.55\ my^{0.64}}$$
$$\mathrm{V_0 = 0.857\ m/s}$$

Now Area, $\mathrm{A = Q/V_0}$

$$\mathrm{A = 276.7333/0.857 = 322.90\,m^2}$$
$$\mathrm{A = y(b + y/2)}$$
$$\mathrm{322.90 = 2(b + 1), \{y = 2\}}$$
$$\mathrm{b = 160.45\,m}$$
$$\mathrm{P = 164.92\,m}$$
$$\mathrm{R = A/P = 1.95\,m}$$

Now calculation for V

$$V = 1/n\left(R^{2/3}\right)\left(S^{1/2}\right)$$
$$V = 1/0.023(1.95)^{2/3}(1/6300)^{1/2}$$
$$V = 0.8569 = 0.857$$
$$\text{So, } V_0 = V = 0.857$$

(ii) **Using Lacey's Theory**

$$Q = 276.7333,\ f = 1$$
$$V = \left[Qf^2/140\right]^{1/6} = [276.7333 * 12/140]^{1/6} = 1.120$$

Now calculating the area and wetted perimeter

$$\mathrm{A = Q/V = 247.08\,m^2}$$
$$\mathrm{R = 2.5\left(V^2/f\right) = 3.136\,m}$$
$$\mathrm{P = 4.75(Q)^{1/2} = 79.01\,m}$$

$$P = b + (5)^{1/2}y \quad \text{(i)}$$
$$A = (b + y/2)y \quad \text{(ii)}$$

By solving the above equation, we get the values of y and b, so they are
So, $y = 3.38$ and the value of $b = 239.50$ m.
So now, using the formula for the value of S

$$S = \left(f^{5/3}\right)/3340 * (Q)^{1/6}$$

$$S = 1/8523$$

4 Results

There are many missing values in the data, so first of all, the missing value has been filled with the help of the arithmetic mean, a method that is shown in Table 1. When all missing values find out then we calculated standard deviation, variance, skewness, and kurtosis to study the nature of the data. This is represented by Table 2. From this, Weibull distribution is found to be the most appropriate one to find out the return period.

When all the necessary parameter calculated then flood discharge, return period, and probability were calculated that shown in Table 3.

By using Gumbel and Weibull method's formulae for a particular return period (10 years, 15 years, and 20 years), the values of Y_T, K, and X_T are calculated and discharge (Q) also calculated. The values of Y_T, K, and X_T are shown below in Table 4.

By using Gumbel's and Weibull's discharge and the help of Kennedy's and Lacey's theory we design the channel section. So, all necessary data that relates to the channel section has been calculated and shown in Table 5.

Table 1 Missing data value treatment

Missing values in 1991–92	Missing values in 1994–95
34.9600	37.3000
105.6200	112.6000
146.6900	156.3900
88.6360	94.0800
35.5100	37.8100
20.4700	21.8000

Table 2 Preliminary statistical data analysis

Number of years	28
Standard deviation	44.7368
Variance	2001.38
Skewness	0.81546
Kurtosis	2.33233

Table 3 Weibull method analysis for return period calculation

m (Order No.)	Flood discharge	*Tp* (return period)	*P* (exceedance probability)
1	276.7333	29.0000	0.034483
2	208.5367	14.5000	0.068966
3	188.0068	9.6667	0.103448
4	177.16	7.2500	0.137931
5	173.7968	5.8000	0.172414
6	172.2733	4.8333	0.206897
7	166.5968	4.1429	0.241379
8	156.39	3.6250	0.275862
9	154.8533	3.222	0.310345
10	152.64	2.9000	0.344828
11	150.54	2.6364	0.379310
12	149.6933	2.4167	0.413793
13	146.69	2.2308	0.448276
14	142.75	2.0714	0.482759
15	140.6167	1.9333	0.517241
16	124.6333	1.8125	0.551724
17	121.2833	1.7059	0.586207
18	119.5933	1.6111	0.620690
19	117.0733	1.5263	0.655172
20	116.1333	1.4500	0.689655
21	109.8568	1.3810	0.724138
22	109.7167	1.3182	0.758621
23	108.0633	1.2609	0.793103
24	101.0333	1.2083	0.827586
25	100.8667	1.1600	0.862069
26	84.0833	1.1154	0.896552
27	82.7733	1.0741	0.931034
28	46.29	1.0357	0.965517

Table 4 Return period calculated from the Gumbel method

T (years)	Y_T	*K*	X_T
10	2.25037	1.5539	208.74
15	2.67375	1.9398	226.006
20	2.9702	2.2049	237.86

Table 5 Summary of the irrigation steps

	Q	$1/S$	n	m	Y	A	P	V_0	R	V
Kennedy's method										
Using discharge from Weibull's method	337.733	6300	0.023	1	2	394.87	200.516	0.857	1.97	0.86
Using discharge from Gumbel's method	276.7333	6300	0.023	1	2	322.90	164.92	0.857	1.95	0.856
Lacey's method										
	Q	$1/S$	n	f	y	A	P	b	R	V
Using discharge from Weibull's method	337.733	8814	6300	1	10	291.652	87.293	64.89	3.352	1.158
Using discharge from Gumbel's method	276.7333	2523	6300	1	3.38	247.8	79.01	71.43	3.136	1.12

5 Conclusion

In the whole process of channel design, we have used the arithmetic mean method for supplementing the missing data and for the return period, we used the Weibull distribution method. Along with for extreme value of discharge, we have used Gumbel's method and with the help of the discharge, we designed the channel section.

The advantage of using Lacey's theory over over Keneddy's theory is that-Due to the high amount of sediments, the true and final regime is followed. Kennedy's theory neglects eddies generated therefore the channels designed using it do not have silt supporting power. That's why we choose Lacey's theory. Lacey established a fixed relationship between bed width and depth while Kennedy's theory ignores them.

References

1. Bhagat, N. (2017). Flood frequency analysis using Gumbel's distribution method: A case study of lower Mahi Basin, India. *Journal of Water Resources and Ocean Science, 6*(4).
2. https://www.reliawiki.com/index.php/The_Weibull_Distribution
3. Cooray, K. (2010). Generalized gumbel distribution. *Journal of Applied Statistics, 37*(1), 171–179.
4. https://www.yourarticlelibrary.com/irrigation/canals/kennedys-theory-of-canal-system-concept-limitations-and-design/60740
5. https://www.yourarticlelibrary.com/irrigation/canals/laceys-theory-concept-equations-and-limitations/60773#:~:text=Thefullsupply/discharge
6. https://www.uttarakhandirrigation.com/hydropower-projects
7. https://indiawris.gov.in/wris/#/RiverMonitoring

Utility Damage Prevention Measures During Excavation: A Review

Bikarama Prasad Yadav, N. A. Siddiqui, Soumya Jain, and Dibendu V. Nayar

1 Introduction

Excavation is the common practice carried out in all the construction projects, and one cannot deny the fact that most of the utilities in particular sewage line, gas pipelines and electrical cables are situated underground. Underground utilities can be referred to as the powerhouse of transfer of various forms of energy in the present era. It can be defined as a collection of various media that are used for transfer of any material, information or energy beneath the ground level such as pipelines, live electric cables, data wires and ethernet cables. As days pass the demand for these types of utilities are increasing at a real fast pace with these utilities playing a key role in the development of a nation. Utilities that are underground or underwater are usually not noticeable by the general public and usually lie invisible. This utility's existence knowledge only comes to attention, when some damage has occurred to the utility and a concerned authority identifies it and does maintenance work. Statistical data indicates that the frequency of occurrence of utility damage via excavation and the resulting injury or fatality is very high, much more than the ALARP levels [1]. There are also indirect consequences caused due to utility damage. Consider an underground utility damaged near to or in a busy roadway. For the maintenance of such a roadway, it needs to be closed down which indirectly results in further time loss, and deviation from the normal cycle of many. In other terms, this results in a huge financial or economic loss as well indirectly [2]. While a construction project begins, be it any type of construction, if there is excavation involved, there should be proper planning and the layout of the underground utilities should be known or identified using suitable mechanisms.

B. P. Yadav (✉) · S. Jain · D. V. Nayar
University of Petroleum and Energy Studies, Dehradun, India
e-mail: bikarama@gmail.com

N. A. Siddiqui
GD Goenka University, Gurgaon, India

N. A. Siddiqui et al. (eds.), *Advances in Construction Safety*,
https://doi.org/10.1007/978-981-19-4001-9_4

The criticality of this issue is important. According to Utility strike avoidance group, damage caused to underground utility in UK are; there occurred more than 2500 utility strikes in 2018–19, 3000 utility strikes in 2017–18 and 2700 utility strikes in 2015–16 [3]. Out of these, most of the strikes occurred due to negligence or unavailability of data. Considering the United States from a period of 2005–2016, there occurred around 875 major utility strike accidents from which there was an estimated fatality to 40 people and leaving 166 others with injuries. Pipeline and hazardous materials safety administration (PHMSA) study revealed that there were around 85,896 gas pipeline utility strikes in the United States in the year 2016 [4]. The 2017 damage information reporting tool (DIRT) report found that the number of incidents related to excavation and utility damage in Canada is estimated to be around 10,644 incidents in Canada (CGA, [5]). Service strikes not only result in immediate consequences leading to fatalities but also impacts environment and the normal work cycle. As an illustration, an explosion in a gas pipeline due to service strike has a potential for numerous fatalities either due to explosion or by inhalation. Another example involves the damage of the sewage line which can lead to the infiltration of harmful chemicals in other water lines, causing severe diseases to the flora and fauna [2]. There is no hard science to understand why these strikes occur. Damage to the underground cable is the result of poor on-site behaviour, due to poor supervision the workers are bit unaware of the procedure to be carried out while working with projects that involve excavation, not using advanced tools and techniques to detect the utilities [6]. There are countless accidents that have been documented in the wake accidental damage to underground cables. Undocumented ones may be more than expected. Coupled with the huge loss of assets and entities, economic loss is another dimension. Cost of repairing the damaged lines may vary from hundreds to millions of dollars [7].

The structure of underground utility infrastructure is imperative in urban culture as the maintenance and installation have an impact on the economic structure, as the poor design may lead to congestion in traffic during repair, delays in the normal work processes, resulting in the increase in losses whether they are direct or the indirect ones [8].

Although different countries have introduced certain methods to reduce or mitigate such instances from occurring, even till date, the utility damage situated underground and underwater occurs. According to the purpose that the utility serves, there are different types of underground utilities. Additionally, to identify the purpose of such a utility, they will also have different colour coding. Different mechanisms are used to identify underground utilities to avoid a collision with them while excavating. This paper has compared available literature in utility strike and damage prevention and a comparative study was done.

In this paper, we will briefly understand and classify the different types of utilities according to the purpose they serve and thereby their colour coding, familiarise the dangers that are associated with utility strikes, identify the common causes that lead to utility damage, identify techniques that can be used to prevent or to understand utility damage prevention techniques, and finally, conclusions are to be drawn from this study.

2 Types of Utilities

Underground utilities are used for different purposes, and according to the change in purposes, they are usually categorized into different types. The most common type of classification is as follows [9].

Information transfer/television cables, as the name indicates this type of utility is used for the transfer of data. Internet data transfer cables, telephonic landline cables, etc., falls under this category. Electrical or power cables are electrical utilities that are commonly used for the transfer of power or in other terms electricity. These are not just confined to underground power cables; and it is also applicable in submarine conditions. Pressurized water pipelines are utilities that transfer water (drinking as well as other purposes) from one location to another. Majority of this type of utility falls under the jurisdiction of respective state authorities. Pressurized gas or fluid pipelines are any utility that is used to transfer gases or fluids under pressure which may be flammable in nature and which poses a high certainty of risk if utility strike occurs falls under this category. Sewer or drainage pipelines are any utility that is used for the transfer of waste material or fluids is referred to as sewer or drainage pipeline. The final classification is any other auxiliary lines that may be used for any other purpose that is not described above (Fig. 1).

This classification can vary nationally. Most countries now make use of colour coding systems to identify which type of utility is present. The colour coding followed by the United States for underground utilities as prescribed by the American Public Works Association for easy identification is described further. The colours consist of white, pink, blue, purple, green, yellow, red and orange. White colour is used for indicating excavation limit or it signifies that beyond the point white colour is encountered, there is the presence of utility. Pink colour indicates markings made during survey and other unidentifiable or undefined utility. Blue colour denotes utilities that transport water which is commonly used for drinking purposes. Purple colour indicates utilities that again transports water, but the difference being the water transported is used for other purposes such as for cultivation and slurry lines. Green colour utility signifies drainage system. Yellow colour indicates utilities that is used for transferring materials that are flammable such as LPG and natural gas. Red colour is used to indicate utilities that live and conduct electricity. Finally, orange colour utility is used to indicate utilities that contain information transfer cables such as telephonic and Internet cable [10]. As earlier discussed, this classification can differ with region. The purpose of this colour coding is to understand if at all during excavation, if they come across any such utility, the content of that utility can be known. If, for example, while excavation proceeds, they come across a white marking, then they can identify the presence of utility beneath that point and hence prevent a direct strike with the utility. Nowadays, cable markers or utility markers are situated above the utility lines to identify the type of utility which is underground, and this marker usually also contains the necessary authorities number to contact in case of emergency.

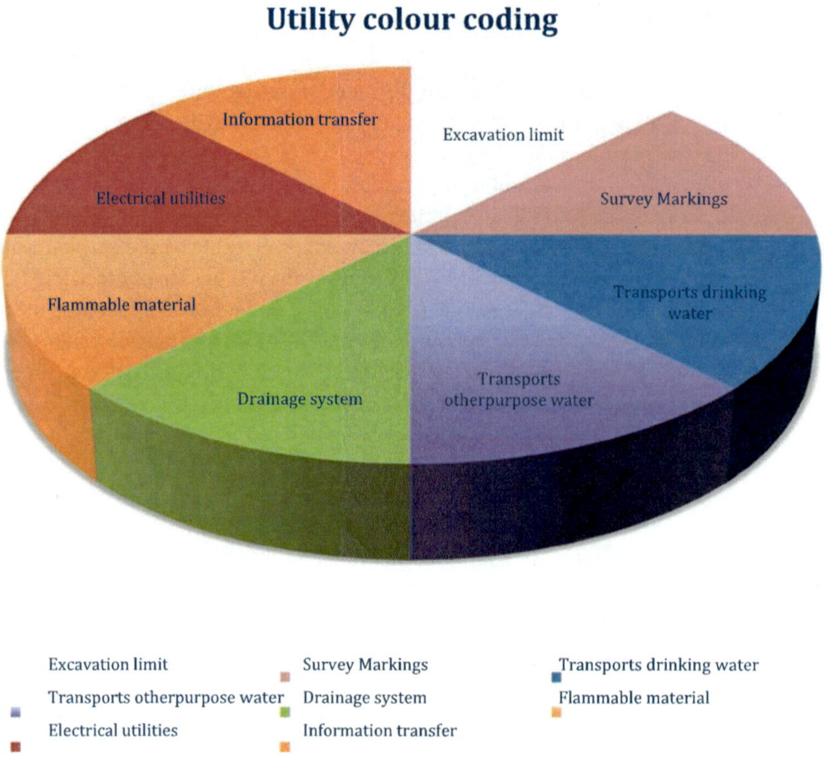

Fig. 1 Colour coding of different utilities

3 Dangers Associated with Underground Utilities

Dangers associated with an underground utility strike can prove to be fatal, and some of these dangers are as follows [11].

Gas pipelines strikes: Striking of gas pipeline will result in the sudden leakage of gas followed by fire and explosion if an ignition source is present.

Petroleum and oil pipelines: In addition to leakage of volatiles and gases, damage in the petroleum pipeline also results in contamination of soil and water ways causing harm to flora and fauna.

Electric cables and lines: Coming in contact with or striking a live electric cable/line can result in electrocution/ fatality. Additionally, it could act as a good ignition source if there are any flammable gases present in the vicinity of the excavation area and could lead to a fire or explosion.

Water pipe: Rupturing of water pipe can release the water with high pressure along with the debris. Sometimes, the area may get flooded and several times its been observed that workers are electrocuted due to electricity line flowing in the nearby vicinity.

Table 1 Percentage of strikes by location in 2019 [3]

Strike location	Percentage strikes
Footpath	54
Private drive way	3
Private land/field	4
Verge	8
Garden	3
Car park	0
Unmade ground	2
Watercourse	0
Railway	0
No location recorded	2
Carriageway	24

Table 2 Percentage of strikes by location in 2018 [3]

Footpath	30
Private drive way	1
Private land/field	3
Verge	9
Garden	1
Car park	0
Unmade ground	6
Watercourse	0
Railway	0
No location recorded	33
Carriageway	17

Sewer lines: Damage to sewer lines results in leakage of wastewater to water supply and can affect the health due to the release of toxic gases present in it.

Telecom: As optical cable comprises infrared radiations, ruptures in it may cause injury to eyes and exposure of workers to radiations. Moreover, if a communication line is damaged, it would be difficult to connect to emergency service and the entire area will be isolated.

Tables 1 and 2 depicts the utility strikes occurred in the year 2019 considering general areas in a location;

4 Causes of Utility Damage

The majority of people are unaware of underground utilities because they are not accessible from the surface. While this is attributed to be a common cause, the actual causes are underlying which points out to factors such as unavailability of locating mechanisms or data, economic constraints within excavating companies, time constraint to finish a project and so on. Some of the causes for damage of underground utility that could be identified were studied and are as follows;

4.1 Location of Utility

The existence of the utility marks, which are temporary and not apparent until the top soil is stripped during excavation, is the main cause of these strikes. Utility strikes are also caused due to inadequate depth information or incomplete utility-location data, due to dependence upon outdated utility maps or a utility locating firm that lacks the experience or ability to have reliable results [1].

4.2 Cost Cutting by Companies

Even though workers are aware of the hazards associated with the underground utilities, the limited budget spent on work practices and availability of utility detecting equipment lead to the disruption of utilities. Another factor is that many companies provide performance-based payments which lead workers to focus more on rushing their work in order to get bonus amount, thereby compromising safety [12].

4.3 Lack of Communication

Most of the time, it's been observed that there is a barrier between the management and site operators. Due to lack of communication between the management and site workers, management is unaware of the actual site data and thereby fails to provide working equipment's as per requirement. On the other hand, the near miss data is not revisited by the site workers requested by the management. Thus, due to improper communication, the alternate measures are not provided even after the identification of the hazards of utility [12].

4.4 Incompetent Workers

As the job is temporary and due to involvement of a lot of contractors, many times work is being carried out by the operatives who have not received adequate training regarding interpreting utilities based on the layouts, using software providing information related to utilities. At times, even after training, a person is not competent enough, but due to poor work to dig the system, they are allowed to carry out operations. This leads to poor assessment of the actual site condition giving rise to utility strike.

4.5 Poor Inspection

Ignorance of factors including extreme weather conditions, type of soil, utility encased in ground during installation of project, quality and material of pipe acts as a contributing factor for the service strike. There are chances of the presence of another utility in nearby proximity. Irregular surveys and planning leads to outdated and inaccurate layouts and improper generic operation carried out as per the inaccurate statutory drawings [13].

4.6 Improper Excavation Techniques

Another main reason that leads to utility strikes is the improper use of excavation tools or the unavailability or ignorance to use precise tools. Even after understanding the utility layout in an area, it has been observed that very often, the excavation proceeds using drilling machines or other excavation tools which are not suitable for precisely digging up to a certain depth, and this in turn causes utility strikes. Sometimes workers are even seen using hand-held manual tools [2].

5 Utility Damage Prevention Techniques

Damage to utility or utility strikes can be mitigated if certain methods are adopted. Some of these methods are discussed below.

5.1 *Location of Utility*

First step to prevent the potential danger of service striking while excavating is to identify the location of utilities.

Types of Utility Locator

The types of utility locators are as follows [14],

1. **Active utility locator**: In order to detect the unimpaired metallic utility from one end to other, active utility locating is proved to be effective. This method of locating utility adds a signal (electromagnetic energy) to the buried utility by a signal transmitter.

   There are two ways by which utilities can be located real time;

   Inductive method: Electromagnetic energy is directly introduced in a line when a transmitter is positioned just above the ground surface of utility.

   Conductive method: Electromagnetic field is generated outside by direct attachment of utility with the transmitter.
2. **Passive utility locator**: In case of broker pipelines, tracing of utilities becomes cumbersome as frequency cannot be transmitted through the entire length of wire, another problem is the lines made up of plastic material does not possess any electrical property. For this purpose, the passive utility locator detects the radio waves already present in utilities [15].
3. **Public utility locator**: Utilities that comes under the services provided to the public such as gas, electricity, water, transport etc., are to be located by public utility locator [15].
4. **Private utility locator**: Utilities not covered by public utility locating are detected by private utility locators [15].

Utility Locating Equipment's

To serve the purpose various equipment are used in detecting the location of utility. Commonly used equipment's involve.

Mobile laser scanner and optical cameras are used to provide 3D utility maps of underground services. Thermal imaging detectors that utilizes the concept of infrared thermography by determining the variations in wavelength are used to determine leaks in pipelines. The operating temperatures of utilities like sanitary sewers and steam pipes are different from the temperature of the surrounding soil. Detection of temperature difference is detected using infrared sensors [16]. Electromagnetic and radiofrequency detection devices are one of the other common devices that are used. These devices provide the information about the depth, physical features, and size of the line by transmitting electromagnetic pulses to ground and tracking the signal through an underground metal pipe. Ground penetrating radar utilizes microwaves to detect the structures below the surface. For locating live electrical utility, a cable locator can be used, provided the utility is conducting. Another instrument, a signal transmitter, can be used along with a cable locator to detect utility more efficiently. This phenomenon is often referred to as active locating and reduces utility striking drastically [17].

An alternative option is to adopt soft digging techniques in order to identify utilities, for example, vacuum and hydro excavation that allows the soil to move away from underground utilities without coming in direct contact with the line. Vacuum excavation is also utilised by utility potholing which determines horizontal location, depth, and any sign of damage in a buried line. Among the above-mentioned techniques, the most important and efficient is the GPR system [15]. A detailed explanation of its use is as follows.

Ground penetrating radar, GPR is a non-invasive, high-resolution technique based on the principle of scattering of electromagnetic waves using Maxwell's Equation. It consists of transmitting antenna, receiving antenna, control unit and a display unit. The link between the properties of material and wave velocity is the crux of using GPR to detect utilities. Simply put, the materials having different electrical properties tend to have different velocities, so whenever a signal is passed through two materials having different electrical properties over the same distance, due to different velocities, the signal will arrive at different times [18].

Initially, a transmitting antenna transmits electromagnetic energy into the soil, this electromagnetic waves travels with a velocity relative to permittivity of material. The wave keeps on travelling until and unless it strikes the buried material having different electrical properties. A part of the wave is reflected back and remaining continues to go further and strike another buried utility in nearby proximity until the energy of the wave reduces to zero [18].

The echoes from the subsurface utility are recorded by a receiving antenna and captured by display unit (usually a digital storage unit).

The advantages of GPR are as follows [19],

1. As it is a non-invasive technique, GPR can be used conveniently in public spaces without causing any damage to nearby landscapes or structures
2. It detects both metallic and non-metallic objects, and measures depth, dimensions and thickness of the buried utilities
3. Without any digging or ground disturbance, this technique provides immediate and accurate data by scanning only a side of the surface
4. Cost-effective and a range of resolution and depth of penetration can be obtained by regulating frequency.

5.2 *Safe Work Practices*

Certain safe work practices can be followed before the beginning of an excavation work that could prevent a utility strike. They are as follows (Fig. 2).

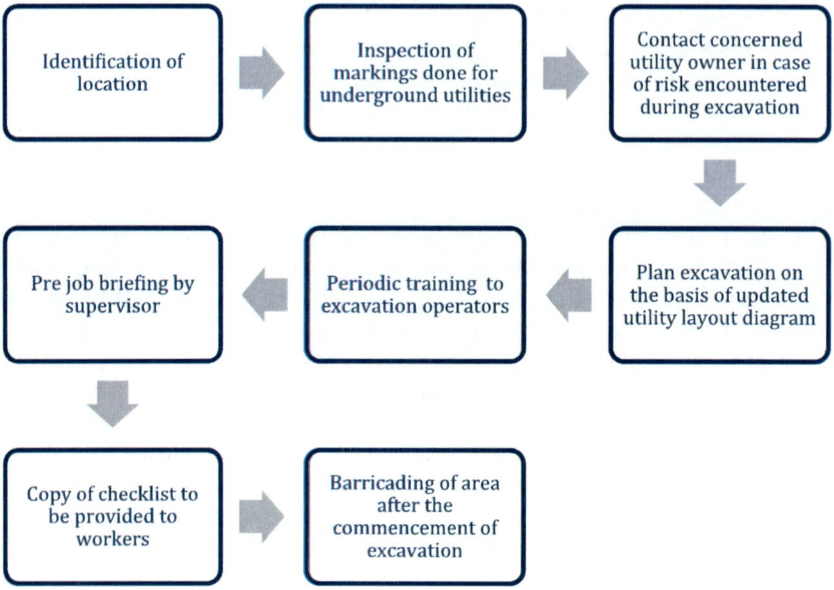

Fig. 2 Safe work practices for excavation [20, 21]

5.3 Continuous Review

Regular monitoring of the site, reporting damaged underground cables, pipelines, potential near miss, should be done before carrying out the excavation process before the beginning of any excavation work. For this purpose, the utility inspector must be appointed by the company. The company should thereby invest more in improving the safety of their excavation works which in turn will not cause any damage to underground utility as well as injury to workers. This should be done and reviewed periodically so that no mishaps may happen in future.

5.4 Training and Competency

In order to achieve safe excavation and competency in detection, training regarding radio detection approved by CAT, and safe digging practices should be made mandatory for workers involved in digging. Training should also cover communication techniques for proper clarification while digging. One of the techniques is the phonetic alphabet which is of great use in noisy environments [20]

5.5 *Statutory or Legal Requirements*

If legal norms, acts or rules are put into place separately for utility damage prevention and legal procedures to be followed while excavation work is being implemented supervised under a national or state regulatory body utility strikes can be reduced to a huge extent. Certain countries have legal norms for preventing utility strikes. The Tennessee state of the U.S has the Underground Utility Damage Prevention Act which puts forward a Underground Utility Damage Enforcement Board to prevent utility strikes [22]). Another state in the U.S is Kansas which puts forward Kansas Underground Utility Damage Prevention Act which puts forward definitions and different statutes to be followed while excavating to prevent utility strikes. Similarly, if all countries or states put forward similar regulations, utility strikes can be mitigated to a certain extent [23].

6 Conclusion

Many methods have been adopted to identify and control the damage to utility. As earlier mentioned, utility damage also results in major economic losses not just for the maintenance of the utility but losses that are incurred during the down time. Damage to utility or utility strike does not only damage the type of utility which comes into contact while excavation but also will affect the workforce adversely; for e.g., direct excavation using hand-held tools and striking a power cable or a flammable gas pipeline could result in fatality of the workforce. The major causes for utility strikes are unavailability of data to identify or locate underground utility, cost cutting by companies which do excavation, lack of communication, incompetent work force, poor inspection techniques and improper excavation techniques. Most of these causes can be mitigated with necessary control measures such as locating utility before beginning excavation using different techniques, following certain safe work practices while excavating, by conducting training programs for excavating employees and contractors, regular review and inspection of existing safety measures for excavation and introducing, implementing or strengthening statutory rules or norms as earlier discussed.

This review paper has done an extensive study on various factors such as the different types of utilities according to the variation in their attributes and the purpose or the medium that is being transferred using the utility, the different dangers that could arise out from it according to what they carry, the major causes that leads to utility strike and damage and all other factors associated with it, and the different terminologies and methods that are used to locate utilities that are located underground and thereby prevent damage to the utilities. The detailed study in this paper has covered the major causes that leads to utility strikes, and it is clearly understandable that excavation is the major cause that lead to utility strikes when compared to any type of other disaster. Even though there are many control measures for detection of utilities that are located underground, utility strikes till remain active and prominent to date. If the control measures suggested in identifying utilities can be invoked in an adept manner then utility strikes can be reduced or mitigated to a major extent.

References

1. Tanoli, W. A., Sharafat, A., Park, J., & Seo, J. W. (2019). Damage prevention for underground utilities using machine guidance. *Automation in Construction, 107*. https://doi.org/10.1016/j.autcon.2019.102893
2. Metje, N., Ahmad, B., & Crossland, S. M. (2015). Causes, impacts and costs of strikes on buried utility assets. *Proceedings of the Institution of Civil Engineers - Municipal Engineer, 168*(3), 165–174. https://doi.org/10.1680/jmuen.14.00035
3. Utility Strike Avoidance Group. (USAG). (2021). *Strike damages report*. https://www.utilitystrikeavoidancegroup.org/strike-damages-report.html
4. National Conference of State Legislatures. (2017). *How states protect pipelines from excavation damage*. http://www.ncsl.org/research/energy/how-states-protect-pipelines-from-excavation-damage.aspx
5. (CGA), C. G. A. (2017). *Damage information reporting tool, analysis and recommendations*. https://commongroundalliance.com/sites/default/files/publications/2017DIRT Report inside final_corrected11-7-2018.pdf
6. Radiodetection Limited. (2021). *Reducing damage to underground pipes and cables*. https://www.radiodetection.com/en/reducing-damage-underground-pipes-and-cables
7. Dockter, L. J. (2008). Development of an underground utility damage prevention plan (UUDPP). *Style (DeKalb, IL)*, 139.
8. Hojjati, A., Jefferson, I., Metje, N., & Rogers, C. D. F. (2018). Sustainability assessment for urban underground utility infrastructure projects. *Proceedings of the Institution of Civil Engineers—Engineering Sustainability, 171*(2), 68–80. https://doi.org/10.1680/jensu.16.00050
9. *What utilities are commonly found underground?* (2018).
10. *Underground Utilities; what you need to know before you dig*. (2021). https://pipespy.com/blog/underground-utility-replacement/
11. Armstrong, J. (2019). *Hit an underground utility line? Here's what you need to do*. https://www.geoscan.com.au/blog/hit-an-underground-utility-line
12. Patel, M., Sherratt, F., & Farrell, P. (2012). Exploring human error through the safety talk of utilities distribution operatives. In *Association of Researchers in Construction Management, ARCOM 2012—Proceedings of the 28th Annual Conference* (Vol. 1, pp. 403–412).
13. Makana, L., Metje, N., Jefferson, I., & Rogers, C. (2016, January). *What do utility strikes really cost?* https://www.researchgate.net/publication/321110173
14. Geosurv. (2019). *Active versus passive utility locating: Which is better?* https://www.geosurv.com.au/active-vs-passive-utility-locating-which-is-better/
15. Tim, H. (2021). *Look it up: Utility locating terminology*. https://www.softdig.com/blog/look-it-up-utility-locating-terminology/
16. Hayes, K., Allouche, E., Sterling, R. L., Anspach, J., Simisevic, J., Rogers, C. D. F., & Weston, K. (2009). Encouraging innovation in locating and characterizing underground utilities. In *Encouraging Innovation in Locating and Characterizing Underground Utilities*. https://doi.org/10.17226/22994
17. Publishers, E. (2017). *Why do utility strikes occur?* https://www.ee.co.za/article/utility-strikes-occur.html
18. Daniels, J. J. (2000). *Ground penetrating radar fundamentals*. https://clu-in.org/download/char/GPR_ohio_stateBASICS.pdf

19. Hoffer, T. (2020). *What is ground penetrating radar (gpr) & how does it work?* https://www.softdig.com/blog/gpr-ground-penetrating-radar-work/
20. Coffey, D. (2016). *Training the next generation of electric utility workers.* https://www.powermag.com/training-next-generation-electric-utility-workers/
21. Armstrong, J. (2020). *Essential tips to prevent utility strikes.* https://www.geoscan.com.au/blog/5-essential-tips-to-prevent-utility-strikes
22. Commission, T. P. U. (2021). *Underground utility damage prevention act.* https://www.tn.gov/tpuc/divisions/uudp-underground-utility-damage-prevention.html
23. Kansas Corporation Community. (2021). *Kansas underground utility damage prevention act.* http://kcc.ks.gov

Issues Faced by Women Construction Workers with Respect to Ill Health Effects, Wage Disparity and Unsecured Work Environment

Ashish Aeri, Chandrakant Singh, Abhishek Nandan, N. A. Siddiqui, P. Mondal, and Akshi Kunwar Singh

1 Introduction

Activities related to construction form a very crucial part of any country's infrastructure. Construction sector in India, when it comes to unorganized labor, is the largest employer. At the top of the fastest growing sectors in India, it is the construction industry with a ten percent growth rate annually. After agriculture, it also happens to be the second largest sector employing people [1]. The women workers, who do not only look after their families but also work and provide for them, are largely engaged in the agriculture sector. After the agricultural sector, it is the construction sector where close to half of the thirty-one million workers at construction sites in India are women. Here, women are mostly employed as unskilled laborers [2].

Women primarily constitute almost half of the workforce in the construction industries. However, the practice of gender discrimination continues to occur in the construction sector as well. Women are mostly hired to work as masons but they are still treated as inferior workers as compared to men. This is because the general view is that women are not as skilled as men when it comes to doing tasks. This very thought process of people and employers creates discrimination against women workers, and this therefore prevents them from being trained like the men who are then employed as masons. Men construction workers mostly start as unskilled workers or laborers; however, unlike women, they climb up the ladder and many also become supervisors and contractors later. It therefore cannot be denied that the construction sector in India is by and large dominated by men where women workers are not encouraged to even become masons. Not only this, also when it comes to giving them opportunities

A. Aeri · C. Singh · A. Nandan (✉) · P. Mondal · A. K. Singh
University of Petroleum and Energy Studies, Dehradun, India
e-mail: abhisheknandan24@gmail.com

N. A. Siddiqui
GD Goenka University, Gurgaon, India

N. A. Siddiqui et al. (eds.), *Advances in Construction Safety*,
https://doi.org/10.1007/978-981-19-4001-9_5

in relation to promotion in construction areas and activities, they are denied despite being qualified to avail the same.

Article 14 of the Indian constitution provides that men and women should be provided equal opportunities as the Article aims to ensure the right to equality of every Indian citizen. The construction sector as we know falls under unorganized sector, and issues and problems which women workers here face form major barriers to their growth and empowerment and thereby constituting contemporary social problems which women face in the construction sector. Most women are not allowed to acquire and learn specific skills which may become useful to them later in order to become masons. But, when it comes to men, many companies are ready to provide them training for the same and many a times they are on the basis of their skills upgraded to higher positions as compared to women who are ignored.

Other than being discriminated, they also face other difficulties like sexual harassment, disparity in payment of wagers as compared to men, absence of any social security and low wages as well. Many women working at the construction areas are also illiterate and are therefore not aware about their rights as well and therefore silently suffer the inadequacies of the society in ensuring them a secure work environment and culture [3]. Thus, sexual harassment, gender-based discrimination and pay gap hamper their growth, making it difficult for them to work in the prevalent environment resulting in them remaining at the same level and with the same skills even after working for years.

Moreover, even the employers think that women cannot work as much as men, and as a result, many women laborers are asked to either settle for lower wages or leave. Wage discrimination is also becoming common with men laborers and construction workers continuously being paid more as compared to women. Wages which females here earn are Rs.40 per day less than the male workers with per day wage of women workers being Rs.80 and men workers being Rs.120. Even while constructing houses, women who do the same work as men there also are yet again paid 30% less wages as compared to male workers [4].

2 Review of Literature

Annette Barnabas and Clifford [2] The construction industry employs the biggest number of building workers in India next just to rural part. Females form a large portion of the workforce, and by decision or by plan, they are not permitted to acquire obvious skills that may empower female workers to become skilled. Females join as untrained laborers and stay incompetent till the finish of their working life expectancy. Be that as it may, men get proper training and deliberately redesign their development abilities to graduate as bricklayers, supervisors and temporary workers. There is a characteristic sex inclination against ladies and furthermore the common general conviction that female construction laborer is not fit to be prepared casually like male workers in the construction segment despite the fact that they have the vital abilities, capacity and want to become masons. In spite of the fact that the temporary workers are happy to acknowledge ladies as masons by providing them proper training and position in the development part, it has been discovered that the social powers that have propagated the idea of ladies as substandard laborers are antagonistic to any similar move. The study likewise shows the philosophy of the training given to men in the construction division in our country and proposes another procedure of preparing that would qualify female laborers to become skilled laborer and enable them monetarily.

Devi and Kiran [5] Construction sector gives work opportunities for huge number of unskilled and skilled workforce. The workforces utilized in this sector need to confront several issues at the workplace. A few issues identified with wellbeing, work pressure and accidents at site are the significant concern of numerous studies among the researchers. The main objective of the research is to distinguish the key variables identified with respect to status of females working in the construction sector. Ladies function as incompetent work and face a few different challenges in contrast with guys. Wage discrimination and biasness based on gender are the main consideration because of which the workplace gets hard for them in the business and female stays at similar level even after subsequently working for number of years.

Kumar [6] Discrimination based on gender is a wide-ranging phenomenon. Ladies involve half of the total populace and perform sixty-six percent of the work and however earn just thirty-three percent of the absolute salary and claim not exactly a one-tenth of the assets. The most kept apart individuals in the world are typically the ones who need financial support. Construction sector is the biggest employer in India. When men construction laborers have profile-raising opportunities, female workers have no chances to secure skills and become supervisors and masons. They should be empowered to develop in their profession.

Royuela et al. [7] described about work life balance to depict practices in accomplishing a harmony among the employee's ' family and work life. The requests and weights of work make hard to extend time for adjusting work life exercises. Work life balance challenges affect their progression. The position and status of female in any general public are a marker of the financial what's more, social accomplishment of that society. Construction industry is the spine of the economy as it fabricates the foundation important for modern development. Around thirty-three percent of these laborers are women with low degree of aptitudes and training and they face significant issues related to work, viz., wage segregation, sex and obscene behavior, undesirable occupation relationship, lower compensation, in present examination the current writing in worried about work life adjusts among the development also, laborers was explored. Devi and Kiran [8] There have been changes both minor and major in the position of females in India throughout the years; however, these progressions are definitely not uniform.

Bonet et al. [9] The conditions at work in the construction business have been improved significantly during the previous decades, and endeavors have been made to diminish the measure of truly difficult work and conveying, yet physical work, static work and impact of climate are still there to consider. Women decide to delay labor or to stay childless so as to seek after a working.

Masood and Jahan [10] From the study in Okhla region, it is clear that women are lagging far behind the men; despite being equally contributing from years, they are always considered as secondary option. The study emphasizes basically on the type of work they are engaged in, differences in the wages although they work for equal number of hours and facilities being provided to them at the site location. One-on-one interview was taken to find out the issues faced by them. The result clearly states that most of the workers belong to lower section of society migrated from central part of India for the job opportunities. Women are basically engaged in transporting the material such as stones, bricks, concrete or mortar mixture and in curing process which gives clear vision about the job status given to them at the construction sector.

Madhok [11] The study was conducted at some selected location of Punjab, and it was observed that women are highly prone to musculoskeletal disorder. The root cause behind the scenario is lack of training provided to them and the basic job awareness [12]. No personal protective equipment is provided to them; as a result, they are directly exposed to the hazardous condition which leads to the occupation health-related diseases. It has been found that facilities of clean drinking water and clean and separate washroom for women which is a basic requirement are ignored.

Dave [13] The present investigation was done with female construction laborers. An effort to comprehend the socio-economic state of female workers, type of work, the prevailing working conditions, wage design, wage distinction and different challenges faced by them at their workplace has been made in this paper. Information was gathered from 350 lady workers from urban and rustic zones of 3 locale of Haryana: Kaithal, Panipat and Kurukshetra. Result from the study shows that greater part of the migrant females were locked in the construction business and were just utilized in incompetent and low paying occupations, workers furthermore assistants.

Anand [14] The women were abused to a more prominent degree as they were paid quite less contrasted with male workers for same degree of work, furthermore hours spent on work. Also, working conditions were unacceptable, and the issues stood up to by them were intense. Also, their inability to read, poverty and obligation constrained them to work for lower compensation and under awful conditions.

Lakhani [15] In his study highlighted occupational health of female workers in the construction site by computing occurrences of occupational health ailments. The sample space taken for the study was 1052 workers by random sampling, and they were medically examined and subjected to interviews, investigations and examination. Out of 1052 selected workers, 520 were female. According to the study, the major finding was the number of female workers suffering from musculoskeletal disorders was much higher and it was about sixty-three percent. 90% of women stated that they were not getting any worker's benefit and social security.

Rasheedha [16] The experts focused uniquely on the female workers who are employed in the construction sites and how they are confronting troubles at the site. Most of the female construction laborers are confronting bunches of troubles like low wages, sexual harassment, inappropriate behavior, gender disparity and many more.

3 Methodology

The purpose of the study is to describe the issues faced by female workers in the construction sector. The study was conducted for Dehradun district of Uttarakhand state. Uttarakhand state lies between 28° 44′ and 31° 28′ N latitude and 77° 35′ and 81° 01′ east longitude, with a total geographical area of 53,483 km^2. The state comprises two regions Garhwal region and Kumaon region. The state capital is Dehradun situated at an altitude of 640 m above mean sea level and is located in Garhwal region of the state. 220 female workers were selected, and the assessment of working conditions for female workers was carried out through questionnaire. All the female workers were interviewed in the local language (Hindi).

4 Result and Discussion

1. **Working conditions for women construction workers**

A majority of India's work power is in unorganized sector. Without financial opportunities in their own states, numerous laborers relocate over different conditions of India to look for employment. Construction sector depends on the migrant laborers, dominant part of which are female. These female laborers have an intense life [17]. The work power employed in this business needs to confront a few challenges at the workplace, viz. wage difference and inappropriate behavior, unhealthy employment relationship, lower compensation and so forth. Rajanna [18] Majority of the female workers are affected by the common workplace hazard existing at the construction site which can be broadly mentioned as slip/trip/fall, swelling of body parts, burns, minor cuts and abrasion of skin, foot injury by pointed objects, respiratory disorders, eye injury, sprains due manual material handling and while waking on unstructured ramps and so on [12].

From the various studies conducted on the health status of female construction workers, it can be clearly stated that the middle-aged women workers were mostly affected by the workplace stress and hazards. The conditions existing at the construction site put women at high risk when it comes to health and hygiene [19]. The pitiable conditions of the female laborers in the construction business need to be addressed; despite of the existing legislation, there exist a huge difference when the actual working conditions at the various construction sites throughout the country are to be addressed. The existing legislation is violated when it comes to providing the basic facilities within the establishment like separate latrines and urinals and availability of drinking water matching the quality parameters [20].

2. **Ill-health effects due to improper working conditions**

For our study, 220 female workers were subjected to interview and majority of them were from the age of 20–35 with no formal education or educated till primary level only (Fig. 1a, b). They were asked health-related questions regarding any persisting

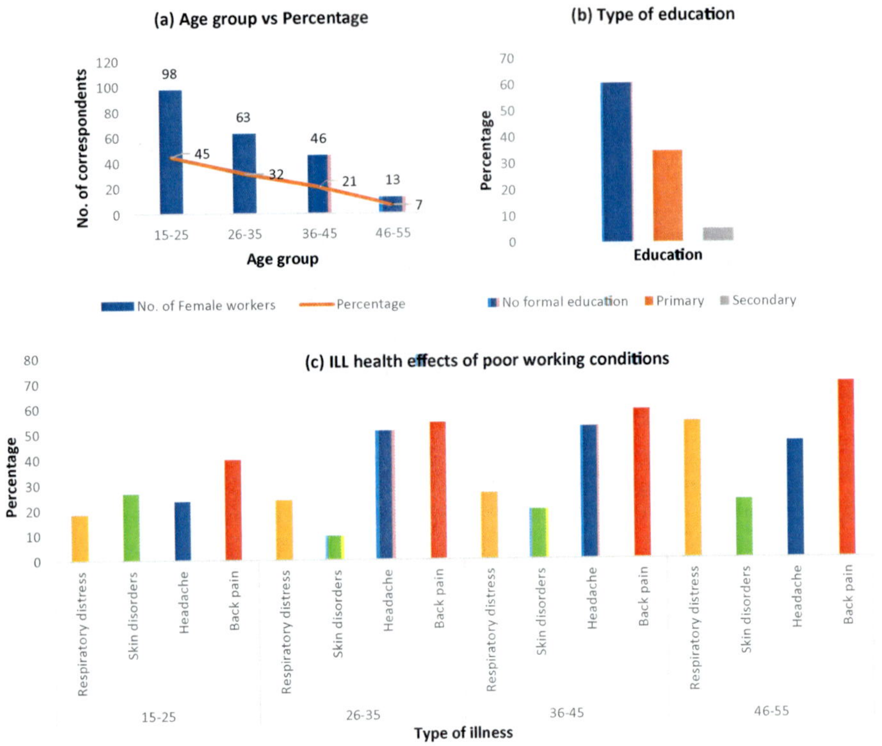

Fig. 1 Assessment of poor working conditions for female workers

back pain, skin disorders like fungal infections and skin irritation issues due to silica and cement dust; they were also asked if they are having any respiratory disorders such as chronic coughing and persistent sore throat. Head ache and back pain were common for female workers from all the age groups.

Most of them suffered from headache on daily basis due to long working hours on average, they have to work for more than 10 h, and activities included brick carrying, brick breaking, manually carrying other construction material and preparing mortar and concrete. Because of carrying material manually without adopting proper lifting posture, majority of female workers were having backpain (Fig. 1c), and no equipment to handle the heavy materials was provided to them by the contractors/site owners. No female worker was involved in brick laying activity, and even after working for the same number of hours, they were paid less than all the male workers.

3. Wage discrimination

The disparity in wages between the men and women construction worker is because of the fact that female workers are always deprived from the opportunities to advance their skills. Hence, women work primarily as untrained workers. Mohapatra [21] The space for unprejudiced consideration and sexual orientation-based comparison is constricted by information insufficiencies as well as is invalidated because of the

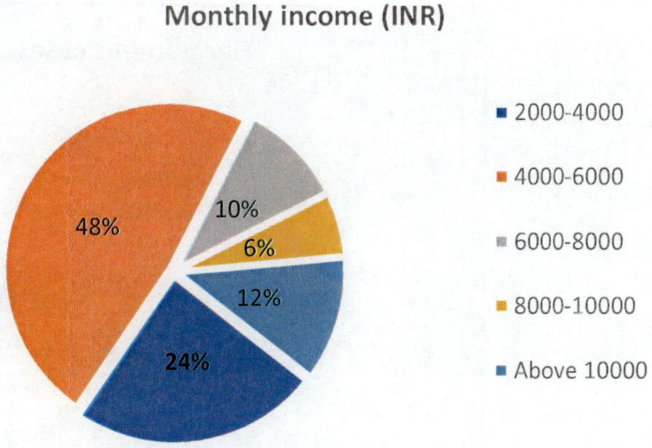

Fig. 2 Monthly income of female workers

perceptions grown from the male centric job stereotyping that goes before any estimation on women's commitment to the economy, requiring the acknowledgment of these components as indications to such investigation on sexual orientation disparities [22]. A majority of workers are facing several health issues, and there exist economic problems which they have to face as they are not paid equally as men worker and there are many women construction laborers who contribute their overall monthly income to the family [23]. Today, financial liberation is the prime reason for refining the status of female in India. It is important that female work support is crucial in upgrading social and economic status just as improving their roles in the family decision-making. In construction practices, wellbeing is the significant element of employment, and it is the crumbling factors consider to laborers particularly the women.

In the research conducted by P. Maneesh, P. T. Jasna in the year 2017, it was found that out of the 50 female workers 82% of the laborers were facing financial problems. The monthly income of most of the female workers was found to be between 4000 and 6000 rupees (Fig. 2). Further, it was also found that about thirty-eight percent of the workers took loan for various purposes and were in debt. The reason behind is there is insufficiency of earning and the growing expenses [24].

4. **Harassment of female workers at work site**

The harassment of women at workplace is a result of society which is following the patriarchal lines, and the construction industry is mainly male-driven sector where men hold the power and the females are excluded from it. Thus, females in male-dominant workplace are treated distinctively [25].

The following data is taken as secondary data from the research conducted by Dr. K. A. Rajanna in the year 2015; the study was conducted on 300 female construction workers, and ninety percent out of the total respondents were stated to be scolded or abused by the mason or male workers or contractors (Fig. 3). It is majorly because of the gender disparity in the construction industry, illiteracy and absence of the knowledge of women education [18].

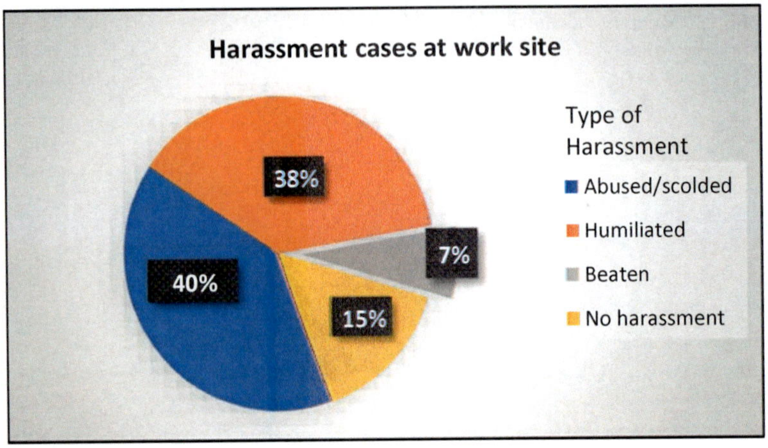

Fig. 3 Harassment of female workers at construction site

5. Improving Conditions of Women Workers

It is high time that issues revolving around plight of women construction workers are not ignored and policy reforms and institutional changes be brought up as soon as possible by the legislature for empowerment of women construction workers. Empowering women and a stricter check on companies will result in change in the power dynamics with workers being favored and paid as equally as men.

There has been large growth over time in the number of women workers employed in India, and therefore, the force of women laborers is also increasing. However, with this there is also growth of worsening conditions of women making them more vulnerable to the ills of the construction industry. Companies must aim at providing women workers with a safe and secure environment as women workers are no less than men and are largely beneficial for any organization as they are disciplined, focused and work oriented. Women also effectively add to a country's economy, and so the need is to prevent practices like hiring them but not paying them at all or even adequately. Women also look after their families and children, and it therefore needs to be understood that providing them weekly holidays is a basic human right which construction companies must cater to as there is a double burden on them as compared to most of the men out there in the country. The society needs to do away with these problems that women face in order to create a better and developed society which gives equal respect to women workers as well. Job security as well as wages should be adequately paid to the workers in the unorganized sector and on failure of the same; penalty/fines and punishments need to be inflicted on the employers or construction companies. Moreover, even male construction workers need to motivate the women in their families who are also working in the construction sites by giving them the same informal training that they receive. Only when the women workers are imparted the same skills as men in construction activities and other related skills, can they in turn further become mentors to many other women workers.

5 Conclusion

Female construction laborers face provocation at home and work spot, they are segregated in wages and advancement, the significant reasons why female workers are not advanced as bricklayers are the sex inclination which people have, and women construction laborers are not allowed a chance to be prepared casually like men in the development industry. The findings from the study show that more than half of the female workers are uneducated and were facing multiple health-related issues due to the poor working conditions at construction sites. The Indian constitution gives equal chances and rights to both of the genders. The issues of female construction workers in the work environment are one of the significant issues in the present social scenario. Larger part of the female construction laborers are confronting loads of troubles like low wages, musculoskeletal disorders, social security, gender discrimination, sexual harassment and so on. Main thing is unawareness and absence of education. They are working in an unbound condition or work culture.

References

1. Baruah, B. J. (2010). Gender and globalization: Opportunities and constraints faced by women in the construction industry in India. *Labor Studies Journal, 35*(2), 198–221.
2. Annette Barnabas, D. J. A., & Clifford, P. S. (2009). A study on the empowerment of women construction workers as Masons in Tamil Nadu, India.
3. Banu, S. R. (2017). Problems of women construction workers with special reference to Mannachanallur Taluk, Trichirappalli District in Tamilnadu. *International Journal of Trend in Research and Development, 4*.
4. Chitra, M. N. (2015). A descriptive study on problems of women workers in construction industry at Tiruchirappalli 2279-0837.
5. Devi, K., & Kiran, U. V. (2013). Status of female workers in construction industry in India: A review. *International Journal of Humanities and Social Science, 14*, 27–30.
6. Kumar, B. R. (2013). Gender discrimination among construction workers with reference to Vijayawada. *Journal of Sociology and Social Work, 1*(1), 42–53.
7. Royuela, V., López-Tamayo, J., & Suriñach, J. (2008). Results of a quality of work life index in Spain. A comparison of survey results and aggregate social indicators. *Social Indicators Research, 90*(2), 225–241.
8. Devi, K., & Kiran, U. V. (2014). Work life balance of women workers in construction industry *2*(4), 4932–4946.
9. Bonet, R., Cruz, C., Kranz, D. F., & Justo R. J. I. R. (2013). Temporary contracts and work—Family balance in a dual labor market. *66*(1), 55–87.
10. Masood, H., & Jahan, Q. (2005). Women workers in unorganized sector in India: Problems and prospects abstract.
11. Madhok, S. (2005). *Report on the status of women workers in the construction industry*. National Commission for Women.
12. Bharara, K., Sandhu, P., & Sidhu, M. (2012). Issues of occupational health and injuries among unskilled female labourers in construction industry: A scenario of Punjab State *6*(1), 1–6.
13. Dave, V. J. (2012) Women workers in unorganized sector *18*(3), 9–12.
14. Anand, V. J. (1998). Advocating for the rights of construction workers: Nirman's experience. *The Indian Journal of Social Work, 59*, 847–863.

15. Lakhani, R. J. (2004). Occupational health of women construction workers in the unorganised sector. *The Journal of Health Management, 6*(2), 187–200.
16. Rasheedha, B. (2017). Problems of women construction workers with special reference to Mannachanallur Taluk, Trichirappalli District in Tamilnadu. *International Journal of Trend in Research and Development, 4*(4), 1–3.
17. Nandal, S. (2006). Women workers in unorganized sector: A study on construction industry in Haryana. *International Journal of Development Issues, 5*(2), 119–132.
18. Rajanna, K. A. (2015). Nature of work, working conditions and problems of women construction workers: A case study. *The International Journal of Business Quantitative Economics and Applied Management Research, 1.*
19. Manhas, S. J. (2014). Assessment of physical health status of female construction workers of Kathau district, J and K. *IOSR Journal of Humanities and Social Science, 19*, 19–24.
20. Sultana, N. (2014). Health problems among women building construction workers.
21. Mohapatra, K. J. (2012). Women workers in informal sector in India: Understanding the occupational vulnerability. *International Journal of Information and Computing Science, 2*(21), 197–207.
22. Rustagi, P. (2005). Understanding gender inequalities in wages and incomes in India. *The Indian Journal of Labour Economics, 48*(2), 319–334.
23. Prasanna, P. (2016). Socio economic empowerment of women construction workers in Tamil Nadu. *International Journal of Current Engineering and Scientific Research (IJCESR), 3.*
24. Maneesh, P., & Jasna, P. T. (2017). Socio-economic condition of women construction workers in Kannur district, Kerala. *Indian Journal of Economics and Development, 5.*
25. Gutek, B. A., & Morasch, B. (1982). Sex-ratios, sex-role spillover, and sexual harassment of women at work. *Journal of Social Issues, 38*(4), 55–74.

Experimental Analysis of Pitting Corrosion in Offshore Structures

Parth Patel, Vikram Garaniya, Rouzbeh Abbassi, Til Baalisampang, and Vahid Aryai

1 Introduction

Marine conditions are considered an exceptionally complex and harsh corrosive environment. The marine environment has a low temperature, oxygen concentration, and varied pH, creating an ideal condition for corrosion of steel structures [1–6]. The higher density of ocean water causes a lower temperature [5–7]. When the metal comes into contact with highly aggressive chloride concentrations, localized corrosion occurs, and pitting starts to initiate and then follow the rest of the material's surface [8]. Moreover, Wang [9], Wang [10] proved that the chloride environment is the most regularly experienced aggressive agent that causes pitting corrosion. Pitting corrosion can be considered a deceptive type of localized corrosion and versatile system failure. Pitting corrosion causes a very high production loss due to equipment failure and shut down of plant for maintenance. According to Frankel [11], pitting corrosion could be controlled by changing several factors such as material and alloy composition, metallurgical, biochemical, and physical factors. Pitting corrosion is one of the most severe corrosion forms and is the cause of notably widespread failure mechanisms for marine and offshore structures [12, 13]. Sharland [14] states that pitting corrosion could be considered one of the most harmful mechanisms in marine and offshore structures, marked by the appearance of several small holes on the exposed metal surface. The pitting corrosion consistently begins when passive film breakdown, which is incited by the heterogeneity [15]. To withstand these environments, metals such as aluminum, stainless steel, and titanium are used in offshore structures. Stainless steel has proven to be one of the most pitting

P. Patel · V. Garaniya (✉) · T. Baalisampang · V. Aryai
Australian Maritime College, University of Tasmania, Launceston, TAS 7250, Australia
e-mail: v.garaniya@utas.edu.au

R. Abbassi
Faculty of Science and Engineering, School of Engineering, Macquarie University, Sydney, NSW, Australia

N. A. Siddiqui et al. (eds.), *Advances in Construction Safety*,
https://doi.org/10.1007/978-981-19-4001-9_6

resistant materials for offshore structures [16]. Several experimental researches have been conducted on stainless steel in terms of pitting corrosion. Steel composites are utilized in different marine offshore structures [17], including stainless steel AISI 304 and AISI 316L materials, which are ordinarily used in offshore structures due to their corrosion resistant nature [12]. These types of steel can withstand chloride solution because of having a passive layer [18–20]. Bhandari [12, 17] believed that good corrosion resistant capacity in the ocean environment is the most widely recognized reason for using stainless steel.

Do Nascimento [21] and Pardo [22] proved that high chromium content stainless steel has more pitting potential than those with lower content. Moreover, the pitting corrosion resistance of stainless steel is primarily influenced by metallurgical factors [23–25]. These include cold working, alloy organization, ion particles, heat treatment, grain size, and chemical properties such as pH, temperature, salinity, and flow speed [23]. According to Dastgerdi [16], utilization of stainless steel in petrochemical, oil, and gas, and offshore industry is increasing these days due to having exceptional corrosion resistance characteristics and passive layer. Hence, the American Iron and Steel Institute (AISI) listed these materials for marine and offshore structures. For example, AISI 304 and AISI 316L are mostly used for piping. However, with the presence of oxygen, fluctuation in temperature, and salinity, even stainless steels are less protectable to pitting corrosion in a longer time [16]. The pit's characteristics rely on many factors, which are alloy composition, temperature, chloride environment, and the orientation of the surface [13]. Klapper [26] studied passive layers, and they believed that the stability of the passive layer relies firmly upon temperature, the concentration of chloride ions, flow speed, and ph. Bhandari [12] found that the temperature can be considered a critical factor in the pitting corrosion phenomenon since it significantly impacts'steels' corrosion phenomenon in ocean water. Generally, the corrosion rate of steel increases with lower pH level [27]. A rough surface completion gives a lower pitting potential [28].

The American Standard of Testing Material (ASTM) G48 provides accelerated pitting corrosion laboratory test guidelines based on some critical factors such as salinity, temperature, and exposure time. Numerous industries (e.g., oil and gas industries [29]) have institutionalized the ASTM G48 as a pre-qualification tool for pitting corrosion tests to save time and cost so that they can make necessary changes before implementation. The suggestion to use ferric chloride by the ASTM G48 is to permit a quick evaluation of the pitting corrosion resistance of stainless steel in a short period. Due to the extremely aggressive nature of ferric chloride ($FeCl_3$), there are similarities between this solution and the actual condition in which pitting corrosion may occur during service in seawater [30, 31]. Many researchers [28, 30–35] have used the ASTM G48 for their corrosion studies; however, ASTM G48 does not suggest how to choose the material for testing. Corbett [35] found difficulty in preparing the solution, exposure time, and testing temperature, whereas Bhandari [31] found problems in sample size and surface preparation. Before and during experiments, they faced surface preparation challenges to mirror polish, maintain a constant temperature, and solve sequentially.

Pitting susceptibility relies on the surface condition of the material. Surface condition is an essential factor since it influences the microstructure and topography of the material as the surface comes in direct contact with the corrosive media [32–34, 36]. If the surface homogeneity is more chemically (by alloy composition) and physically (by surface homogeneous) both, then there is a more chance of having a more pitting resistant capacity [31]. Accordingly, the resistance capacity of metal toward corrosion would be improved, and the number of pits could be diminished if the surface level of the material is even. If the surface condition is not homogenous, and there is any weak point, chloride (Cl^-) ions start to enter that point, which initiates the corrosion process [37]. Thus, to avoid mitigate the risk associated with the pitting initiation site, the mirror polish is considered on all the samples used in the current study.

The temperature is considered a critical factor [38] as it can greatly affect corrosion behaviour. As a result, the material loses its pitting resistant capability significantly. Dastgerdi [16] believed that temperature has the most notable Influence on AISI 304 steel corrosion, especially between the range of 20–40 °C. Dastgerdi [16] found that the corrosion rate increases and pitting resistant capacity declines with increased temperature. Numerous research [39, 40] have considered different exposure times, such as 24 h, 48 h, 30 days, and 196 days. The ASTM G48 suggests 72 h to be considered for the exposure time in the pitting corrosion experiments. Byrne [40] illustrated that at 50 °C and 24 h of exposure time, pitting does not occur in numerous stainless steel standard grades. The ASTM G48 guidelines are generally used for corrosion in materials as a pre-qualification in marine and offshore structures and oil and gas industries [29]. Otero [39] and Pardo [22] proved that by increasing temperature and chloride ion concentrations, pitting corrosion rate increases in AISI 304 and AISI 316L. They followed ASTM G48 guidelines in conducting their experiments. Prawoto [41] followed the exact procedures in the ASTM G48 guidelines without any modification to identify stainless steel's pitting corrosion rate. They found that the corrosion rate increased by increasing the temperature, salinity, and exposure time. Buscail [42] revealed that AISI 304 has a lower pitting potential compared to AISI 316L. The surface finish at the mirror stage was highly recommended by Bhandari [31].

This study aims to determine the effects of salinity, temperature, and exposure time on the pits' depth and diameter of AISI 304 and AISI 316L in corrosive environments. It can be seen from the abovementioned literature that most of the previous studies conducted on pitting corrosion for stainless steel were based on the ASTM G-48 guidelines. Very few investigations have been done by modifying the ASTM G-48 standards for pitting corrosion, so there is significantly less data on pitting immersion laboratory tests. To fill this gap, several experiments are conducted by considering different salinity, temperature, and exposure time by changing ASTM G48 guidelines to meet real ocean conditions.

Furthermore, the test environment was created by modifying the guidelines of the ASTM G48 for accelerated pitting corrosion tests. The evolution of pit depth, diameter, and the total number of pits on each sample was studied using an optical microscope to analyze how each factor influences the corrosion phenomenon on both materials. This study will help better understand the pitting corrosion phenomena, which in turn helps enhance the safety and reliability of marine and off shore structures.

2 Materials and Methods

Mirror polish was carried out on all 136 samples physically by utilizing wet and dry sandpaper from 80 to 2000 coarseness to maintain consistency in test and analysis. The sample size was 50 × 25 × 5 mm (length × width × thickness). The effects of different environmental factors such as temperature (25, 30, 40, and 50 °C), salinity (1, 3, 6, and 9%), and the exposure time (24, 48, and 72 h) on the depth and diameter of the pits were experimentally assessed. The ASTM G48 guidelines for pitting corrosion tests were modified to create real ocean conditions in the laboratory. Changes were made in the defined temperature, salinity, exposure time, and sample orientation. The experiments were designed by applying a Response Surface Methodology (RSM). A quadratic model was used to identify the influence of temperature, salinity, and exposure time on corrosion on both AISI 304 and AISI 316L steels.

2.1 *Sample Preparation*

In this study, two types of stainless steel materials (AISI 304 & AISI 316L) in the form of a flat bar were used. The surface of the materials was very rough at the beginning. It is widely acknowledged by Chang [34] that surface preparation could play a vital role in removing manganese sulfide (MnS) inclusions, which can reduce steel's vulnerability to pitting corrosion.

Thus, the surface polish must be reached at the mirror polish stage to prevent corrosion. In these experiments, we have utilized wet and dry sandpaper, ranging 80, 180, 240, 350, 400, 600, 800, 1000, 1200, 1400, 1500, 1600, 1800, and 2000 to finish the mirror polish of locally purchased materials AISI 304 and 316L. Both types of materials were cut by using a water jet machine. Factorial analysis was conducted to design the number of experiments. To perform the experiment, some mechanical procedures (e.g., cutting, grinding, and buffing) were accepted. For example, 125 strengthened grinding wheels were used to expel burs, wet and dry sandpaper were used for mirror polishing, and lastly, all the samples were buffed through stainless steel sisal buff and compounded to make it smooth. As illustrated in Fig. 1, the left-hand side figure shows a sample before polishing, and the right-hand side model is the one after mirror polishing. All the test samples were degreased with

Fig. 1 Surface preparation (as received and after polishing)

acetone and washed with ethanol before and after the experiment. After the immersion experiment, the tested samples were immediately kept into the desiccators to avoid environmental contact and a further increase in the corrosion rate. The chemical composition for the pre-owned flat bars is given in Table 1.

2.2 Experimental Setup

The primary purpose of this study is to investigate how different environmental factors, such as salinity, temperature, and exposure time, can affect the corrosion properties in two stainless steel materials. Thus, accelerated pitting corrosion tests were conducted in the lab environment. The ASTM G48 has provided only two types of temperatures, namely 25 and 50 °C. Salinity is 6% only. These temperatures and salinity are not enough to meet real ocean conditions. So, to replicate the testing environment same as ocean condition in terms of salinity and temperature, this study has selected salinity as 1, 3, 6, and 9%, and temperature as 25, 30, 40, and 50 °C. Three exposure times 24, 48, and 72 h are considered.

All the experiments were designed using RSM and adopting D-optimal custom design in design expert software (version 12) [43]. The replicate points were considered as 20 to increase the accuracy of the experiment and the software suggested a quadratic design model. The total number of designed runs was 68 for each material. For the surface orientation, G48-11 [30] suggested a rubber band to hang the testing sample, but Deforce [44] stated that it is a risk if the test sample falls during the test. So, by using a PVC pipe to support the sample, this study has modified the guideline of G48-11 for surface orientation.

As shown in Fig. 2, part A, the test samples were placed on top of the PVC pipe to minimize direct contact with the sample. This setup is placed into the standard

Table 1 Chemical composition of AISI 304 and AISI 316L

Grade		Carbon (C)	Manganese (Mn)	Silicon (Si)	Phosphorus (P)	Sulfur (S)	Chromium (Cr)	Molybdenum (Mo)	Nickel (Ni)	Nitrogen (N)
AISI 304	Min.	–	–	–	–	–	17.5	–	8.0	–
	Max.	0.07	2.0	0.75	0.045	0.030	19.5	–	10.5	0.10
AISI 316L	Min.	–	–	–	–	–	16.0	2.00	10.0	–
	Max.	0.030	2.0	0.75	0.045	0.030	–	3.00	14.0	0.10

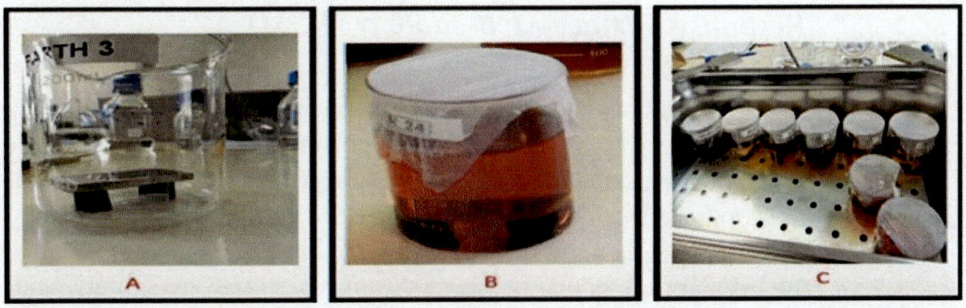

Fig. 2 Experimental setup, sample setup (**a**), immersed sample in glass beaker (**b**), samples are in water bath (**c**)

lab measuring glass beaker. A 200 ml beaker was used for the experiment, which was filled up to 200 ml with the ferric chloride solution, as shown in Fig. 2, part B. All the experiments were conducted using a water bath to maintain the desired temperature, shown in Fig. 2, part C. The solution was poured into the bottle and kept in a water bath until the desired temperature was reached. At this point, the specimen was immersed in the solution.

Surface orientation plays a crucial role in an accelerated pitting corrosion test. We have modified ASTM G48 guidelines for sample orientation in this experiment. Instead of pivoting the test piece through glass wire, we placed the sample on two PVC-corrosion safe columns, as shown in Fig. 2, part A. A preliminary trial was conveyed by pivoting the sample. As this study was focused on one side of the sample to observe the pits, the opposite side was not prepared and analyzed.

Even though pH is an essential factor in the pitting corrosion technique, ASTM G48 has not given any procedure to prepare the $FeCl_3$ solution. Numerous research studies ([22, 45]) have been conducted to assess the impacts of pH on pitting corrosion in marine environments, and results show that a more aggressive condition that usually has lower pH (high salinity) could increase the pitting corrosion rate. The ASTM G48 guideline proposes pH for test solution to be between 1 and 2. In these experiments, pH values were 2.03, 1.53, 1.32, and 1.13 observed for 1, 3, 6, and 9% salinity accordingly. Before and after each experiment, the pH was measured. The variation of pH before and after the experiments were ± 0.2.

After a specified emersion time, each sample was carefully taken out, washed with water, cleaned with a nylon brush, and dipped into the acetone before drying at normal room temperature. This procedure can help to remove surface corrosion products and make samples clean. Thus, the subsequent analysis of the samples can be conducted appropriately.

2.3 *Response Surface Methodology (RSM)*

The RSM is the highly recommended method for many engineering fields such as materials engineering, food engineering, chemical engineering, bioprocess engineering, and pharmaceutical engineering to analyze the influence of each parameter and interactions between them [46]. Using RSM in the quadratic model can limit the number of experiments, reduce process variability and duration, and save overall cost. The key parameters that influence the corrosion rate are salinity, temperature, and exposure time. A quadratic model was adopted to study the combined effects of these parameters.

Analysis of variance (ANOVA) was applied to evaluate the factors that significantly affect pitting corrosion. Its *p*-value can determine the influence of each parameter. Statistics from ANOVA (Fig. 3) elucidate that among all the factors, pH has the highest impact on pit depth as its *p*-value is less than 0.05. Other factors, such as temperature, have a *p*-value of 0.07, and exposure time has a *p*-value of 0.39 for AISI 304. Moreover, for the AISI 316L samples, pH, temperature, and exposure time have *p*-values of 0.0001, 0.306, and 0.804, respectively.

The relationship between pH and exposure time has the strongest influence than those of pH and temperature, temperature and exposure time for AISI 304. Because pH and exposure time have the lowest *p*-value, whereas the relationship between temperature and exposure time has the least influence. On the other hand, the relationship between pH and temperature dramatically influences the corrosion rate on AISI 316L compared to temperature, exposure time, and pH, and exposure time. The

Source	Sum of Squares	Df	Mean Square	F- Value	p- Value	
Model	8.01	9	0.8895	5	<0.0001	significant
A- pH	5.32	1	5.32	29.9	<0.0001	
B- Temperature	0.5722	1	0.5722	3.21	0.0782	
C- Exposure Time	0.1322	1	0.1322	0.7428	0.3923	
AB	0.2458	1	0.2458	1.38	0.2448	
AC	0.6884	1	0.6884	3.87	0.054	
BC	0.0202	1	0.0202	0.1135	0.7375	
Residual	10.33	58	0.178			
Lack of fit	7.83	38	0.2061	1.65	0.1154	Not significant
Pure Error	2.49	20	0.1246			
Cor Total	18.33	67				

Std. Dev.	0.4219	R^2	0.43367	ANOVA test for Quadratic Model and Fit statistics for AISI 304 (Maximum pit depth)
Mean	0.5697	Adjusted R^2	0.3493	
C.V. %	74.07	Predicted R^2	0.2695	
		Adequate Precision	8.6315	

Source	Sum of Squares	Df	Mean Square	F- Value	p- Value	
Model	24.55	9	2.73	8.07	<0.0001	significant
A- pH	19.24	1	19.24	56.96	<0.0001	
B- Temperature	0.3589	1	0.3589	1.06	0.3069	
C- Exposure Time	0.0208	1	0.0208	0.0616	0.8048	
AB	0.3539	1	0.3539	1.05	0.3103	
AC	0.0071	1	0.0071	0.021	0.8854	
BC	0.0658	1	0.0658	0.1947	0.6607	
Residual	19.59	58	0.3377			
Lack of fit	11.91	38	0.3135	0.8166	0.7121	Not significant
Pure Error	7.68	20	0.3839			
Cor Total	44.13	67				

Std. Dev.	0.5812	R^2	0.5561	ANOVA test for Quadratic Model and Fit statistics for AISI 316L (Maximum pit depth)
Mean	1	Adjusted R^2	0.4873	
C.V. %	57.98	Predicted R^2	0.4115	
		Adequate Precision	9.7305	

Fig. 3 ANOVA tests for AISI 304 and AISI 316L in design expert software [43]

relationship between pH and exposure time has the lowest impact on corrosion for AISI 316L.

The information on each factor's influence and interaction effects could not be predicted if this study had used the conventional method to design experiments. The analysis of the coefficient of determination (R^2) depicts that the RSM approach was appropriate for this study. The coefficient of determination (R^2) was 0.4367 and 0.5561 for AISI 304 and AISI 316L, respectively, representing the correlation between observed and predicted values.

2.4 Pit Observation

A metallurgical microscope (ME300TZB-2L-14 M) was utilized to quantify pitting attributes such as depth, diameter, and the number of pits. To quantify pitting depth, the fine-focus method was adopted. Perusing the top surface of the sample and the base of pitting were recorded using a computerized dial gauge. The contrast between these two peruses was considered a pitting depth. All the images were taken at 50× magnification, and the diameter was determined by utilizing circular, two-point, or three-point circle tools of metallurgical microscope (ME300TZB-2L-14 M).

As demonstrated in Fig. 4, section A, firstly surface of pits was captured and noted the reading in dial gauge. After that, pit depth was captured, as shown in Fig. 4, section B. This technique is called "Fine Focus." Section C shows a close look inside the pit.

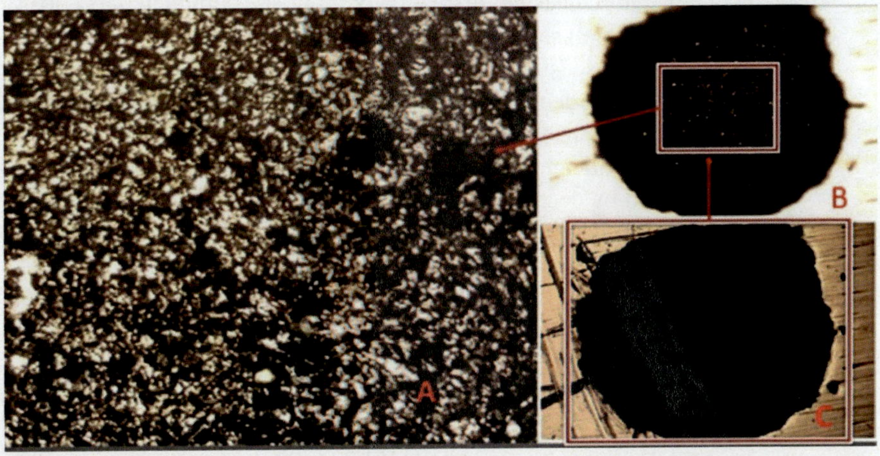

Fig. 4 Microscopic pit observation, surface of depth (**a**), depth of pit (**b**), and close look of depth (**c**)

3 Results and Discussion

3.1 Effect of Interactions Relationship Among the Factors on Pit Depth and Diameter

From the experimental data and results given in Table 2, it is evident that pitting resistant capacity is higher in AISI 316L samples than in AISI 304 in terms of the number of pits per sample due to having a good corrosion resistant capacity, as acknowledged by Ge [47].

However, the maximum pit depth is 3.265 mm in AISI 316L with an exposure time of 24 h, a temperature of 30 °C, and a 9% salinity. On the other hand, with the similar environment, AISI 304 has 0.075 mm depth (as given in Table 2). The uneven surface could be a reason for an unexpected result. Moreover, AISI 304 samples have lesser pit depth and diameter in almost all the samples than AISI 316L. However, the number of pits in all the samples are higher in AISI 304 than those of AISI 316L.

It should be noted that surface preparation plays a significant role in pitting potential, which was explained by Burstein [28] and G48-11 [30].

Figure 5 illustrates the exposure time and pH relationship impact on pit depth and AISI 304 material diameter. Pit depth is 1 mm and diameter is 1.2 mm at the highest exposure time. It is observed from Fig. 5 that pH does not significantly impact pit depth and diameter, while depth and diameter both steadily increase by increasing exposure time.

Figures 6 and 9 depict the relationship between exposure time and temperature for AISI 304 and AISI 316L. It is visible that the interaction between exposure time and the temperature does not dramatically influence both materials corrosion property. ANOVA (Fig. 3) shows that this interaction does have a high *p*-value for both materials. As we discussed in RSM, a high *p*-value means less impact on results.

Figures 7 and 10 demonstrate the interactions between temperature and pH for both materials. Figures 6 and 9 show that the pit diameter grew continuously in both materials. While pit depth is also increasing by raising temperature and salinity and is decreased by decreasing the salinity level. As a result, it could be concluded that temperature and pH significantly impact the pitting depth and diameter in both considered materials. According to Kolawole [48] low pH (high salinity) means an acidic environment, which significantly causes the pitting than high pH (less salinity). By reducing pH value, there would be a high likelihood of increasing the corrosion rate.

Figure 8 illustrates a similar relationship for the AISI 316L as that of AISI 304. Exposure time and pH both significantly influence the pitting depth and diameter for the AISI 316L than those of the AISI 304. Pit depth and diameter both increase with the increase of exposure time and pH. It could be concluded from Fig. 8 that the interaction between exposure time and pH highly influences the AISI 316L corrosion.

Besides, G48-11 [30] and Hoseinpoor [49] mentioned the minimum depth at 40 °C and 24 h of exposure time should be 0.025 mm, while in this experiment, it

Table 2 Experimental data and results

S. No.	pH (%)	Temp. (°C)	Exposure time (h)	AISI 304			AISI 316L		
				No. of pits	Max. pit diameter (mm)	Max. pit depth (mm)	No. of pits	Max. pit diameter (mm)	Max. pit depth (mm)
1	9	25	24	6	0.877	0.411	5	3.000	1.199
2	9	25	24	4	1.736	0.816	9	2.399	1.728
3	3	25	24	1	0.147	0.154	1	0.342	0.667
4	6	25	24	2	1.404	0.716	5	1.514	0.995
5	1	25	24	1	0.327	0.084	2	0.855	0.936
6	1	25	48	3	0.157	0.065	1	0.553	0.038
7	6	25	48	6	0.273	0.075	6	3.430	2.391
8	1	25	48	2	0.290	0.074	1	0.074	0.011
9	3	25	48	8	0.206	0.110	1	0.095	0.035
10	6	25	48	7	1.827	0.846	2	2.115	1.908
11	9	25	48	6	1.201	0.795	3	1.397	2.732
12	3	25	72	2	0.349	0.079	1	0.439	0.915
13	1	25	72	5	0.043	0.068	1	0.505	0.067
14	1	25	72	1	0.622	0.054	1	0.700	0.77
15	9	25	72	6	1.170	1.275	1	1.850	1.189
16	9	25	72	6	1.120	0.977	4	2.585	1.919
17	6	25	72	7	0.661	0.805	1	2.367	1.587
18	1	30	24	1	0.123	0.032	1	0.528	0.528
19	3	30	24	2	0.364	0.027	1	0.446	1.341
20	1	30	24	1	0.195	0.026	1	0.219	0.016

(continued)

Table 2 (continued)

S. No.	pH (%)	Temp. (°C)	Exposure time (h)	AISI 304			AISI 316L		
				No. of pits	Max. pit diameter (mm)	Max. pit depth (mm)	No. of pits	Max. pit diameter (mm)	Max. pit depth (mm)
21	9	30	24	3	0.170	0.075	1	1.647	3.265
22	3	30	24	1	0.110	0.012	1	0.223	0.012
23	6	30	24	3	0.608	0.542	1	0.909	0.349
24	6	30	48	2	1.994	0.617	2	2.130	1.598
25	1	30	48	1	0.393	0.387	1	0.195	0.137
26	1	30	48	1	0.395	0.389	1	0.531	0.01
27	9	30	48	13	1.508	0.813	5	3.291	1.821
28	6	30	48	3	1.708	0.735	2	2.201	1.428
29	3	30	48	2	0.101	2.173	1	0.931	1.287
30	6	30	72	1	0.666	0.311	5	2.345	2.237
31	9	30	72	18	1.740	0.829	4	3.216	1.971
32	3	30	72	10	0.158	0.041	2	1.943	1.459
33	1	30	72	1	0.564	0.023	1	0.166	0.025
34	1	40	24	2	0.436	0.301	1	0.355	0.014
35	1	40	24	1	0.440	0.341	1	0.375	0.064
36	3	40	24	3	0.847	0.986	1	0.686	0.791
37	3	40	24	3	0.936	0.813	1	0.102	0.067
38	6	40	24	3	1.318	0.643	3	2.014	0.946
39	9	40	24	25	1.362	0.637	4	2.109	1.153
40	3	40	48	5	0.660	1.196	4	1.402	2.808
41	9	40	48	10	1.664	0.241	4	2.280	1.492

(continued)

Table 2 (continued)

S. No.	pH (%)	Temp. (°C)	Exposure time (h)	AISI 304			AISI 316L		
				No. of pits	Max. pit diameter (mm)	Max. pit depth (mm)	No. of pits	Max. pit diameter (mm)	Max. pit depth (mm)
42	6	40	48	5	0.836	1.189	2	1.402	1.036
43	9	40	48	18	1.511	0.964	5	3.477	1.682
44	3	40	48	3	1.292	0.765	1	1.055	0.1
45	1	40	48	2	0.212	0.023	1	0.213	0.015
46	9	40	72	16	1.131	0.816	3	1.113	0.745
47	1	40	72	7	0.133	0.040	1	0.117	0.042
48	6	40	72	5	2.230	1.193	2	0.616	0.381
49	9	40	72	13	1.644	0.901	5	2.697	1.731
50	1	40	72	1	0.073	0.002	1	0.183	0.017
51	3	40	72	1	0.412	0.002	2	0.441	0.177
52	9	50	24	14	1.695	0.936	2	1.803	1.316
53	9	50	24	4	1.597	0.748	4	2.508	1.585
54	6	50	24	4	1.415	0.149	4	1.588	0.912
55	6	50	24	4	1.269	1.169	3	1.618	1.189
56	1	50	24	1	0.264	0.139	1	0.125	0.025
57	3	50	24	6	1.060	0.992	2	1.582	0.653
58	9	50	48	19	2.603	1.511	2	1.730	0.944
59	1	50	48	3	0.450	0.065	1	0.796	0.37
60	3	50	48	3	0.487	0.112	1	0.832	1.246
61	3	50	48	5	0.920	0.162	1	0.728	2.029
62	9	50	48	14	1.539	0.774	3	2.073	2.394

(continued)

Table 2 (continued)

S. No.	pH (%)	Temp. (°C)	Exposure time (h)	AISI 304			AISI 316L		
				No. of pits	Max. pit diameter (mm)	Max. pit depth (mm)	No. of pits	Max. pit diameter (mm)	Max. pit depth (mm)
63	6	50	48	6	3.062	0.914	3	1.347	1.238
64	6	50	72	9	0.814	2.262	2	1.571	1.185
65	6	50	72	11	1.049	0.938	2	1.148	1.269
66	3	50	72	8	1.435	0.729	1	1.731	0.901
67	1	50	72	2	0.287	0.032	1	0.279	0.039
68	9	50	72	25	1.330	1.617	3	1.970	1.03

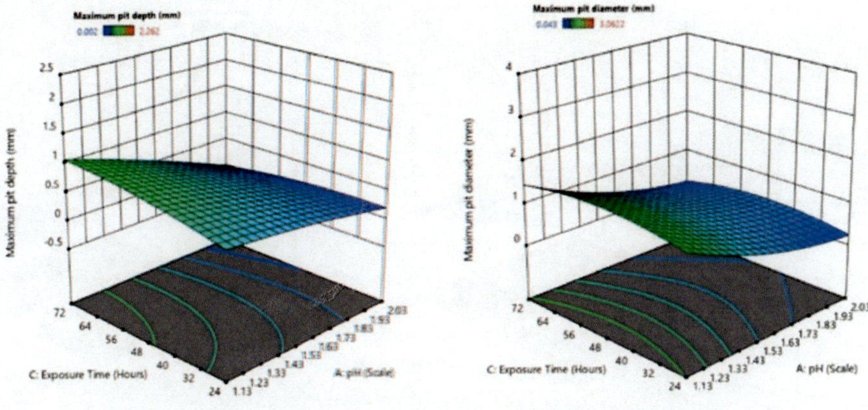

Fig. 5 The variation of exposure time verses pH (pitting depth and diameter) (AISI 304)

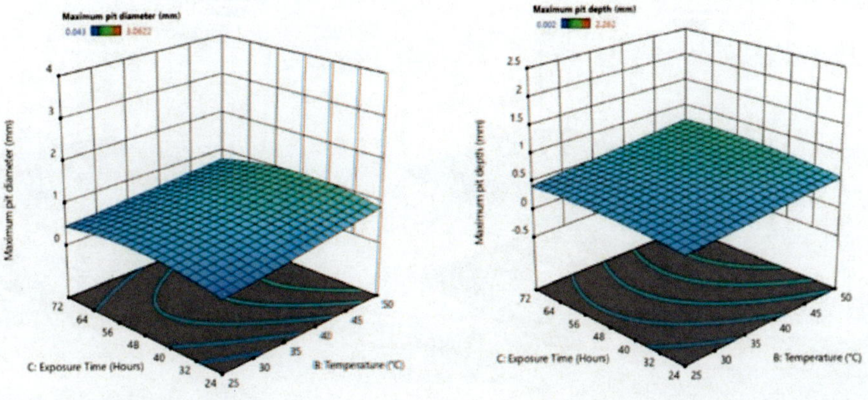

Fig. 6 The variation of exposure time verses temperature (Pitting depth and diameter) (AISI 304)

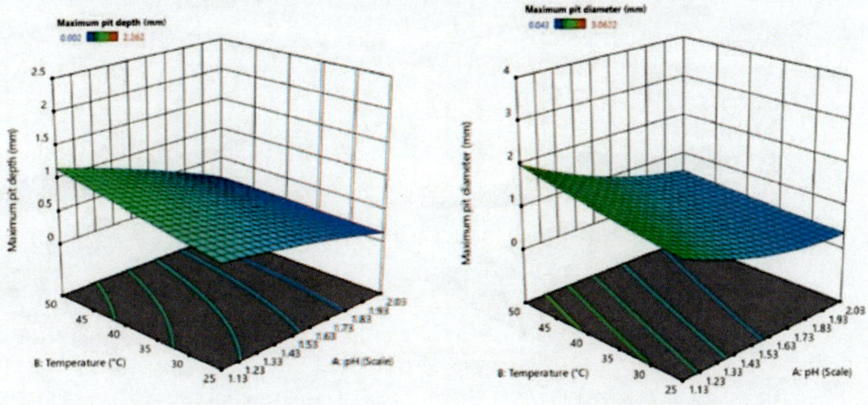

Fig. 7 The variation of temperature verses pH (pitting depth and diameter) (AISI 304)

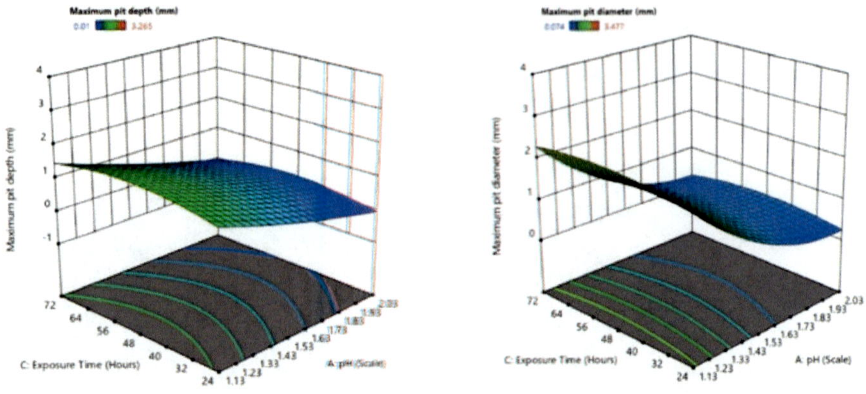

Fig. 8 The variation of exposure time verses pH (pitting depth and diameter) (AISI 316L)

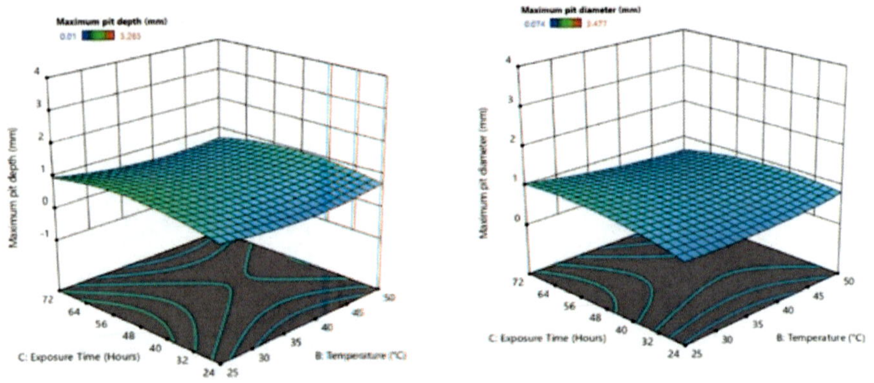

Fig. 9 The variation of exposure time verses temperature (pitting depth and diameter) (AISI 316L)

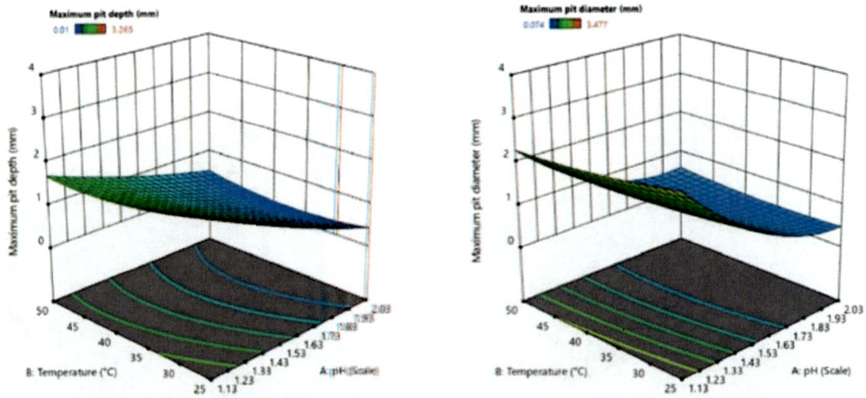

Fig. 10 The variation of temperature versus pH (pitting depth and diameter) (AISI 316L)

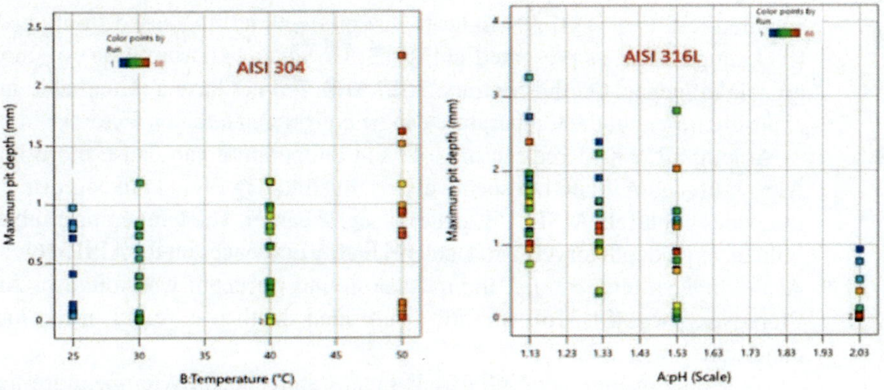

Fig. 11 The variation of temperature verses depth (AISI 304) and pH verses depth (AISI 316L)

was 0.643 mm for AISI 304 and 0.946 mm for AISI 316L, which is much higher than expected. Surface roughness could be a reason to have different results.

The number of pit is more in AISI 304 than that of AISI 316L. Hence, it can be concluded that AISI 316L has higher pitting resistant capacity than AISI 304 in all environments considered in this study.

3.2 *Effect of an Individual Factor on Pit Depth and Diameter*

According to the acquired results, exposure time does not have a remarkable impact on pitting depth and diameter than temperature and salinity. Temperature and salinity significantly influence the depth of pits in AISI 304 and AISI 316L, as illustrated in Fig. 11. Moreover, surface homogeneity, pitting potential, and resistance of materials and passivity can also influence the corrosion rate. Results elucidate that both materials have the highest number of pits at 50 °C and lowest at 25 °C, as illustrated in Fig. 11. This means passivity breakdown is strongly influenced by pH and temperature compared to exposure time, especially between 20 and 40 °C, which agrees well with Dastgerdi [16].

All the experimental factors considered in this study have an impact on pitting corrosion. Pit depth was increased when the salinity level was increased (shown in Fig. 11). A 0.5 mm increment in depth could be noted from a low level to a high level of salinity. For 1, 3, 6, and 9% salinity, pH was 2.03, 1.53, 1.32, and 1.13 sequentially. As a result, low pH has more Cl-ions and starts the penetrating process after breaking the passivity layer, which causes pitting corrosion. According to Figs. 5 and 8, exposure time did not significantly affect pit depth even the salinity level was increased for both AISI 304 and AISI 316L. A low pH level (1.13) causes deeper pit depth in AISI 316L than that of AISI 304, as demonstrated in Fig. 11. The acquired results of AISI 304 indicate that pit depth has increased from low to high-level temperature, as presented in Fig. 11. However, exposure time does not have a notable influence. On the contrary, AISI 316L did not have a remarkable impact of exposure time, but the temperature can be considered a major factor.

Among all considered factors, pH and temperature can cause more impact on AISI 316L and AISI 304, respectively, as illustrated in Fig. 11. In contrast, exposure time does not affect AISI 316L material significantly. The temperature substantially influences pit depth for AISI 304, and pH highly influences pit for AISI 316L material. At the highest temperature, the most profound pit depth was found in AISI 304. Moreover, for AISI 316L, the most profound depth was found in the lowest pH condition.

As shown in Fig. 12, AISI 304 is greatly influenced by temperature than AISI 316L. From 25 to 50 °C, pitting depth is gradually increasing in AISI 304 except at 40 °C, while salinity was varied. It could be clear from Fig. 12 that at each considered temperature, the AISI 316L is more pitting resistant than AISI 304. However, as shown in Table 3, both materials' deepest pit was found at 50 °C. It has been explained by Dastgerdi [16] that temperature can easily accelerate pit initiation.

As illustrated in Fig. 13, AISI 316L is significantly impacted by salinity (confirmed by Fig. 11), whereas there was no notable impact on AISI 304 samples. With increasing salinity, pit depth is significantly enhanced in the AISI 316L. As explained

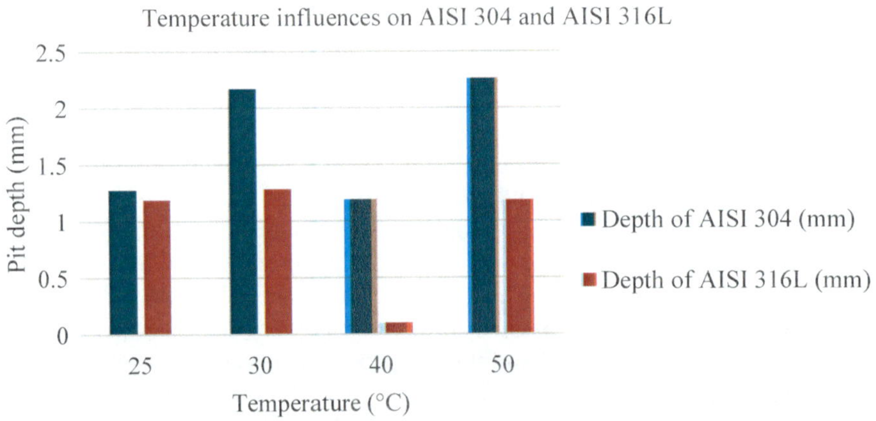

Fig. 12 Temperature influences on AISI 304 and AISI 316L

Table 3 Temperature influences on AISI 304 and AISI 316L

Temperature °C	Salinity (%)	Exposure time (h)	Depth of AISI 304	Depth of AISI 316L
25	9	72	1.275	1.189
30	6	48	2.173	1.287
40	6	48	1.196	0.101
50	3	72	2.262	1.185

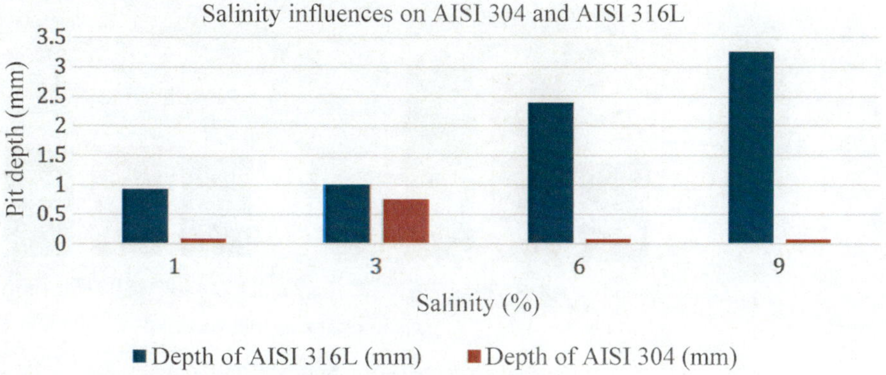

Fig. 13 Salinity influences on AISI 304 and AISI 316L

Table 4 Salinity influences on AISI 304 and AISI 316L

Salinity (%)	Temperature °C	Exposure time (h)	Depth of AISI 316L	Depth of AISI 304
1	25	24	0.936	0.084
3	40	48	1.01	0.765
6	25	48	2.391	0.075
9	30	24	3.265	0.075

in Table 4, the maximum pit depth is 3.265 mm in AISI 316L, where salinity is 9%. However, the maximum pit depth in AISI 304 is 0.765 mm, where salinity is 3%, and the temperature is 40 °C. So, it is concluded that temperature can highly influence AISI 304 (demonstrated in Figs. 11, 12, and Table 3), and salinity has an enormous impact on AISI 316L (seen in Figs. 11, 13, and Table 3) in terms of pitting depth. Results agreed with Malik [50], in which they revealed that AISI 316L has the lowest number of pits at low pH and the highest number of pits at high pH. The study concluded that an environment with low pH and high Cl^- is the most favorable condition to initiate and propagate pits on AISI 316L. Wang [51] stated that an increase in temperature could increase the porosity of passive film, and the chemical composition and physical structure of the passive film are strongly affected by the increasing temperature of AISI 304.

As shown in Fig. 14, exposure time significantly influences the pitting depth on AISI 304 than that of AISI 316L. According to Table 5, the deepest pit depth has been found to be 2.262 mm for AISI 304 at 72 h of exposure time. However, AISI 316L has substantially the same depth in different exposure times.

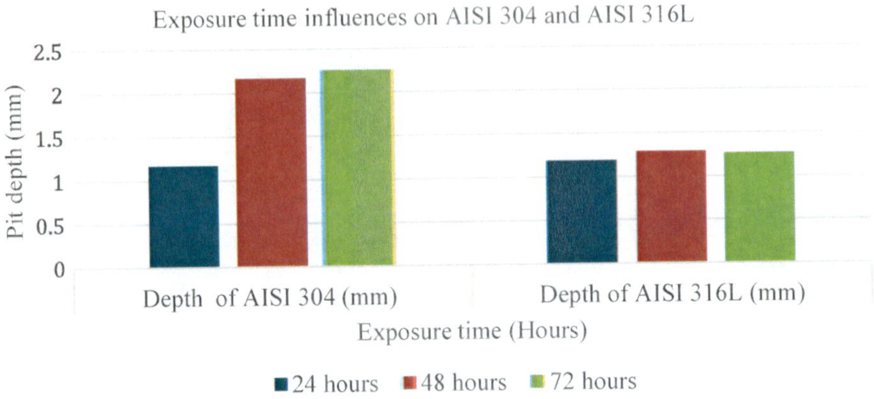

Fig. 14 Exposure time influences on AISI 304 and AISI 316L

Table 5 Exposure time influences on AISI 304 and AISI 316L

Exposure time (h)	Temperature °C	Salinity (%)	Depth of AISI 304	Depth of AISI 316L
24	50	3	1.169	1.189
48	30	6	2.173	1.287
72	50	6	2.262	1.269

4 Conclusions

To demonstrate the pitting susceptibility of marine offshore structural materials (AISI 304 and AISI 316L), accelerated laboratory tests were conducted by following ASTM G48 guidelines with modification related to orientation, salinity, temperature, and exposure time. This study compares AISI 304 and AISI 316L materials under the same salinity conditions, surface preparation, and temperature. Response surface methodology (RSM) study, particularly the quadratic model, was used to analyze the effect of each parameter on corrosion and interactions between them. Results were analyzed by utilizing the quadratic model. From the results achieved in this study, the following findings are highlighted.

(1) The corrosion rate is high for AISI 316L in a high salinity environment. However, AISI 304 has fluctuating salinity impacts as the maximum pit depth is found at 3% salinity.

(2) According to the ANOVA test and experimental results, the temperature is the second major factor that influences the corrosion rate. Especially for AISI 304, the number of pits significantly was increased by increasing the temperature from 25 to 50 °C. The numbers of pits were high at 25 and 30 °C, but low at 40 and 50 °C for AISI 316L. Thus, AISI 316L steel has more pitting potential than AISI 304 at higher temperatures.

(3) The temperature did not significantly influence the corrosion rate on AISI 316L compared to AISI 304. The most profound depth was found at 72 h of exposure time and the lowest at 24 h for AISI 304; however, the temperature does not have a remarkable impact on AISI 316L.

Overall, AISI 316L has more pitting resistant capacity than AISI 304 in terms of the number of pits in every environment considered in the current study. Influence of surface roughness and hydrostatic pressure on pitting corrosion rate can be regarded as the future direction of this study.

Acknowledgements The authors would like to acknowledge the financial and technical support received from the Australian Maritime College (AMC) of the University of Tasmania for the project.

References

1. Duan, T., Peng, W., Ding, K., Guo, W., Hou, J., Cheng, W., Liu, S., & Xu, L. (2019). Long-term field exposure corrosion behavior investigation of 316L stainless steel in the deep sea environment. *Ocean Engineering, 189*, 106405.
2. Canepa, E., Stifanese, R., Merotto, L., & Traverso, P. (2018). Corrosion behaviour of aluminium alloys in deep-sea environment: A review and the KM3NeT test results. *Marine Structures, 59*, 271–284.
3. Luciano, G., Letardi, P., Traverso, P., & Belsanti, L. (2013). *Corrosion behaviour of Al, Cu, and Fe alloys in deep sea environment.* La Metallurgia Italiana.
4. Venkatesan, R., Venkatasamy, M. A., Bhaskaran, T. A., Dwarakadasa, E. S., & Ravindran, M. (2002). Corrosion of ferrous alloys in deep sea environments. *British Corrosion Journal, 37*(4), 257–266.
5. Melchers, R. E. (1999). Corrosion uncertainty modelling for steel structures. *Journal of Constructional Steel Research, 52*(1), 3–19.
6. Melchers, R. E. (2020). Long-term durability of marine reinforced concrete structures. *Journal of Marine Science and Engineering, 8*(4), 290.
7. Rohling, E. J., Abu-Zied, R. H., Casford, J. S. L., Hayes, A., & Hoogakker, B. A. A. (2009). The marine environment: Present and past. In *The physical geography of the Mediterranean* (pp. 33–67).
8. Burstein, G. T., Liu, C., Souto, R. M., & Vines, S. P. (2004). Origins of pitting corrosion. *Corrosion Engineering, Science and Technology, 39*(1), 25–30.
9. Wang, Y., Cheng, G., Wu, W., Qiao, Q., Li, Y., & Li, X. (2015). effect of pH and chloride on the micro-mechanism of pitting corrosion for high strength pipeline steel in aerated NaCl solutions. *Applied Surface Science, 349*, 746–756.
10. Wang, Y., Cheng, G., Wu, W., & Li, Y. (2018). Role of inclusions in the pitting initiation of pipeline steel and the effect of electron irradiation in SEM. *Corrosion Science, 130*, 252–260.
11. Frankel, G. (1998). Pitting corrosion of metals a review of the critical factors. *Journal of the Electrochemical Society, 145*(6), 2186–2198.
12. Bhandari, J., Khan, F., Abbassi, R., Garaniya, V., & Ojeda, R. (2015). Modelling of pitting corrosion in marine and offshore steel structures—A technical review. *Journal of Loss Prevention in the Process Industries, 37*, 39–62.
13. Burstein, G. T., Liu, C., Souto, R. M., & Vines, S. P. (2013). Origins of pitting corrosion. *Corrosion Engineering, Science and Technology, 39*(1), 25–30.
14. Sharland, S. (1987). A review of the theoretical modelling of crevice and pitting corrosion. *Corrosion Science, 27*(3), 289–323.

15. Zhang, B., & Ma, X. (2019). A review-pitting corrosion initiation investigated by TEM. *Journal of Materials Science & Technology*.
16. Dastgerdi, A. A., Brenna, A., Ormellese, M., Pedeferri, M., & Bolzoni, F. (2019). Experimental design to study the influence of temperature, pH, and chloride concentration on the pitting and crevice corrosion of UNS S30403 stainless steel. *Corrosion Science, 159*, 108160.
17. Ryan, M. P., Williams, D. E., Chater, R. J., Hutton, B. M., & McPhail, D. S. (2002). Why stainless steel corrodes. *Nature, 415*(6873), 770–774.
18. Nishimoto, K., & Ogawa, K. (1999). Corrosion properties in weldments of stainless steels (1). Metallurgical factors affecting corrosion properties. *Welding International, 13*(11), 845–854.
19. Gooch, T. (1996). Corrosion behavior of welded stainless steel. *Welding Journal-Including Welding Research Supplement, 75*(5), 135s.
20. Lu, B. T., Chen, Z. K., Luo, J. L., Patchett, B. M., & Xu, Z. H. (2005). Pitting and stress corrosion cracking behavior in welded austenitic stainless steel. *Electrochimica Acta, 50*(6), 1391–1403.
21. Do Nascimento, A., Ierardi, M. C. F., Kina, A. Y., & Tavares, S. S. M. (2008). Pitting corrosion resistance of cast duplex stainless steels in 3.5% NaCl solution. *Materials Characterization, 59*(12), 1736–1740.
22. Pardo, A., Otero, E., Merino, M. C., López, M. D., Utrilla, M. V., & Moreno, F. (2000). Influence of pH and chloride concentration on the pitting and crevice corrosion behavior of high-alloy stainless steels. *Corrosion, 56*(4), 411–418.
23. Kamachi Mudali, U., Shankar, P., Ningshen, S., Dayal, R. K., Khatak, H. S., & Raj, B. (2002). On the pitting corrosion resistance of nitrogen alloyed cold worked austenitic stainless steels. *Corrosion Science, 44*(10), 2183–2198.
24. Szklarska-Smialowska, Z. (1971). Review of literature on pitting corrosion published since 1960. *Corrosion, 27*(6), 223–233.
25. Sedriks, A., Schultz, J., & Cordovi, M. (1979). Inconel alloy 690-A new corrosion resistant material. *防食技術, 28*(2), 82–95.
26. Klapper, H. S., Goellner, J., Burkert, A., & Heyn, A. (2013). Environmental factors affecting pitting corrosion of type 304 stainless steel investigated by electrochemical noise measurements under potentiostatic control. *Corrosion Science, 75*, 239–247.
27. Köliö, A., Pakkala, T. A., Annila, P. J., Lahdensivu, J., & Pentti, M. (2014). Possibilities to validate design models for corrosion in carbonated concrete using condition assessment data. *Engineering Structures, 75*, 539–549.
28. Burstein, G., & Pistorius, P. (1995). Surface roughness and the metastable pitting of stainless steel in chloride solutions. *Corrosion, 51*(5), 380–385.
29. Mathiesen, T., & Anderko, A. (2014). *Challenges in pre-qualification corrosion testing of CRAs based on ASTM G48*. NACE Corrosion.
30. G48-11, A. (2009). *Standard test methods for pitting and crevice corrosion resistance of stainless steels and related alloys by use of ferric chloride solution*.
31. Bhandari, J., Lau, S., Abbassi, R., Garaniya, V., Ojeda, R., Lisson, D., & Khan, F. (2017). Accelerated pitting corrosion test of 304 stainless steel using ASTM G48: Experimental investigation and concomitant challenges. *Journal of Loss Prevention in the Process Industries, 47*, 10–21.
32. Chang, L., Jonathan Duff, M., Burke, G., & Scenini, F. (2019). SCC initiation in the machined austenitic stainless steel 316L in simulated PWR primary water. In *Proceedings of the 18th International Conference on Environmental Degradation of Materials in Nuclear Power Systems–Water Reactors*. Springer.
33. Cissé, S., Laffont, L., Tanguy, B., Lafont, M.-C., & Andrieu, E. (2012). Effect of surface preparation on the corrosion of austenitic stainless steel 304L in high temperature steam and simulated PWR primary water. *Corrosion Science, 56*, 209–216.
34. Chang, L., Burke, M. G., & Scenini, F. (2019). Understanding the effect of surface finish on stress corrosion crack initiation in warm-forged stainless steel 304L in high-temperature water. *Scripta Materialia, 164*, 1–5.

35. Corbett, R. (1992). *Problems in utilizing ASTM G 48 to evaluate high-alloy stainless steels*. R. A. Corbett, Paper (298).
36. Lozano-Perez, S., Lindsay, J., Chang, L., Wang, Y. L., Burke, M. G., Pimentel, G., Tice, D., Mottershead, K., & Addepalli, V. (2017). *Oxidation and SCC initiation studies of type 304L SS in PWR primary water*.
37. Szklarska-Smialowska, Z. (1986). *Pitting corrosion of metals*.
38. Zhang, Z., Tan, J., Wu, X., Han, E.-H., Ke, W., & Rao, J. (2019). Effects of temperature on corrosion fatigue behavior of 316LN stainless steel in high-temperature pressurized water. *Corrosion Science, 146*, 80–89.
39. Otero, E., Pardo, A., Utrilla, M. V., Sáenz, E., & Perez, F. J. (1995). Influence of microstructure on the corrosion resistance of AISI type 304L and type 316L sintered stainless steels exposed to ferric chloride solution. *Materials Characterization, 35*(3), 145–151.
40. Byrne, G., Warburton, G., Wilson, J., & Francis, R. (2011). Fabrication of superduplex stainless steel for optimum seawater corrosion resistance. In *World Congress/Perth Convention and Exhibition Centre (PCEC), Perth, Western Australia September*.
41. Prawoto, Y., Ibrahim, K., & Wan Nik, W. (2009). Effect of pH and chloride concentration on the corrosion of duplex stainless steel. *Arabian Journal for Science and Engineering, 34*(2), 115.
42. Buscail, H., El Messki, S., Riffard, F., Perrier, S., & Issartel, C. (2011). effect of pre-oxidation at 800 °C on the pitting corrosion resistance of the AISI 316L stainless steel. *Oxidation of Metals, 75*(1–2), 27–39.
43. Statease. (2020). *Design expert*.
44. DeForce, B. S. (2016). *Comments on ASTM G48-Standard Test Methods for Pitting and Crevice Corrosion Resistance of Stainless Steels and Related Alloys by Use of Ferric Chloride Solution*. in *NACE International Corrosion Conference Proceedings*. NACE International.
45. Malik, A. U., Ahmad, S., Andijani, I., & Al-Fouzan, S. (1999). Corrosion behavior of steels in Gulf seawater environment. *Desalination, 123*(2–3), 205–213.
46. Goh, K.-H., Lim, T.-T., & Chui, P.-C. (2008). Evaluation of the effect of dosage, pH and contact time on high-dose phosphate inhibition for copper corrosion control using response surface methodology (RSM). *Corrosion Science, 50*(4), 918–927.
47. Ge, H., Zhou, G., & Xie, Q. (2005). Comparison of anti-corrosion behavior to chloride ions and sulfide ions between 304 and 316L stainless steel. *East China Electric Power, 9*.
48. Kolawole, S. K., Kolawole, F. O., Enegela, O. P., Adewoye, O. O., Soboyejo, A. B. O., & Soboyejo, W. O. (2016). Pitting corrosion of a low carbon steel in corrosive environments: Experiments and models. *Advanced Materials Research*.
49. Hoseinpoor, M., Momeni, M., Moayed, M. H., & Davoodi, A. (2014). EIS assessment of critical pitting temperature of 2205 duplex stainless steel in acidified ferric chloride solution. *Corrosion Science, 80*, 197–204.
50. Malik, A. U., Mayan Kutty, P. C., Siddiqi, N. A., Andijani, I. N., & Ahmed, S. (1992). The Influence of pH and chloride concentration on the corrosion behaviour of AISI 316L steel in aqueous solutions. *Corrosion Science, 33*(11), 1809–1827.
51. Wang, J., Su, C., & Szklarska-Smialowska, Z. (1988). Effects of Cl− concentration and temperature on pitting of AISI 304 stainless steel. *Corrosion, 44*(10), 732–737.

Prioritisation of Safety Factors in a Refinery by Intuitionistic Fuzzy Analytic Hierarchy Process

G. Suresh, V. R. Renjth, and A. B. Bhasi

1 Introduction

Accident prevention is the main activity of all workplace safety programs. Analysing the cause of the accident is the first step in preventing their reoccurrences. There are several accident causation theories like "The Domino theory, Human factor theory, Accident/Incident theory, Systems theory, Combination theory". When we look into the various factors which cause accidents/incidents in an industry with time, we can understand that the approaches in industrial safety move from loss prevention of assets, including human to organisational and human activities [1] (Fig. 1). It is beneficial to categorise the factors under different heads like human factors, organisational factors and technical or engineering factors.

In almost all industries, there are reportable as well as non-reportable incidents. Analysing past incidents in an organisation highlights the areas of weakness in that particular industry. This paper analyses incidents at different refinery plants to identify the weak areas or factors.

The multi-criteria decision-making process (MCDM) can be effectively utilised for prioritising safety factors. The analytical hierarchy process (AHP) possesses the unique advantage of comparing parameters with no units or scale of measurements. The intuitionistic fuzzy analytical hierarchy process (IF-AHP) further improves the AHP in terms of vagueness, uncertainty and handles the imprecise data. The IF-AHP is mainly applied to prioritise an alternative through highly complex criteria. The IF-AHP demonstrates strong decision-making capability in highly complex domains.

G. Suresh (✉) · V. R. Renjth
Safety and Fire Engineering, School of Engineering, Cochin University of Science and Technology, Kochi, India
e-mail: sureshkankhal@gmail.com

A. B. Bhasi
Mechanical Engineering, School of Engineering, Cochin University of Science and Technology, Kochi, India

N. A. Siddiqui et al. (eds.), *Advances in Construction Safety*,
https://doi.org/10.1007/978-981-19-4001-9_7

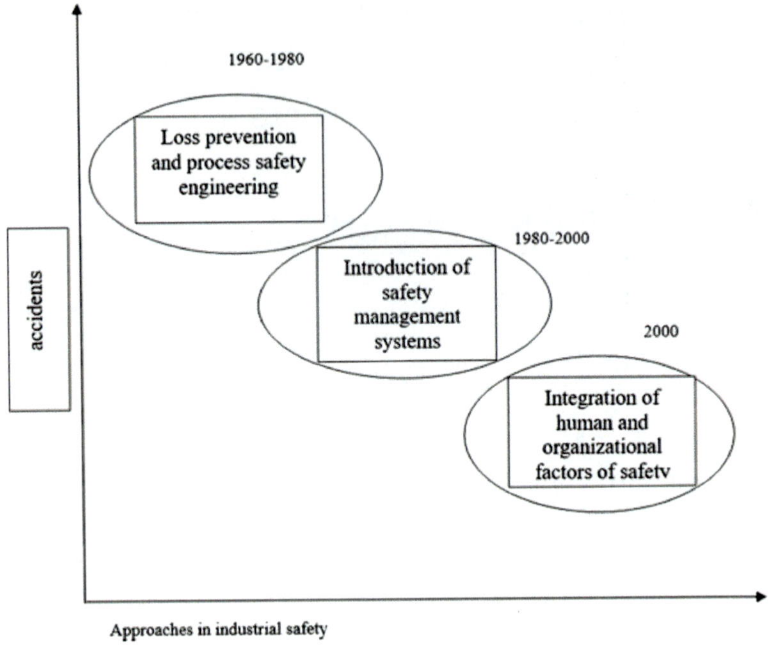

Fig. 1 Approaches in industrial safety

2 Literature Review

The IF-AHP is utilised in various domains such as supply chain management, engineering, e-commerce, banking and risk assessment. Some of the applications of IF-AHP are described below.

2.1 IF-AHP in Prioritising Alternatives

Sadiq [2] applied to select best drilling fluid in environmental decision-making process. It utilised a similarity measure to reduce the number of alternatives by grouping different alternatives into a single class or cluster. Also, for ranking generalised mean and standard deviations are followed.

Wu [3] applied interval-valued IF-AHP in e-commerce domain, here a score judgement matrices and obtained its associated interval multiplicative matrix to calculate the priority vector.

Kahraman [4] applied intuitionistic fuzzy originated interval type-2 FAHP in the application of dam less hydroelectric power plants. The triangular IF linguistic evaluation scale is utilised in pairwise comparison, which is transferred into triangular type-2 fuzzy (TT2F) pairwise comparison matrix. Applied classical AHP to find best alternatives and finally TT2F is defuzzied.

Nirmala [5] proposed triangular IF-AHP with location index number and fuzziness index function for represent the TIFN in the application to select best computer.

Xu [6] proposed an algorithm to repair inconsistencies in his article for global supplier selection.

Yu [7] described to rank the risk factors in transnational public–private partnership projects based on IF-AHP. The consistency is checked by the distances between given intuitionistic preference relation and its perfect multiplicative consistent intuitionistic preference relations were used. An algorithm is proposed for repair the inconsistent relations.

2.2 IF-AHP in Risk Estimation

Nguyen [8] Introduced IF-AHP in ship system risk estimation. The priority vector of consequences is found out and a membership knowledge measure of IFV's is utilised for final ranking.

Cebi [9] is proposed IVIF-AHP for warehouse risk estimation, score judgement matrix and possibility degree matrix were utilised for prioritisation.

3 Basic Concepts

3.1 Intuitionistic Fuzzy Sets

The fuzzy sets concepts given by Zadeh is generalised by Atannasov [10]. The intuitionistic fuzzy set consists of membership function and non-membership function for those elements of the universe of discourse. An Intuitionistic Fuzzy Set '*F*' in the Universe of Discourse '*X*' is defined as an object of the following form

$$F = \{[x, \mu F(x), \upsilon F(x)]/x \in X\} \quad (1)$$

where the function

$\mu F : X \to [0, 1]$

$\vartheta F : X \to [0, 1]$

$\mu F\ (x)$—degree of membership

$\upsilon F\ (x)$—degree of non-membership.

such that $0 \leq \mu\ F\ (x) + \upsilon\ F\ (x) \leq 1$.

Obviously, each ordinary fuzzy set may be written as

$$\{[x, \mu F(x), (1 - \mu F(x))]/x \in X\}$$

The value of $\pi\ F\ (x) = 1 - \mu\ F\ (x) - \upsilon\ F\ (x)$ is called the degree of non-determinacy or uncertainty or hesitancy.

3.2 Intuitionistic Fuzzy Set Operations [6]

Let $f_{pq} = (\mu_{pq}, \upsilon_{pq})$; $f_{rs} = (\mu_{rs}, \upsilon_{rs})$

$$1.\quad f_{pq} + \text{ars} = (\mu_{pq} + \mu_{rs} - \mu_{pq}\mu_{rs}, \upsilon_{pq}\upsilon_{rs}) \tag{2}$$

$$2.\quad f_{pq} \times \text{arsars} = (\mu_{pq}\mu_{rs}, \upsilon_{pq} + \upsilon_{rs} - \upsilon_{pq}\upsilon_{rs}) \tag{3}$$

$$3.\quad \lambda f_{pq}apq = \left(1 - (1 - \mu_{pq})^{\lambda}\upsilon_{pq}^{\lambda}\right), \lambda > 0 \tag{4}$$

$$4.\quad apq\lambda^{\lambda} = \left(\mu_{pq}^{\lambda}, (1 - (1 - \upsilon_{pq})^{\lambda}\right), \lambda > 0 \tag{5}$$

4 Intuitionistic Fuzzy AHP Prioritisation Methods

4.1 Hierarchical Structuring

The human mind structures complex realities into their constituent parts and can perceive things and ideas, identify them and communicate what they observe. By breaking down the reality into homogeneous clusters and subdividing these clusters into smaller ones, we can integrate a large amount of information to structure a problem and form a complete picture of the whole system. The mind's natural tendency to sort elements of a system into different levels and group the like elements in each level is known as hierarchical structuring [11].

4.2 Setting Priorities

Humans can perceive relationships amongst the things they observe, compare similar things against specific criteria, and discriminate between them by judging the intensity of their preference for one over the other. The result of the discrimination process is a vector of priority that describes the relative importance of each element at a level to the element of the next higher level. The latter elements serve as a criterion and are called property. This pairwise comparison is repeated for all the elements in each group. The final step is to come down the hierarchy by weighing each vector by the priority of its property. This synthesis results in a net priority weight for the bottom level. The elements with the highest weight are the one that merits the most serious consideration for action, although the others are not ruled out entirely.

4.3 Establish Priorities

The first step in establishing priorities [12] is the pairwise comparisons of each element in a cluster. That is, to compare elements in pairs against given criteria. For pairwise comparison, a matrix is the preferred form. The matrix is a simple, well-established tool that offers a framework for testing consistency, obtaining additional information through making all possible comparisons, and analysing the sensitivity of overall priorities to change in judgements.

To begin with, the pairwise comparison process, adopt the following steps:

1. Start at the top of the hierarchy to select the 'C' criterion. That will be used for making the first comparison.
2. Then, from the level immediately below, take the elements to be compared A1, A2. … An.
3. In the matrix, compare the element A1 in the column on the left with elements A1, A2 … An. in the row on the top concerning the property 'C' in the upper left-hand corner.
4. Then repeat the comparison in column elements A2 and so on.

To compare elements, ask how much more strongly do these elements or activities possess or contribute to, dominate, influence, satisfy or benefit the property than does the element with which it is being compared [13]. To form the matrix for pairwise comparison, we use numbers to represent the relative importance of one element over another concerning the property.

The fundamental scale for pairwise comparisons is described below (Table 1).

Experience has confirmed that a scale of nine units [12] is reasonable and reflects how one can discriminate the intensity of the relationship between elements. Using the scale in a social, psychological, or political context, express the verbal judgement first and then translate it to numerical values. The numerically translated judgements are approximations, and a consistency test can evaluate their validity.

Table 1 Fundamental scale for pairwise comparisons

Intensity	Definition
1	Equal importance
3	Moderate importance
5	Strong importance
7	Very strong importance
9	Extreme importance
2, 4, 6, 8	For comparisons between the above values
Reciprocals	If activity '*i*' has one of the above non zero numbers assigned to it, when compared with activity '*j*', then '*j*' has the reciprocal value when compared with '*i*'
1.1–1.9	When elements are close and nearly indistinguishable
1.3	Moderate importance
1.9	Extreme importance

Table 2 Pairwise comparison preferences

Relativity in linguistic term	Relativity in intuitionistic fuzzy sets
Extremely not related	(0.8, 0.1)
very strongly not related	(0.7, 0.2)
strongly not related	(0.6, 0.3)
moderately not related	(0.55, 0.35)
equally related	(0.5, 0.5)
moderately related	(0.35, 0.55)
strongly related	(0.3, 0.6)
very strongly related	(0.2, 0.7)
Extremely related	(0.1, 0.8)

4.4 Intuitionistic Preference Relations

Intuitionistic fuzzy values represented the pairwise comparisons. The Intuitionistic preference relations F_{pq} are conveniently represented as

$$f_{pq} = (\mu_{pq}, \upsilon_{pq}),$$

where μ_{pq} represents the degree to which the object is preferred over the object x_q, υ_{pq} represents the degree to whic the object x_p is not preferred over the object x_q and the $\pi = 1 - \mu_{pq} - \upsilon_{pq}$ is represented as a hesitancy function with the conditions.

$$\mu_{pq}, \upsilon_{pq} \in [0, 1], \mu_{pq} + \upsilon_{pq} \leq 1, 1, \mu_{pq} = \upsilon_{qp}, \mu_{qp} = \upsilon_{qp},$$
$$\mu_{pq} = \upsilon_{qp} = 0.5 \quad \text{for all } p, q = 1, 2, \ldots, n.$$

4.5 Intuitionistic Fuzzy Preferences

The pairwise comparison of alternatives can be represented in a suitable form. In this study, intuitionistic fuzzy numbers are utilised. Comparing two alternatives is more effortless than assessing a single alternative with a measurand or scale in crisp values. The scaling method for priorities in a hierarchical structure is extremely, very strongly, strongly, moderately and equally in both directions of agreeable and disagreeable. The pairwise preference of x_i over X_u is expressed in the table. The x_i values over x_j are inversely related to the preference of x_j over x_i (Table 2).

4.6 Consistency Checking

Consistency means establishing relationships among objects or ideas to coherent or relate well to each other [6]. It is uncertain that our judgements in the pairwise comparison matrix are perfect reciprocals of the transpose position because we integrate new experiences into our consciousness so that the relationships may change and some consistency may lose. The latest ideas that affect our lives tend to maintain coherence among the objects of our experiences, which causes us to rearrange some of our previous commitments. If we decided never to change our minds, we would be afraid to accept new ideas. All knowledge must be admitted into our narrow corridor between tolerable and perfect consistency. In this work, we used the algorithm by Xia and Xu [14] to construct a perfect multiplicative consistent intuitionistic preference relation.

$$\overline{F} = (f_{pq})_{n\times n}$$

For $q = p + 1$, let $(\overline{f}_{pq} = (\overline{\mu}_{pq}, \overline{\upsilon}_{pq})$, where a

$$\overline{\mu}_{pq} = \frac{\sqrt[q-p-1]{\prod_{r=p+1}^{q-1} \mu_{pr}\mu_{rq}}}{\sqrt[q-p-1]{\prod_{r=p+1}^{q-1} \mu_{pr}\mu_{rq}} + \sqrt[q-p-1]{\prod_{r=p+1}^{q-1}(1-\mu_{pr})(1-\mu_{rq})}}, \quad q > p+1 \tag{6}$$

$$\overline{\upsilon}_{pq} = \frac{\sqrt[q-p-1]{\prod_{r=p+1}^{q-1} \upsilon_{pr}\upsilon_{rq}}}{\sqrt[q-p-1]{\prod_{r=p+1}^{q-1} \upsilon_{pr}\upsilon_{rq}} + \sqrt[q-p-1]{\prod_{r=p+1}^{q-1}(1-\upsilon_{pr})(1-\upsilon_{rq})}} q > p+1 \tag{7}$$

$\overline{f}_{pq} = a_{pq}, q = p + 1;$

$\overline{f}_{pq} = (\overline{\upsilon}_{pq}, \overline{\mu}_{pq}),\ q < p$

By applying these equations, we can only update less than half of the elements in the original intuitionistic preference relations to construct the perfect multiplicative consistent intuitionistic preference relations

$\overline{F} = (\overline{f}_{pq})_{n\times n}$for F

The intuitionistic preference relations F is an acceptable multiplicative consistent intuitionistic preference relation if $d(F, \overline{F}) < \tau$.

where $d(F, \overline{F})$is distance measure between the intuitionistic preference relations F and its corresponding perfect multiplicative consistent intuitionistic preference relations $\overline{F}$ [15]. Which can be calculated by

$$d(\overline{F}, F) = \frac{1}{2(n-1)(n-2)} \sum_{p=1}^{n}\sum_{q=1}^{n} \left(\left|\tilde{\mu}_{pq} - \mu_{pq}\right| + \left|\tilde{\upsilon}_{pq} - \upsilon_{pq}\right| + \left|\tilde{\pi}_{pq} - \pi_{pq}\right| \right) \tag{8}$$

"τ" is the consistency threshold.

The inconsistent intuitionistic preference relations $F = (f_{pq})_{n\times n}$ can be transformed into corresponding perfect multiplicative consistent intuitionistic preference relations $\overline{F} = ((f_{pq})_{n\times n}$.

The distance $d(\overline{F}, F)$ between intuitionistic preference relations and the transformed one can be calculated. If distance $d((\overline{F}, F)$ is too larger, then the transformed intuitionistic preference relations $\overline{F}$ cannot represent the initial preference of the decision-maker. The modified intuitionistic preference relations should have acceptable multiplicative consistency and maintain the original preference information of the decision-maker as much as possible.

Hence, it is beneficial to combine initial intuitionistic preference relations F and the corresponding perfect multiplicative intuitionistic preference relations $\overline{F}$ into a joined intuitionistic preference relation.

$\tilde{F} = (\tilde{f}_{pq})apq_{n\times n,}$ where each element is defined as

$$\tilde{\mu}_{pq}^{(t)} = \frac{(\mu_{pq}^{(t)})^{1-\sigma}\ (\tilde{\mu}_{pq}^{(t)})^{\sigma}}{(\mu_{pq}^{(t)})^{1-\sigma}(\tilde{\mu}_{pq}^{(t)})^{\sigma}+(1-(\mu_{pq}^{(t)})^{1-\sigma}(1-(\tilde{\mu}_{pq}^{(t)})^{\sigma}} \tag{9}$$

$$\tilde{\upsilon}_{pq}^{(t)} = \frac{(\upsilon_{pq}^{(t)})^{1-\sigma}(\tilde{\upsilon}_{pq}^{(t)})^{\sigma}}{(\upsilon_{pq}^{(t)})^{1-\sigma}\ (\tilde{\upsilon}_{pq}^{(t)})^{\sigma}+(1-(\upsilon_{pq}^{(t)})^{1-\sigma}(1-(\tilde{\upsilon}_{pq}^{(t)})^{\sigma}} \tag{10}$$

where t is the number of iterations, σ is a controlling parameter, that is determined by the decision-maker the smaller the value of σ, $\tilde{F} = F$; if $\sigma = 0$, $\tilde{F} = F$; if $\sigma = 1$, $\tilde{F} = \overline{F}$ also $\tilde{F}$ is an intuitionistic preference relation.

Generally, the combined intuitionistic preference relations contain the original preference information of the intuitionistic preference relations and its corresponding perfect multiplicative consistent preference relations. The controlling parameters σ also represent the decision-maker's preference to some extent. Based on the analyses mentioned above, the equation to repair inconsistent intuitionistic preference relations is described.

This equation can improve the consistency level of any intuitionistic preference relation without losing much original information. By this iterative method, we can save much time of the decision-maker.

4.7 Composite Priorities and Ranking

The intuitionistic preference relations do not directly give the priorities. According to Saaty's concepts, n-dimensional vectors $\omega = (\omega_1, \omega_2 \ldots \omega_n)$ is estimated from the multiplicative preference relations. The ω_p is the weight that accurately represents the relative dominance of the alternative F_p among the alternatives in F.

The intuitionistic preference relations $F = (f_{pq})_{n\times n}$,

where $f_{pq} = (\mu_{pq}, \upsilon_{pq})$, since $\mu_{pq}, \upsilon_{pq} \in [0, 1]$,$\mu_{pq} + \upsilon_{pq} \leq 1$, then $\mu_{pq} = 1 - \upsilon_{pq}$.

So, the intuitionistic preference relation's membership and non-membership pairs which can be transformed into $(\mu_{pq}, 1 - \upsilon_{pq})$ the interval-valued preference relations. So that, the intuitionistic preference relations $A = (f_{pq})_{n\times n}$ is transformed into an interval-valued preference relation $A' = \left(a'_{\text{pq}}\right)_{n\times n} = (\mu_{pq}, 1 - \upsilon_{pq})_{n\times n}$ based on the operational laws of intervals, the priority weights are estimated using the formulae.

$$\omega_p = \left(\frac{\sum_{q=1}^{n} \mu_{pq}}{\sum_{p=1}^{n} \sum_{q=1}^{n} (1 - \upsilon_{pq})}, 1 - \frac{\sum_{q=1}^{n} (1 - \upsilon_{pq})}{\sum_{p=1}^{n} \sum_{q=1}^{n} \mu_{pq}} \right) \tag{11}$$

For an intuitionistic fuzzy set $\pi\ F(x) = 1 - \mu\ F(x) + \upsilon F(x)$, the degree of uncertainty of the membership of element $x \in X$ to the set F. In ordinary fuzzy set π $F\ (x) = 0$ for every $x \in X$.

$\pi\ F(x) = [0,1]$ for all $x \in X$, Let $\alpha = (\mu_\alpha, \upsilon_\alpha, \pi_\alpha)$ be an intuitionistic fuzzy value.

Szmidt and Kaeprzyk [16] proposed a relation to rank the intuitionistic fuzzy value

$$\rho(\alpha) = 0.5\,(1 + \pi_\alpha)(1 - \mu_\alpha) \tag{12}$$

The smaller the value of $\rho(\alpha)$, the more significant the intuitionistic fuzzy value α in a sense, the amount of positive information included and the reliability of the data.

5 Methodology

A step-by-step process for the intuitionistic fuzzy AHP is described in Fig. 2. The application of IFAHP in prioritising safety factors in a crude refining industry based on past incidents consists of two levels of factors: primary level and sub levels. In each sub-level, the preference is based on experts' opinions. The sequence of the process is described below.

5.1 Development of Hierarchal Structure

In this stage, the objective, safety factors and sub-factors are identified based on the incidents that happened in the organisation. The hierarchal structures are formed at different levels. The first level represents primary factors, and the next level constitutes Sub factors. It describes the hierarchy of each factor, sub-factors, and its weightage in the overall goal.

5.2 Formation of IF Pairwise Comparison Matrix

The intuitionistic fuzzy set numbers represent the pairwise comparison of experts' opinions. The experts can specify the degree of belief and non-belief, represented by the functions μx and υx, respectively. At this stage, evaluation of intuitionistic preference relation (IPR) is done by pairwise comparison between each factor and

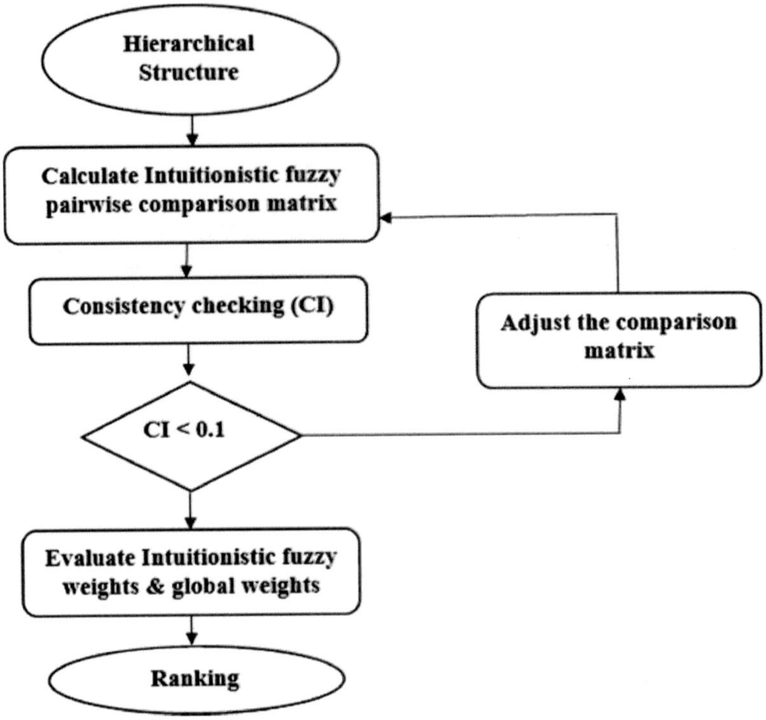

Fig. 2 The general flow chart of IF-AHP

sub-factors against the respective criterion. From the IPR, the perfect multiplicative consist preference relation is calculated.

5.3 Consistency Check

The pairwise comparison of experts is subjected to inconsistency. The inconsistency in any judgement matrix is identified by measuring the distance. The threshold of distance must be less than ten per cent for acceptance. If it is inconsistent ($\tau > 0.1$), repair the inconsistent IPR by iteratively applying the auto-correction formula until $\tau < 0.1$.

5.4 Estimation of Intuitionistic Fuzzy Set Priority Weights

The priority vector for each consistent intuitionistic preference relation is calculated and find out the overall weights from the lowest level to the highest level by intuitionistic fuzzy (IF) operation.

5.5 *Defuzzification and Ranking of iF Priority Factors*

The ranking of intuitionistic fuzzy numbers is done by using the relation.

$$\rho(\alpha) = 0.5(1 + \pi_\alpha)(1 - \mu_\alpha).$$

6 Numerical Analysis

This analysis is based on the past incidents that occurred at different production plants in a refinery by the subject experts. The factors are described below.

6.1 *Basic Factors*

The basic factors are again divided into sub-factors, such as human, organisational and technical factors. Each sub-factors are discussed below in detail.

6.2 *Human Factors*

i. **Slips and lapses**: are errors that result from some failure to execute a sequence of actions. Lapses occur when an action is omitted or not carried out. Slips are generally called when appropriate action is not carried out correctly.

ii. **Laxness:** Laxness scribes the undesired quality of negligence or slackness to follow established rules, procedures or practices.

iii. **Communication or Perception:** Communication and perception are overlooked and crucial in workplace safety. Ineffective communication has many negative consequences on workplace safety, especially in rotating shift workers and relay workers. The job status in the former shift needs to be updated in the continuous process plant. The perception of process problems or other safety issues is critical for the right solutions. Understanding many safety-related incidents in process plants will conclude that perception significantly impacts the incidents. So, communication and perception play a vital role in ensuring safety.

iv. **Collaboration or Teamwork:** Collaboration includes sharing information and instructions from multiple departments/sections for the smooth execution of the jobs in the plant. Teamwork is mainly beneficial in handling complex work requiring simultaneous activities at different locations.

v. **Psychological Stresses:** In the case of psychological stress, the operators or workmen may execute jobs under stress, leading to accidents. The stress may be due to job insecurity, work overload, bullying and harassment or shift work.

6.3 Organisational Factors

i. **Improper Standard Operating Procedures (SOP) :** Standard operating procedures are the set of instructions that need to be followed in word and spirit by the employees strictly and without any deviations. Through adherence to SOP, an organisation can ensure safety. However, it has often been seen that SOPs are found improper in various situations and may lead to accidents.
ii. **Inadequate time/Work overload :** It has been seen that there are organisations that assign responsibilities or works to employees that cannot be done within the stipulated time. This culture leads to laxness.
iii. **Planning & Controlling:** Improper or poor planning and controlling can lead to accidents. Planning plays an increasing role, especially during the execution of time-bound and critical activities during breakdown maintenance plant shutdown.
iv. **Handling of Job Complexities:** It is often found that organisations causally handle the work complexities. Work complexities include significant variables and interfaces, ambiguity, unpredictability, dynamics (rapid rate of change), social structure and interrelationships.
v. **Training and Competence:** Competencies are particular qualities that a company has identified for its employees to perform a specific or identified task. Persons assigned to perform critical tasks must be competent enough and sufficiently experienced to perform the assigned tasks. Young, untrained and new workers generally sustain injuries more frequently than older, trained and experienced employees.

6.4 Technical Factors

i. **Engineering/Design Constraints:** Engineering/design constraints include physical design constraints and social design constraints. The physical design constraints include the factors like natural laws and properties of materials. The example enlightens the concept that an object will fly only if its shape allows air current to lift it; otherwise, gravity will keep it on the ground. The social design constraints include ease of use and ease of access.
ii. **Cost/Time Trade-off:** In the industry, we may compromise safety and take a calculated risk in executing some works. If we cannot take calculated risks, we may have to bear the excess cost or compromise time.
iii. **Systems Failure:** The technological failure in a broad sense includes failure due to cracking, corrosion, improper working of safety systems and other safety equipment, inability to sustain process parameters, variations including process fluids hammering, malfunction of sensors, indicators, controllers and also includes software faults.

In almost all industries, there are reportable as well as non-reportable incidents. Analysing past incidents in an organisation enlightens the actual areas of weakness in that industry. In this paper, all the reportable and non-reportable incidents in a crude distillation unit of a refinery are analysed to identify the weak areas or factors. The multi-criteria decision-making process (MCDM) can be effectively utilised for prioritising safety factors. The analytical hierarchy process (AHP) possesses the unique advantage of comparing parameters with no units or scale of measurements. The interval-valued intuitionistic fuzzy analytical hierarchy process (IF-AHP) further improves the AHP in terms of vagueness, uncertainty and handles the imprecise data. The IF-AHP is mainly applied to prioritise an alternative through highly complex criteria. The IF-AHP demonstrates strong decision-making capability in highly complex domains. The main drawbacks are that it is a perception-based assessment.

7 Sample Calculations

The intuitionistic preference relation of human sub-factors is given in Table 3.

The relations to evaluate the perfect multiplicative consistent intuitionistic preference relation is given by

$$\overline{\mu}_{pq} = \frac{\sqrt[q-p-1]{\prod_{r=p+1}^{q-1} \mu_{pr}\mu_{rq}}}{\sqrt[q-p-1]{\prod_{r=p+1}^{q-1} \mu_{pr}\mu_{rq}} + \sqrt[q-p-1]{\prod_{r=p+1}^{q-1} (1-\mu_{pr})(1-\mu_{rq})}}, q > p+1$$

Let us consider $\bar{a}_{15,}$ its perfect multiplicative consistent intuitionistic preference membership and non-membership functions are

$\overline{\mu}_{15} = 0.378856.$
$\overline{\upsilon}_{pq} = 0.480795$ (Table 4).

Table 3 Intuitionistic preference relation of human sub-factors

Human factors	Slips/lapses	Laxness	Communication/ perception	Collaboration/ team work	Psychological stress
Slips/lapses	(0.5, 0.5)	(0.55, 0.35)	(0.3, 0.6)	(0.55, 0.35)	(0.35, 0.55)
Laxness	(0.35, 0.55)	(0.5, 0.5)	(0.3, 0.6)	(0.35, 0.55)	(0.55, 0.35)
Communication/ perception	(0.6, 0.3)	(0.6, 0.3)	(0.5, 0.5)	(0.55, 0.35)	(0.35, 0.55)
Collaboration/ team work	(0.35, 0.55)	(0.35, 0.55)	(0.35, 0.55)	(0.5, 0.5)	(0.35, 0.55)
Psychological stress	(0.55, 0.35)	(0.55, 0.35)	(0.55, 0.35)	(0.55, 0.35)	(0.5, 0.5)

Table 4 Perfect multiplicative consistent intuitionistic preference relation for human factors

Human factors	Slips/lapses	Laxness	Communication/ perception	Collaboration/ team work	Psychological stress
Slips/lapses	(0.5, 0.5)	(0.55, 0.35)	(0.343, 0.446)	(0.369, 0.421)	(0.378, 0.480)
Laxness	(0.35, 0.55)	(0.5, 0.5)	(0.3, 0.6)	(0.343, 0.446)	(0.205, 0.623)
Communication/ perception	(0.446, 0.343)	(0.6, 0.3)	(0.5, 0.5)	(0.55, 0.35)	(0.396, 0.396)
Collaboration/ team work	(0.421, 0.369)	(0.446, 0.343)	(0.35, 0.55)	(0.5, 0.5)	(0.35, 0.55)
Psychological stress	(0.480, 0.378)	(0.623, 0.205)	(0.396, 0.396)	(0.55, 0.35)	(0.5, 0.5)

Table 5 Perfect multiplicative consistent intuitionistic preference relation for human factors (first iteration)

Human factors	Slips/lapses	Laxness	Communication/ perception	Collaboration/ team work	Psychological stress
Slips/lapses	(0.5, 0.5)	(0.55, 0.35)	(0.335, 0.478)	(0.405, 0.407)	(0.373, 0.495)
Laxness	(0.35, 0.55)	(0.5, 0.5)	(0.3, 0.6)	(0.345, 0.467)	(0.261, 0.569)
Communication/ perception	(0.478, 0.335)	(0.6, 0.3)	(0.5, 0.5)	(0.55, 0.35)	(0.387, 0.427)
Collaboration / team work	(0.407, 0.405)	(0.427, 0.383)	(0.35, 0.55)	(0.5, 0.5)	(0.35, 0.55)
Psychological stress	(0.495, 0.373)	(0.609, 0.230)	(0.427, 0.387)	(0.55, 0.35)	(0.5, 0.5)

$$d(\overline{A}, A) = \frac{1}{2(n-1)(n-2)} \sum_{p=1}^{n} \sum_{q=1}^{n} \left(\left| \tilde{\mu}_{pq} - \mu_{pq} \right| + \left| \tilde{\upsilon}_{pq} - \upsilon_{pq} \right| + \left| \tilde{\pi}_{pq} - \pi_{pq} \right| \right)$$

$d(\bar{A}, A) = 0.23 > 0.1$ so apply auto-correction relations

$$\tilde{\mu}_{pq}^{(t)} = \frac{(\mu_{pq}^{(t)})^{1-\sigma} (\tilde{\mu}_{pq}^{(t)})^{\sigma}}{(\mu_{pq}^{(t)})^{1-\sigma} (\tilde{\mu}_{pq}^{(t)})^{\sigma} + (1 - (\mu_{pq}^{(t)})^{1-\sigma} (1 - (\tilde{\mu}_{pq}^{(t)})^{\sigma}}$$

Auto-correction first iteration (Table 5).

$$\mathrm{d}\,(\overline{A}, \mathrm{A}) = 0.10947 > 0.1$$

Auto-correction second iteration (Table 6).

$$\mathrm{d}(\overline{A}, \mathrm{A}) = 0.085534 < 0.1$$

Weight vectors for primary factors under severity calculated by (Table 7)

Table 6 Perfect multiplicative consistent intuitionistic preference relation for human factors (second iteration)

Human factors	Slips/lapses	Laxness	Communication/ perception	Collaboration/ team work	Psychological stress
Slips/lapses	(0.5, 0.5)	(0.55, 0.35)	(0.337, 0.471)	(0.398, 0.410)	(0.374, 0.492)
Laxness	(0.35, 0.55)	(0.5, 0.5)	(0.3, 0.6)	(0.345, 0.463)	(0.249, 0.580)
Communication/ perception	(0.471, 0.337)	(0.6, 0.3)	(0.5, 0.5)	(0.55, 0.35)	(0.389, 0.421)
Collaboration/ team work	(0.410, 0.398)	(0.431, 0.375)	(0.35, 0.55)	(0.5, 0.5)	(0.35, 0.55)
Psychological stress	(0.492, 0.374)	(0.612, 0.225)	(0.421, 0.389)	(0.55, 0.35)	(0.5, 0.5)

Table 7 Weight vectors for human factors

Sub-factors (human) priority weights		μ	υ
Slips or lapses	$\omega 1$	0.1667	0.7696
Laxness	$\omega 2$	0.1518	0.7869
Communication or perception	$\omega 3$	0.1925	0.7391
Collaboration/team work	$\omega 4$	0.1407	0.8
Psychological stress	$\omega 5$	0.2	0.7304

$$\omega_p = \left(\frac{\sum_{q=1}^{n} \mu_{pq}}{\sum_{p=1}^{n} \sum_{q=1}^{n} \left(1 - \upsilon_{pq}\right)}, 1 - \frac{\sum_{q=1}^{n} \left(1 - \upsilon_{pq}\right)}{\sum_{p=1}^{n} \sum_{q=1}^{n} \mu_{pq}} \right)$$

Overall priority of slip or lapses = weight of human (primary) factor X weight of slips or lapses (sub factor)

$$\begin{aligned} &= (0.072,\ 0.119) X (0.1667, 0.7696) \\ &= (0.012019, 0.7969) \end{aligned}$$

Rank the intuitionistic fuzzy value = $\rho(\alpha) = 0.5\ (1 + \pi_\alpha)(1 - \mu_\alpha) = 0.588334$ (Table 8).

Table 8 Pairwise comparison of primary factors and sub-factors

Primary factors	Human factors	Organisational factors	Technical factors
Human factors	(0.5, 0.5)	(0.3, 0.6)	(0.7, 0.2)
Organisational factors	(0.6, 0.3)	(0.5, 0.5)	(0.7, 0.2)
Technical factors	(0.2, 0.7)	(0.2, 0.7)	(0.5, 0.5)

Human factors	Slips/lapses	Laxness	Communication/ perception	Collaboration/ team work	Psychological stress
Slips/lapses	(0.5, 0.5)	(0.55, 0.35)	(0.3, 0.6)	(0.55, 0.35)	(0.35, 0.55)
Laxness	(0.35, 0.55)	(0.5, 0.5)	(0.3, 0.6)	(0.35, 0.55)	(0.55, 0.35)
Communication/ Perception	(0.6, 0.3)	(0.6, 0.3)	(0.5, 0.5)	(0.55, 0.35)	(0.35, 0.55)
Collaboration / Team work	(0.35, 0.55)	(0.35, 0.55)	(0.35, 0.55)	(0.5, 0.5)	(0.35, 0.55)
Psychological stress	(0.55, 0.35)	(0.55, 0.35)	(0.55, 0.35)	(0.55, 0.35)	(0.5, 0.5)

Organisational factors	Improper SOPs	Inadequate time/workload	Planning and controlling	Job complexity	Training and competency
Improper SOPs	(0.5, 0.5)	(0.35, 0.55)	(0.55, 0.35)	(0.55, 0.35)	(0.35, 0.55)
Inadequate time/workload	(0.55, 0.35)	(0.5, 0.5)	(0.6, 0.3)	(0.6, 0.3)	(0.6, 0.3)
Planning and Controlling	(0.35, 0.55)	(0.3, 0.6)	(0.5, 0.5)	(0.35, 0.55)	(0.35, 0.55)
Job complexity	(0.35, 0.55)	(0.3, 0.6)	(0.55, 0.35)	(0.5, 0.5)	(0.35, 0.55)
Training and Competency	(0.35, 0.55)	(0.3, 0.6)	(0.55, 0.35)	(0.55, 0.35)	(0.5, 0.5)

Technical factors	Engineering/design constraints	Cost/time trade-off	System failure
Engineering/design constraints	(0.5, 0.5)	(0.3, 0.6)	(0.35, 0.55)
Cost/time trade-off	(0.6, 0.3)	(0.5, 0.5)	(0.6, 0.3)
System failure	(0.55, 0.35)	(0.3, 0.6)	(0.5, 0.5)

7.1 *Final Ranking*

The final ranking of the safety factors is tabulated below (Table 9).

Table 9 Ranking of safety factors

	Human factors	μ	υ	ρ	Rank
1	Slips or Lapses	0.00986129	0.77612461	0.601021	7
2	Laxness	0.00796705	0.81404931	0.584299	10
3	Communication or Perception	0.01147070	0.75066551	0.611832	5
4	Collaboration/team work	0.00932366	0.78817341	0.595646	8
5	Work stress	0.01176296	0.74513180	0.614241	3
	Organisational factors				
6	Improper SOPs	0.01344510	0.74500390	0.612429	4
7	Inadequate time/work overload	0.01591755	0.71717986	0.623368	1
8	Planning and controlling	0.01066519	0.79564783	0.590478	9
9	Handling of job complexity	0.01291218	0.76082346	0.605215	6
10	Training and competency	0.01443921	0.73282077	0.617326	2
	Technical factors				
11	Engineering/design constraints	0.00777712	0.96809204	0.508083	13
12	Cost/time trade-off	0.00961205	0.94066156	0.519818	11
13	Systems failure	0.00788916	0.96601702	0.508999	12

8 Conclusions

This analysis found that the organisational factors represent the most priority category in a refinery process safety. Work overload, competency and work stress are the prominent sub-factors. The intuitionistic fuzzy AHP tackles the uncertainties of preference relations by membership and non-membership functions successfully. The intuitionistic fuzzy AHP structure's transparency makes complex decision problems simpler, and the pairwise comparison between the factors is also easy to understand.

References

1. Daniellou, F., Simard, M., & Boissière, I. (2001). Human and organizational factors of safety: State of the art 119.
2. Sadiq, R., & Tesfamariam, S. (2009). Environmental decision-making under uncertainty using intuitionistic fuzzy analytic hierarchy process (IF-AHP). *Stochastic Environmental Research and Risk Assessment, 23*, 75–91.
3. Wu, J., Huang, H., & Cao, Q. (2013). Research on AHP with interval-valued intuitionistic fuzzy sets and its application in multi-criteria decision-making problems. *Applied Mathematical Modelling, 37*, 9898–9906.
4. Kahraman, C., Öztaysi, B., Onar, S. Ç., & Dogan, O. (2018). Intuitionistic fuzzy originated interval type-2 fuzzy AHP: An application to damless hydroelectric power plants. *International Journal of the Analytic Hierarchy Process, 10*, 266–292.

5. Nirmala, G., & Uthra, G. (2017). Triangular intuitionistic fuzzy Ahp and its application to select best product of notebook computer. *International Journal of Pure and Applied Mathematics, 113*, 253–261.
6. Xu, Z., & Liao, H. (2014). Intuitionistic fuzzy analytic hierarchy process. *IEEE Transactions on Fuzzy Systems, 22*, 749–761.
7. Yu, Y., Darko, A., Chan, A. P. C., Chen, C., & Bao, F. (2018). Evaluation and ranking of risk factors in transnational public-private partnerships projects: Case study based on the intuitionistic fuzzy analytic hierarchy process. *Journal of Infrastructure Systems, 24*, 1–13.
8. Nguyen, H. (2016). An application of intuitionistic fuzzy analytic hierarchy process in ship system risk estimation. *Journal of KONES Powertrain and Transport, 23*.
9. Cebi, S., & Ilbahar, E. (2018). Warehouse risk assessment using interval valued intuitionistic fuzzy AHP. *International Journal of the Analytic Hierarchy Process, 10*, 243–253.
10. Atanassov, K. T. (1999). *Intuitionistic fuzzy sets, theory and applications*. Physica-Verlag.
11. Saaty, T. l. (1995). *Decision making for leaders*. RWS Publication.
12. Saaty, T. l. (1980). *The analytical hierarchy process*. Mc Graw-Hill.
13. Saaty, T. L. (1986). Axiomatic foundation of the analytic hierarchy process. *32*, 841–855. https://doi.org/10.1287/mnsc.32.7.841
14. Xia, M.M., & Xu, Z. (2011). On consensus in group decision making based on fuzzy preference relations. *International Journal of Intelligent Systems, 26*, 787–813.
15. Szmidt, E., & Kacprzyk, J. (2009). Amount of information and its reliability in the ranking of Atanassov's intuitionistic fuzzy alternatives. *Studies in Computational Intelligence, 222*, 7–19.
16. Szmidt, E., & Kacprzyk, J. (2000) Distances between intuitionistic fuzzy sets. *Fuzzy Sets and Systems, 114*, 505–518.

Hydraulic Design of a Water Treatment Plant—A Case Study for Devprayag Town in Uttarakhand

Rahul Silori and S. C. Gupta

1 Introduction

Water is an important natural resource on Earth. It is the second necessary substance after air necessary for survival of all the living organisms on this planet [1]. However, the total amount of freshwater sources is limited [2]. Earth's surface is mostly covered by water (71%), out of which ocean contains about 97% of water which is saline and unsafe for drinking and other purposes. Only 3% of the water is fresh and suitable for drinking and other purposes [1]. One-third of the freshwater is in the form of icebergs and polar ice caps, and only a small amount is stored as groundwater and surface water. The surface water basically available in the rivers, ponds and lakes is the water which we use as a source of water supply for the cities, towns, etc. [1]. Improved lifestyle, industrial development and population growth are the primary reasons led to increase in demand on water supply [3]. But, before facilitating the society with clean and palatable water treatment of the raw water from the source is an essential requirement. Therefore, water treatment plant (WTP) is an essential component for any town/city. The concern area of the study here is Devprayag town located at Tehri Garhwal district, Uttarakhand, India shown in Fig. 1. To serve the demand of the water supply for the town, the project for a WTP construction under Akri-barjula water supply scheme was launched by Uttarakhand Pey Jal Nigam and M/s Unipro Techno Infrastructure Pvt. Ltd, Chandigarh. The WTP has the treatment capacity as 4.1 MLD, in which extra 5% loss during the treatment process is considered, and therefore, the design flow works have been worked out for 4.31 MLD.

R. Silori (✉)
School of Engineering, University of Petroleum and Energy Studies, Dehradun, India
e-mail: rsilori@ddn.upes.ac.in

S. C. Gupta
Avadh Consultancy Services, Dehradun, India

N. A. Siddiqui et al. (eds.), *Advances in Construction Safety*,
https://doi.org/10.1007/978-981-19-4001-9_8

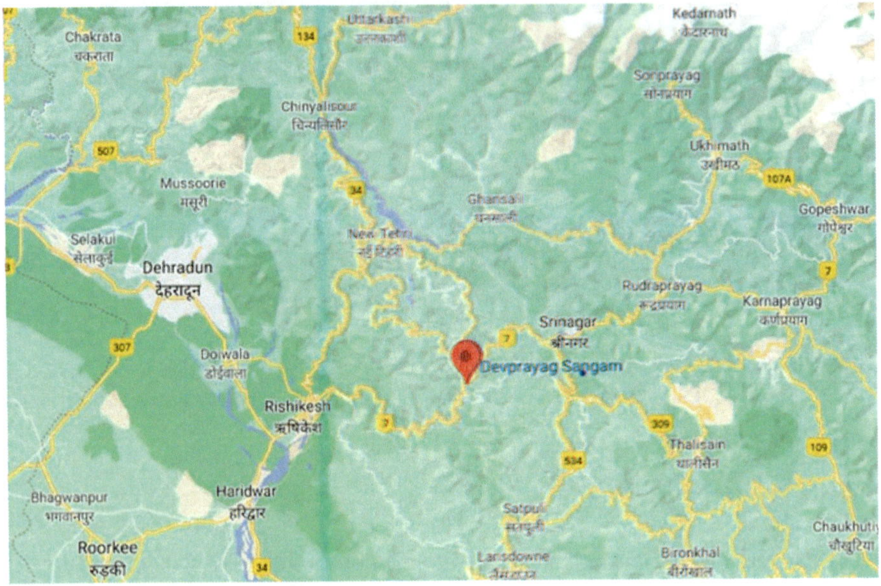

Fig. 1 Google Map showing Devprayag town location in Uttarakhand [4]

2 Proposed Layout for the WTP

Raw water inlet chamber has been provided as the first unit in the plant for measurement of flow and uniform distribution of the flow to further units. Due to the withdrawal of water from river, pre-settling unit is designed for settling of heavier particles afterwards. The other units of treatment include coagulation, flocculation, secondary settling and rapid sand filtration. For the flocculation as gentle mixing is required, it is obtained by diffusing low pressure compressed air through the water held in tank. The rapid sand filters have the air and water wash. Flat RCC roof is provided over the filters. Wash-water storage tank will be put over the operating platforms of the filters.

For storage of chemicals and housing chemical solution tanks, a separate building will be put adjoining to filter house. The solution tank can be housed in a hall made of the roof of the filter units. The chemical storage tanks will have acid proof lining. Alternately, HDPE tanks of suitable size may be used for chemical solutions. Dosing of chemical solution with raw water will be done at the dosing chamber.

To distribute and regulate the flow for the ensuring units, common chambers need to be provided between different stages of treatment. Drain is required to be provided between different stages of treatment plant to carry underflow and overflow from various units. Dual units have been provided to facilitate operational shutdown and maintenance process.

For the design of the water treatment units, guidelines given by CEEPHO manual on water supply and best design practices have been followed.

The proposed and hydraulic designed various units of the WTP are as follows:

1. **Raw water inlet chamber**
2. **Pre-settling chamber**
3. **Overflow, underflow and wastewater drain**
4. **Dosing chamber and mixing chamber**
5. **Coagulation units/chemical house**
6. **Flocculation unit**
7. **Secondary settling tank**
8. **Rapid sand filter**
9. **Chamber between the units (intermediate chambers)**
10. **Clear water reservoir (CWR)**.

3 Hydraulic Design

Capacity: 4.1 MLD.

Considering 5% water loss.

Design Capacity—4.31 MLD or 4310 m^3/d.

Two parallel units (from flocculation unit) have been designed for treatment of 4.31 MLD water.

It means each unit (from flocculation unit) will treat 2.155 MLD water.

3.1 Raw Water Inlet Chamber

These will have twin chambers in series with 90° V-notch in between having 50 cm free fall. This will provide measurement of flow and uniform distribution of the flow to further units [5].

Each of these chambers is $1.0 \times 1.0 \times 2.0$ m^3 SWD. The V-notch may be cut out of a plate. The edges and scale of the V-notch may be of aluminium.

3.2 Pre-settling Chamber/Tank

This single unit is designed for 4.31 MLD capacity.

Assume horizontal flow velocity of 0.2 m/min and detention period of 1 h [5].

Length of the tank required $= 0.2 \times 1 \times 60 = 12$ m.

Capacity of the tank for 1 h $= 179.58$ m^3.

Cross-sectional area of the tank required $= 15.0$ m^2.

Assume water depth in the tank $= 4.0$ m $+ 0.30$ m free board.

Width of the tank $= 3.75$ m.

Providing 0.5 m for sludge zone, overall depth of the tank $= 4.3 + 0.5 = 4.8$ m.

So, **size of pre-settling tank 12.00** m × **3.75** m × **4.8** m is required (with slope of 1:24 in bottom slab for effective collection of settled sludge).

3.3 Overflow, Underflow and Wastewater Drain

An overflow pipe of 30 cm dia will be provided on the first inlet chamber. A scour outlet pipe and valve of 20 cm dia will be provided at bottom of the first chamber.

A 50 cm dia drain of RCC Hume pipe with intermediate chambers be laid parallel to the treatment units at a minimum slope of 1 in 250 to drain out the overflow, wastewater and sludge. Alternately, a RCC/brick masonry drain is put with the removable slabs on the top [5].

3.4 Dosing and Mixing Chamber

It consists of twin chamber with 90° V-notch in between having free fall. The dosing of the alum coagulant to the raw water will occur in dosing chamber, whereas uniform mixing of the same with raw water will occur in mixing chamber.

Both the chambers provided are of sizes 1 × 1 × 3.1 m SWD.

3.5 Coagulation Units

Total flow to be treated = 4.31 MLD = 4310 m^3/d.

Approx. alum dose = 60 mg/l [5].

Conc. of stock solution of the chemicals = 8% [5].

Capacity of the solution storage tank = to be adequate to serve for 8 h flow [5].

No. of solution storage tanks: 2 nos for alum (1 for dissolving chemical and 1 for storage of the solution).

Solution dosing tanks: 2.

Size of the Units

Total flow to be treated = 4.31 MLD = 4310 m^3/d.

Quantity of alum required at a dose of 60 mg/l = 259 kg/d (approx) ~ 87 kg in 8 h. (approx).

Volume of the solution at 8% strength for 8 h working = 1.09 m^3.

Min size of solution tanks to be put = 1 m × 1 m × 1.2 m (considering capacity of 1.2 m^3).

Size of solution dosing tank of HDPE/PVC (2 nos) = 1 × 1 × 0.6 ^{3}.

Type of dosing: Float controlled submerged orifice tube. Alternately, pumps may be used for dosing of solutions.

System of dissolving: By pumped recirculation of the solution to and from the solution tanks.

Capacity of the chemical solution mixing pumps: 1.5HP*10 m head with 2.5 cm dia. pipes.

No. of pumps for mixing of chemical solutions: 2 nos for each chemical plus standbys.

Each solution tank has to be provided with a sludge hopper, pit and sludge valve of brass or SS at the bottom.

3.6 Flocculation Unit

Each flocculation tank will have the following sizes:

1. Assuming the detention time of 10 min [5], capacity for each tank will be 14.96 m^3
2. Provide a tank of size: $4 \times 2 \times 2 = 16$ m^3
3. Available DT = V/Q = 10.69 min
4. Quantity of air required @150 lpm/m^2 = 1200 lpm
5. Pressure of the air required at the point of discharge at tank bottom: 3 m
6. HP of the twin lobe blower to be used for the air agitation: 2.0HP.

So, **two units** with size of each flocculation tank as **4.00** × **2.00** × **2.00** m^3 are provided with 0.50 m deep hopper bottom in side slope of 1:1 for sludge.

3.7 Secondary Settling Tank

It will be an upward flow inclined type of rectangular-shaped tank having sludge hopper at the bottom. The settling will take place on the multiple corrugated AC sheets held at 45° slope in the tank. Standing size of corrugated AC sheet is taken as 1 m wide and 2.5 m long.

Size of Each Unit

Design flow = 2155 m^3/d (half of the total flow).

SLR = 30 m^3/m^2/d [5].

Area of settling tank = Q/SLR = 71.83 m^2.

Area of the sloping plates required = 101.58 m^2.

Size of AC sheets 1.0 m wide × 2.5 long.

Area of sheets = 2.50 m^2.

Effective area of sloping settling with 2.0 m sloping depth = 2.0 m^2.

No. of plates required = 101.58/2 = 50.79 ~ 52.

Provide two tiers in series, then required total row for 52 sheets = 52/2 = 26 rows.

No. of A.C. sheets = 26 × 2 tiers in series.

Provide width of the tank = 1.0 × 2 + 0.60 m for weir = **2.6** m.

At a horizontal gap of 12.5 cm between two rows of sheets, the length of the tank for 26 rows = 26 × 12.5 = 3.25 m.

Adding 0.60 m length for baffle wall, so total length of the tank = 3.25 + 0.60 = 3.85 m, say **4.0** m.

Sloping water depth of the tank = 3.0 m.

So, vertical height of the tank = 2.12 + 0.30 m free board = 2.42 m say 2.50 m.

So, size of each settling tank = 4.0 x 2.6 x 2.5 m **above hopper**.

Size of Bottom Hopper

Two hopper bottoms will be provided at the bottom. With the above dimensions of the settling tank, the top size of the individual bottom hopper will be 2.0 × 2.6 m.

With 1.0 × 1.6 m length flat apron at floor level, a horizontal offset (H) of 0.5 m is left on each side of the hopper.

Provide a vertical depth of 0.5 m in the hopper, side slope = V/H = 0.5/0.5 = 1:1.

Weir Loading

Provide one number of 0.3 × 0.5 m overflow collection troughs on full width of the tank at water surface.

Total weir length = 4.0 × 2 = 8.0 m (edges).

Weir loading = 269.375 m^3/d/m < 300 m^3/d/m [5], O.K.

3.8 *Rapid Sand Filter (RSF) Design*

Capacity of RSF = 4.31 MLD (*desired for the plant*).

Sand Filtering Media

Filtered water required per hour = 0.179 ml/hr.

Assume the rate of filtration be 6000 l/h/m^2 of bed [5].

Area of filter = 29.93 m^2.

Provide two units. Each bed area = 14.96 m^2.

Assume L/B = 1.5;

B = 3.15 m; L = 3.15 × 1.5 = 4.74 m.

Therefore, adopt 2 filter units each of dimensions (effective size) ~ 4.8 mx 3.20 m **(for factor of safety).**

Provide 0.75 m for wide wash-water gutter along 3.20 m wide.

Therefore, total size of the bed will be **4.80** × **3.95** m^2.

Assume depth of sand = 75 cm (*finer at the top and coarser at the bottom and can be laid in layers*).

Effective size (D10) of sand particles = 0.35 − 0.55 mm.

Uniformity coefficient—1.3–1.7.

Below sand layer provide 60 cm gravel support layer (*finer at the top and coarser at the bottom and can be laid in 5 layers).*

Layer no. (from top towards bottom)	1	2	3	4	5
Gravel size (mm)	3–6	6–12	12–20	20–40	40–80
Thickness (mm)	100	120	120	120	140

Depth of the Tank

1. Media depth—1.35 m
2. Depth required for column of standing water—2.15
3. Free board—0.5 m.

Total height = 4.00 m.

Underdrainage System

Total area of holes = 0.2 to 0.5% of bed area [5].

Assume 0.2% of bed area = 0.0307 m^2.

Area of lateral = 2 (area of holes of lateral) (for 10 mm dia holes).

Area of manifold = (area of laterals).

So, area of manifold = 2 × area of holes = 2 × 0.0307 = 0.0061 m^2.

Assume width of manifold = 25 cm.

Depth of manifold = 0.245 m ~ 30 cm.

So, **size of manifold is 25 x 30** cm.

Assume c/c of lateral = 30 cm.

Total numbers = 3.2/0.3 = 10.66 + 1 = 11.66 say 12 no. on either side.

Length of each lateral = (4.8 − 0.25)/2 = 2.275 m.

C.S. area of lateral = 2 × area of perforations per lateral.

Assumed dia. of holes = 10 mm.

Number of holes (n) = 391.08, say 394.

Number of holes per lateral = 394/24 = 16.41 ~ 20.

Using two rows alternate of holes in each lateral at 60°.

Spacing of holes = 2.275/10 = 22.75 m ~ 20 cm.

Area of perforations per lateral = 15.7 cm^2.

C.S. area of each lateral = 2 × area of perforations per lateral = 2 × 15.7 = 31.4 cm^2.

Diameter of each lateral = 6.32 ~ 6.5 cm.

Check: Length of lateral < 60 d = 60 × 6.5 = 3.9 m. provided L = 2.275 m (hence acceptable).

Rising wash-water velocity in bed = 50 cm/min. (= 500 lpm/m^2) [5].

Wash-water discharge per bed = 0.128 m^3/s.

Velocity of flow through lateral = (0.128/ total lateral area) = 3.39 m/s (O.K.).

Manifold velocity = 0.128/(0.25 × 0.3) = 1.71 m/s < 2.4 m/s (max. permissible) (O.K.).

Wash-Water Trough

Discharge of wash water per bed = 0.128 m^3/s. Size of bed = 4.8 × 3.2 m.

Assume 2 troughs running lengthwise.

Discharge of each trough = Q/2 = 0.128/2 = 0.064 m^3/s.

$Q' = 1.376 \times b \times h^{3/2}$.

Assume b = 30 cm or 0.3 m, h = 0.288 m, say 30 cm.

Hence, 2 wash-water troughs of size 30 cm × 30 cm laid in slope.

Air Compressor unit = 600–800 l of Air/min/m^2 Bed Area [5].

Take 600 l/min/m^2:

Air flow rate required = 0.6 × 4.8 × 3.2 = 9.22 m^3/min.

Pressure of air: 0.35 kg/cm^2.

Provide twin lobe blower of 3.0 HP and air pipe 8 cm dia vertically connected on manifold.

Wash-Water Storage Tank

At 500 lpm/m^2 [5], for washing time of 5 min water required = 500 × 4.8 × 3.2 × 5 = 38.40 m^3 (dimensions can be altered if washing time is kept more). Provide a wash-water storage tank of capacity = 4.8 m × 3.2 m × 2.5 m (either dimensions can be altered according to the volume of water required). Apart from that, provide a free board of 0.3 m So, **size of wash-water storage tank is 4.8 x 3.2 x 2.8** m.

Clean Wash-Water Gravity Main

Design flow: 0.128 m^3/s.

Dia. of down flow wash-water pipe (velocity = 3 m/s): 23.31 cm, say 25 cm.

Wash-Water Drainage Pipe

Rate of wash water = 500 lpm/m^2.

Design flow = 500 × 4.8 × 3.2 = 7680 lpm = 0.128 m^3/sec.

Assume velocity of flow = 2 m/sec.

Dia. of outlet pipe for dirty water required = 28.55 cm.

Provide 30 cm pipe.

Inlet and Outlet Pipes to the Filter Unit

For carrying a discharge of 2.155 MLD = 0.0249 m^3/s to each of the two filter units, a MS pipe of 20 cm dia. may be provided. Same for the outlet pipe from the filters.

3.9 Chamber Between the Units (Intermediate Chambers)

Wherever required, 1.0 × 1.0 × 1.7 m chambers will be provided between different stages of treatment to feed the flow to the next units. These chambers will have the outlets at the top for overflow and scour connection at the bottom. The flow to and from the units will be carried under gravity by pipes of suitable sizes with valves.

Clear Water Chamber/Clear Water Reservoir (CWR)

A set of 3 chambers, each of size 1.0 × 1.0 × 2.0 m SWD, will be put to receive effluent of the two filter units.

4 Conclusion

The summary of designed dimensions for all the proposed units for the water treatment plant has been stated below in Table 1. Water treatment plant design/development in the city has been found to be an essential/critical requirement for the society nowadays. The above work exemplifies the hydraulic design for the real site 4.1 MLD water treatment plant in consideration with the key factors. This study will provide a detailed overview to the hydraulic engineers/researchers for hydraulically designing a field water treatment system in accordance with given conditions and design considerations.

Table 1 Summary of dimensions for different designed units of WTP

S. No.	Unit description	Size in 'm'
1	Raw water inlet chamber	$1 \times 1 \times 2$ m^3
2	Pre-settling chamber	$12 \times 3.75 \times 4.8$ m^3
3	Dosing chamber and mixing chamber	$1 \times 1 \times 3.1$ m^3
4	Flocculation unit/tank	$4 \times 2 \times 2.5$ m^3
5	Secondary settling tank	$4 \times 2.6 \times 3$ m^3
6	**Rapid sand filter**	$4.80 \times 3.95 \times 4$ m^3
7	Back wash tank	$4.8 \times 3.2 \times 2.8$ m^3
8	Intermediate chambers	$1 \times 1 \times 1.7$ m^3
9	Clear water chamber	$1 \times 1 \times 2$ m^3

References

1. Water treatment plant and its importance to earth. https://www.earthreminder.com/water-treatment-plant-its-importance/
2. Silori, R., Gupta, C. S., & Hussain, I. (2018). Hydraulic design of water treatment plant-A case study for Srinagar town. *International Journal of Advances Engineering Research Development, 5*(02), 152–158.
3. Farhaoui, M., & Derraz, M. (2016). Review on optimization of drinking water treatment process. *Journal of Water Resource Protection, 08*(08), 777–786.
4. Devprayag—Google Maps. https://goo.gl/maps/r3mvMUVr8JcWrV9V9
5. Manual on water supply and treatment—1999: Central public health and environmental engineering organisation (CPHEEO), Govt of India. http://cpheeo.gov.in/cms/manual-on-water-supply-and-treatment.php

A Review of Ergonomic Risk Assessment Techniques Employed in Construction Industry

Rajneesh Patial, Himani Gusain, Bikarama Prasad Yadav, and N. A. Siddiqui

1 Introduction

The 1/3rd of the total sick leaves in industrialised countries are due to musculoskeletal disorders or complaints which are the result of poor ergonomic design [1]. MSDs or ergonomic injuries affect muscles, tendons, ligaments while performing tasks that involve awkward position, frequent activities and repetitive jobs causing overexertion and repetition of the same activity [2]. Approximately, 1.71 billion people have musculoskeletal disease worldwide and is a leading contribution of disability by limiting the mobility and dexterity of worker and leading to early retirement from their work [3]. Construction industry is one of the most stable and growing sectors in the world, and workers of construction industry are involved in too much physical work rather than using machines, leading to MSD [4]. MSD is the single largest cause of 77% newly reported occupational disease among construction workers all over world which leads to productivity reduction of workers, disability, sick leaves and work days lost [4]. Construction workers are injured 2.5 times more than general work force workers as it involves manual work; leading cause of overexertion injuries was manual lifting, carrying, moving objects, shovelling and digging [5]. Among construction workers, maximum MSDs due to overexertion were found in carpenters(29%) followed by plumbers and fitters(28%) and least by roofers(22%) [5]8.

Repetitive strain injury causes MSDs by repetition of the same work for a longer duration without taking breaks or mechanical aid described by prolonged pain in shoulders, neck or arms and soft tissue injury like spasm, carpal tunnel syndrome, tingling sensation, numbness, stiffing, swelling, weakness, etc. [6]. Musculoskeletal

R. Patial · H. Gusain · B. P. Yadav (✉)
University of Petroleum and Energy Studies, Dehradun, India
e-mail: bikarama@gmail.com

N. A. Siddiqui
GD Goenka University, Gurgaon, India

N. A. Siddiqui et al. (eds.), *Advances in Construction Safety*,
https://doi.org/10.1007/978-981-19-4001-9_9

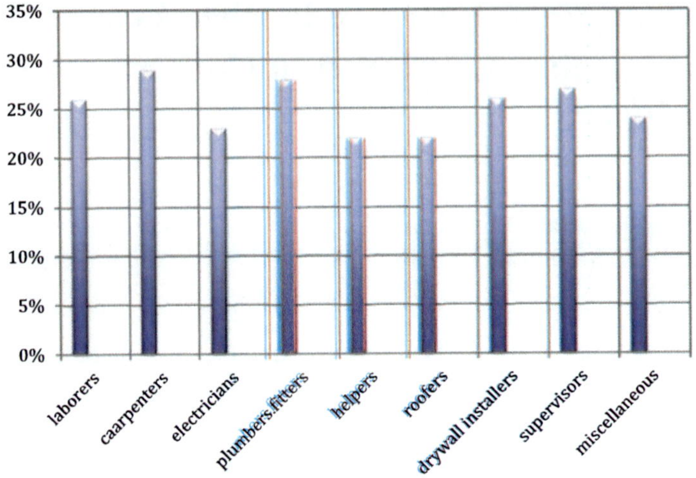

Fig. 1 No. of MSDs in construction industry in different construction tasks [5]

condition affected more than two thirds of person in the age group 51–61 and in all except 10% of the disabilities, prevention will improve their employment opportunity [7].

With the increasing risk and adverse effects of MSDs, there has been new methods introduced to identify the risk factors leading to the disease. These methods have been developed and improvised over the course of time. Aim of the review is to carry out study of the ergonomic risk analysis in various construction operation, physical and manual work that can cause fatigue in muscles and shows chronic effects. Various methods that have been adopted or used are illustrative type like the nordic musculoskeletal questionnaire [8], used to find out musculoskeletal problems in construction workers; Tonometry measurements were made to find out skeletal muscle and fatigue in muscles, grip force as measurement of muscle strength [9] (Fig. 1).

However, the risk factors like different body posture and repetition can be identified but other risk factors like vibration and exertion in workplace are still not quantified. The risk of injury is also related to activity duration, combination of exposure to risk factors, intensity of work and no. of times activity is done. Number of studies have been carried out for better understanding of WMSD in construction industry. Li and Buckle reviewed the methods for the assessment of physical workload stress and reviewed the methods for assessing exposure due to physical workload leading to WMSD risks due to poor ergonomic; also described the already available assessment method for WMSD exposure as not suitable for real life work situations [10]. Definition of ergonomic and risks in construction industry was studied and a link was presented between humans, work environment, machines and design of job [11]. A very little time left for the recovery of the muscles may lead to musculoskeletal disorders due to repetitive awkward posture activities, lifting, pulling, gripping and pushing. A range of methods have been developed for the assessment of risk factors of WRMSDs such as self-report from operators, observational methods and direct measurement methods by using sensors attached to the subject for measurement of exposure to risk factors at work [12]. A study done on WRMDs lost time claims statics to show the losses and adverse social impacts due to current legislation and practices and potential areas of development [13]. In study, a big highway tunnel construction project was observed, and a number of ergonomic hazards were observed. A method was developed to compare risks, identifying process to reduce ergonomic hazard and deciding the most accurate way to lower MSDS of each job [14]. A research focused on analysis of ergonomic risks involved in the construction industry in different operations at work by carrying out the research using objective and subjective methods, including tonometer measurements, nordic musculoskeletal questionnaire and muscle's force determination with dynamometer [14]. The results proved that muscle tone in construction workers within 1 week cycle of work and fist muscle forces also decreases [15]. It was believed that this combination of objective and subjective ergonomic risk analysis methods presents a complete plan and gives trustworthy results [15].

2 Common Risk Factors of Musculoskeletal Disorders

2.1 Vibrations

Vibration is important factor in working environment, which is from the power tools and when used over time, it can cause MSDs.

Vibration acts as an impacting force to the organs at higher frequencies which harms the organs. The vibration applied on hands can cause blood deficiency in the fingers called Raynaud's infection).

2.2 Repetition

When an activity or movement is performed by a muscle tissue, body part or joint over a period of time with very little chance of recovery or rest to that part is called repetition. Repetition leads to higher risk of injury when factors such as awkward posture and heavy force. Repeated motions of similar type involving similar muscle groups can lead to muscle fatigue. Greater degree of risk is associated with greater repetitions. The risk can be reduced by factors such as force, task time, recovery time and way of sitting or operation.

2.3 Force

Force is a mechanical or physical push to carry out exertion or movement. The physical exertion of force is required to operate equipment or tools or perform specific tasks which could overload muscles or tendons. The physical exertion of force is a primary risk factor. Force is generally required to lift, lower, carry, gripping an object, pushing or pulling. The over exertion of muscles causes the muscles to contract harder than normal which further causes stress on joints, muscles and tendons.

2.4 Awkward Posture

The position which generated extra stress on different parts of the body at a certain time is generally called awkward posture. The muscles and tendons work harder than normal when in awkward position, then over a time, it results in stress build-up. The bending and twisting of body excessively out of regular or comfortable or normal range of motion have more risk of injury. Stooping, crouching, twisting and standing on fix position with your hands or arms overhead are considered as awkward posture.

2.5 Contact Stress

Injury to the tissues and nerves under the skin is called contact stress. The injury or impingement is caused due to sharp objects, hard objects, using or grabbing equipment. For example, when muscles hit the sharp edge, it causes muscular impingement. Contact stress generally affects the soft tissues which are on shins, palms, feet, forearms and thighs.

2.6 Extreme Temperature

Temperature has two extremes either it is too cold or too hot. Cold temperature accentuates symptoms such as nerve end impairment and hinders manual dexterity. The human body when exposed to cold causes symptoms such as shivering, dilated pupils and clouded consciousness. Heat stroke is caused when body is exposed to extreme heat. Heat stress can cause less serious conditions such as heat exhaustion and dehydration.

2.7 *Static Loading*

Human body needs movement, remaining still puts stress on it. Maintaining same body position over an extended period of time without changing position can be tiresome and uncomfortable. The body movement allows the joints to dissipate the stress which is developed over time by staying in same position.

3 Risk Assessment Methods of Work-Related Musculoskeletal Disorders

A wide range of section of people has tried to find out measures to find risk factors and treat them as base for reduction and detection of WMSD. People involved in it are researchers, regulating authorities, owners and employee representatives. Researchers and medical technology have been working towards the motive of mitigation and prevention of the disorder. Various interventions have been introduced to provide new instruments and to redesign the job plan. WMSDs are generally developed due to uncomfortable posture, repeating similar activity and high force exertion. Method for risk assessment of MSDs is done according to following categories:

(1) Self-report
(2) Observational methods
(3) Direct measurement
(4) Remote sensing
(5) Smart wearable devices.

3.1 *Self-Report*

Workers diaries, interviews and questionnaire are used to collect information of the physical pain, postural discomfort, exertion and psychosocial factors responsible for MSDs, also includes videos and questionnaire from Internet [12]. Estimation can be made for longer time period by making observation at the workplace, taking large sample size [12]. Workers provide description of the symptoms occurring due to a work by marking on a body diagram, the area of discomfort while working and report can be submitted online [16]. Historically, a number of questionnaire have also been developed to assess musculoskeletal symptoms assessed by Nordic musculoskeletal questionnaire [17]. A study done on self-report method to assess ergonomic exposures concludes that the method has acceptable internal consistency, greater construct and criterion validity compared to observational methods [18]. One of the advantages is that workers can directly report their problems which occur during work and could not be identified by observation; another is that the method is cheaper and provide accurate assessment of the work-related MSDs [12]. The nordic musculoskeletal questionnaire (NMQ) is method used with other techniques to identify MSD, psychosocial risks and assessment of the sustainability of the organization [19].

While its limitation includes the imprecise and unreliable perception of exposure of the worker, worker illiteracy and interpretation of questionnaire; hence, the reliability and validity are too low for this method [12]. In 2017, Lopez et al. researched about the application of the standardised Nordic musculoskeletal questionnaire in sectors such as agriculture, forestry and manufacturing industries which says that the ergonomic NMQ limits the study to a limited body parts such as neck, lower back, shoulders while leaving out extremities, upper back [19].

3.2 *Observational Methods*

Observation is the process of recording the exposure factors in workplace so as to assess the potential risk of MSDs by using different techniques. Observation is carried out by experts and recorded on the sheets [12]. A number of tools were developed for observation-based method to record exposure factors such as posture, load, force, movement frequency, duration, recovery and vibration. These tools were developed to allow experts in evaluating and recording the observation data, e.g. checklist for different body parts, Ovako Working Posture Analysing System (OWAS) and Posture Activity Tools Handling(PATH) [20]. It is defined as posture, activity, tools and handling when compared with OWAS, and it assessed the working postures with workers activity, load handling, grasp and tool use [17].

Rapid Upper Limb Assessment is developed and used for upper limb disorders reported in workplace, RULA is a survey technique which provides quick assessment of the upper limbs, trunk and neck with respect to muscle function, without any need of special equipment [21]. Another tool rapid entire body assessment (REBA) is a specific designed tool sensitive to unpredictable working postures in areas such as service industries and health care [22]. Finnish Institute of Occupational Health in 2014 developed a semi-quantitative and time-based observation method to estimate the physical loads on upper extremities assessing factors such as use of hand force, pinch grip, non-neutral wrist posture, elevation of upper arm, use of hand and local mechanical pressure, and the data was validated against expert observation from the video, continuous recordings of EMG [23].

EAWS is an ergonomic screening tool for physical work load and it is an internationally accepted method, also a European legal requirement [1]. European Assembly Worksheet (EAWS) is an risk assessment tool which assess the work cycle in five different sections: general, postures, action forces, repetitive load of the upper limbs, manual material handling and assigns the final score in traffic light scheme of three levels green, yellow and red with ergonomic intervention recommendations [1, 24].

Postural ergonomic risk assessment (PERA) is method used to assess the postural ergonomic risk of the short cyclic assembly work which provides analysis of every work task in the work cycle [24]. PERA uses Sperling's cube method for assessing the ergonomic risk. The work cycle is divided into different tasks according to postures and job content. Then, every task is classified into three demand levels (low, medium and high risk) for posture, force and duration. Then, work task score is calculated and validated, evaluated against EAWS [24]. PERA gives analysis of every job task as well as overall evaluation of the job which helps in identification of high risk sources. The findings by [25] indicate that there is significant impact of the lack accuracy of inputs on the outputs in ergonomics analysis [25].

3.3 Direct Measurement

It is used to replace the observation by experts and is more accurate than the observation method. In this method, sensors are attached to body or clothes of worker to record body segments and motion of the joints at particular angles, angular velocities and moments at particular body part, leading to a much more precise response of the body to the effect of physical work on that particular portion using a device called inertial motion capture (IMC) [26]. Direct assessment is carried out in laboratory or real situations and the measurement devices include optical markers, electromyography (EMG), goniometers, optical scanners, inclinometers and sonic sensors [26]. EMG is used primarily in studying muscle exertions by attaching the sensors to the target muscle group which is used to record EMG signals from 8 trunk muscles [17]. There are certain postures which are more desirable than the other and inclination of some postures causes discomfort [10]. Data obtained from the sensors is combined with the data from EMG and this combination gives the estimated load on tissues in muscles by biomechanical modelling [27]. Another instrument uses reflective markers to capture the motion of a person working with the help of a software for tracking markers [28]. In 2017, a biomechanical analysis EMG study of WMSDs during repetitive lifting operations is carried out by construction workers done by [29] in an controlled laboratory experiment revealed that repetitive lifting weights significantly increases chances of surface EMG which is directly related to muscle fatigue. Research revealed that fatigue in lumbar erector spinae was the highest compared to other bicep branchii, brachioradialis, medial gastrocnemius and rectus femoris muscles, while lowest EMG activity was recorded on rectus femoris muscle [29]. Main advantage of the software is its flexibility to track human movement in an unsupervised environment; tracking environment was successfully tested in activities like walking, running, breathing, etc. [28]. A carbon nanotube sensor has been developed which is wearable and stretchable to track strains up to 280% with the help of gaps formed in nanotube structure due to work strains, small joints like fingers, upper, lower limbs, etc. [30]. These nanotubes have high durability, low creep and faster response and can be attached to various clothing items like stockings, bandages, gloves, etc. to measure typing, breathing and speech [30]. These instruments require controlled conditions to assemble data related to motion of body and muscles [31]. Main advantage of the process is that direct measurement method provides accurate and detailed information; processing of data after collection is simple [12]. Disadvantage includes the large number of subjects or the long-term data collection, high cost, time requirement, data storage problem also the sensors attached to body makes them uncomfortable and hence alter the result of measurement [17].

3.4 Remote Sensing Techniques

Advanced remote sensing techniques use video feed and image sensors to capture movement of body parts, then risk assessment is done using kinematics motion data that acts as input for existing observation risk assessment methods, and these practical skills do not demand equipment directly hooked to human body [17]. Various studies have been conducted to determine the feasibility of video stream to carry out WMSD-related assessments. This approach provides the ergonomist a cost friendly method to conduct onsite evaluations of the biomechanics of the lifting job using a camcorder without the need of complex equipment [32]. The assessment of lower back moments and loads using an video-based analysis in an occupational environment for lifting tasks is an cheap and easily applicable method [33]. The video-based analysis requires manual identification of joints to obtain kinematic data such as angle of the joints and body part acceleration [32, 33]. A real-time tracking system is developed to keep track of complex human dynamic motion by using an integration of two modules Kinect and augmented reality module, Kinect sensor records the depth of the images while augmented reality module computes distance parameters [34]. SDK identifies the 20 joints in human skeleton based on depth of each pixel, colour data and Kinect provides input to the augmented reality module which processes it into virtual reality experience. Kinect devices have a high sampling rate which accurately determines duration and frequency of exposure with other observational techniques for postural assessment [35]. In 2013, [36] developed a framework of an automated vision-based unsafe posture detection for recording the behaviours of the workers. Microsoft Kinect a widely used motion sensing device may suffer instrument error due to poor lighting conditions or maybe due to target reflections and low resolution of depth measurements [25, 37]. According to accuracy study of Kinect depth sensor, it is found that as the distance of subject increases from the sensor, a random error involved in depth measurement increases and it ranges from 4 cm to millimetres [37] (Table 1).

3.5 Smart Wearable Devices

Wearable devices are tiny electronics device which has capabilities of storage, sensing, communication and processing and is worn by the user to determine awkward working postures, vital signs, kinetics and human movement developed for evaluating or controlling physical loads or posture [46].

Table 1 Comparing risk assessment methods for WMSD

Assessment methods	Example techniques	Advantages	Limitations
Self-report	"Questionnaire/Human body map" [38, 39]	Easy technique, boosts morale, highly applicable, budget friendly, very practical to use, less work disturbance, could be used in both indoors and outdoors	Personal perception, low resolution and accuracy
Observation	"OWAS/PATH" (Buchholz et al. 1996); "RULA" [21] "REBA" [22] EAWS[1] FIOH[23] PERA[24]	Highly applicable, budget friendly, very practical to use, less work disturbance, could be used in both indoors and outdoors	Dependent on the bias of expert, OWAS, RULA, REBA cannot combine risk from multiple sources, very less diagnosing can be done in certain period of time, high input inaccuracy
Direct measurement	"(Wireless) EMG" [40] "Vicon" [41] IMC	Automatic and accurate data collection, variety of data collection, could be used in both indoors and outdoors depending on the technique used	Very time-consuming, makers to be attached very technically to the skin tissue, uncomfortable, work hindrance, limits the movement, movements may be non-uniform
Remote sensing	"Kinect" [36] "Stereo camera System" [32, 33, 36]	Budget friendly, could be used in real-work places, can be automated, could be used in both indoors and outdoors	Reflection from the surfaces, sensitive to light radiation, less range of few metres, data processing, technical, high chances of input inaccuracy
Wearable devices	Insole pressure system [42], (Antwi-Afari, Li, Umer, et al.) Wearable mobile phones [43] Wearable IMU [44, 45]	Non-invasive, cost friendly, highly feasible to use in construction industry, post processing not required once system/algorithm established, can be used both indoors and outdoors, processing power or modern mobile devices could be used	Low output frequency, low battery capacity, proper installation required

In 2018, a wearable insole pressure system was developed by [42] which detects and classifies static awkward working postures (as per ISO11226) involved in construction industry which will help safety managers and researchers to continuously and objectively evaluate bad working postures. Insole pressure system uses OpenGO system (Motion GmbH Munich, Germany) for measuring spatiotemporal foot planar pressure distribution and triaxial acceleration to detect awkward postures [42]. The author performed the study in controlled laboratory environment involving only one legged kneeling, stooping, semi-squatting, squatting and overhead working [42]. The study lacks on a lot of areas such as effect of ambient environment, pulling/pushing load, walking with loads, vibration and individual factors (age, gender, experience). In 2020, [47] published a feasibility study (as per UMass Lowell OSHA) to measure the risk of WMSDs due to overexertion by using data sets of activities such as grip force, lift/lower/carry, pull/push and any other non-risk activity. The limitations are similar to [42] also the number of data sets is less compared to previous study. The current study is limited to selected to task or activities not including sawing, hammering, installing rebar, etc.

An ergonomic study published by [43] in 2017 in which he experimented using wearable mobile phones by making use of their built-in accelerometers, gyroscope and linear acceleration sensors to determine trunk and shoulder flexion during the task. The flexions measured for both trunk and shoulders were within ±3 degrees of the true value, hence, wearable mobile sensors is a low procurement and low maintenance cost, easy to use tool for ergonomic analysis [43].

Ref. [44] used wearable IMU devices to develop a novel system and data processing frame work for posture evaluation experimenting on brick laying task. In his study, integrated IMU devices in a wireless body area network and used machine-based data processing method to determine inadequate working postures according to ISO [44]. Although, the study needs to expand the experiments to other construction activities to develop better machine learning algorithms. A wearable inertial motion sensor (WIMU) that has real-time warning system to aware the workers on bad posture related to back and neck, with 2 IMU sensors placed on the back of the worker's safety helmet and other on middle upper part of the wearer's back connected to smartphone application to receive and process real-time motion data [45], while the disadvantage of the IMU-based system is the proper installation of sensors on the human body, battery life and output frequency.

4 Discussion

This review lets us know about the gap in current techniques for ergonomic risk assessment, the requirement of assessment tools in construction industry and how the techniques evolved. Over the years, as the safety awareness increased, this led to the development of new methods for ergonomic analysis which can be used in real-world setting. The earliest tools were developed was self-report in which workers were asked to fill the questionnaire according to the difficulties or problem faced

Safety and Risk Management in Oil and Gas Industries: A Review

Anilett Benny and V. R. Renjith

1 Introduction

Process industries involve numerous process susceptible to hazard that can cause accidents in the operational sequence. Risk and safety are two significant factors considered inevitable for any process in an industry for safe operation and achievement of desired targets. Risk is the probability of a hazard to cause an accident that could pose a threat to health, property or environment. The key focal components of that provide vital parameters for an effective risk assessment are risk identification, analysis & evaluation [1]. The procedure of risk identification helps to identify, isolate and describe the risks that could influence the target objectives. Risk analysis helps in understanding the nature, sources, causes of the risks that have been identified, and to quantify the level of risk. It is also used to study impacts and consequences. It is also a vital tool to examine the process controls that currently exist. Risk evaluation compares risk analysis results with risk criteria in order to determine whether or not a specified level of risk is acceptable or tolerable. Therefore it is pivotal to examine all possible risks of a condition, assess and formulate the preventive measures to avoid occurrence of an incident or control the intensity of the accident that may occur [2].

Process industries like oil and gas, more specifically offshore oil and gas exploration activities fall under high intensity risk criteria. An accident in offshore oil and gas industry can lead to devastating consequences and catastrophic effects to men, material & environment on top of the multitude of financial burden it can lead to. The accidents in oil and gas rigs have been occurring since hydrocarbon exploration started in mid nineteenth century. History has witnessed several such catastrophes in form of Piper Alpha in 1988 killing 167 personnel to the most recent Deepwater

A. Benny (✉)
Division of Safety and Fire Engineering, SoE, CUSAT, Kochi, India
e-mail: bennyanilett1989@cusat.ac.in

V. R. Renjith
Division of Safety and Fire Engineering, SoE, CUSAT, Kochi, India

N. A. Siddiqui et al. (eds.), *Advances in Construction Safety*,
https://doi.org/10.1007/978-981-19-4001-9_10

Horizon incident on April 20, 2010 that led to the largest accidental oil spill in history. The cause of Piper Alpha and Deepwater Horizon were both manifestation of a series of system or process failure. When insufficient management of change (MOC) procedure is largely speculated to have caused the dominos effect on Piper Alpha platform, failure of critical safety equipment namely subsea blowout preventer (BOP) led to Deepwater Horizon incident. Deviations from SOPs can also be regarded as primary initiating cause of both Piper Alpha and Macondo incident (Deepwater Horizon). It is a pertinent observation that human influence that led to changes in operating practices can be attributed to have played a major role in instigating the preliminary cause of accident. Oilrigs deal with extreme pressures, heavy machinery, constrained space and highly exothermic chemicals all in the vast expanse of seas that create a huge logistical constraint in itself. Hence, safety and risk management play a very crucial role is offshore oil and gas rigs.

2 Oil Well Drilling

The upstream oil and gas industry revolve around exploration and production of hydrocarbons. In spite of quantum leaps in field of seismic science, there is no substitute to drilling to positively determine presence of commercially viable hydrocarbons in the deepness of earth's crust. Oil and gas wells are drilled using drilling rigs that are usually onshore or offshore. The present generation rigs are highly sophisticated machinery that can drill into the earth's crust to depths of up to 30000 ft in water depths of upto 12000 ft. With more than a century since the first oil well was drilled, infusion of technology and thus inherent safety has gained substantial importance on the table. Drilling process involves planning and execution of creating a borehole to a target depth as per a laid out scientific plan. On a deepwater drilling rig, this involves components like station mooring of the rig, physical drilling machinery, medium to clean the cuttings generated & various safety barriers. The synchrony of all the aforesaid systems working in unison is the most desired process scenario [3].

3 Station Mooring

A floater deepwater rig can be broadly classified as Semi-Submersible and Drill Ships. The functionality & utilization is each type is based on several factors that focus on sea/weather conditions of the region. The primary requirement of such a drilling vessel is to ensure stability in position in the sea during almost all weather conditions within design parameters. The design capability of the rig to maintain position is known as station mooring capability. This is achieved by dependency on Global Positioning System (GPS), onboard Mechanical Thrusters & Feedback system.

Once drilling rig position itself and commences drilling well below seabed, under no circumstance can the drilling rig deviate beyond an acceptable design parameter away from the moored station. Once the drilling rig is connected by means of marine riser system to BOP and wellhead the design parameter for deviation reduces drastically. The rigs are also expected to operate Offshore Supply Vessels (OSV) and helicopters for supply & logistics of men and material. In a dynamic sea state, the rig is expected to maintain position to successfully work the vessels and chopper. There exists scenario of transfer of highly combustible fuel oil or oil base mud system from OSVs and hence margin of deviation from moored station further reduces. The complexity of operation and risk further increases as both rig and OSV are expected to maintain their mooring within limits. Failure of station mooring system can lead to catastrophic consequences.

The drilling rigs are equipped with inherent redundancy mechanism to address such scenario of equipment or process failure. Most modern rigs are equipped with Dynamic Positioning-Third level (DP-III) system that has not one but two levels of redundancy. The incidents that led to catastrophic disaster in ONGC owned BHN platform is a grim reminder about the need to DP-III system.

4 Drilling Process

Drilling process associates itself with a background of large scientific study that includes detailed geological prognosis based on seismic study & offset well data. Hydrocarbon wells are designed with safety parameters laid down based on technical analysis and historical experiences. In spite of all preparedness, several complications are experienced in drilling operations. These include stuck ups, downhole tool failures, surface equipment failures etc. The risk associated with drilling operation is substantial due to involvement of several components. The financial implication is very huge but the most dangerous of all these revolve around the human presence in the process or complication mitigation that could lead to personnel injury.

One of the most common procedures adopted by industry is the system of Management of Change (MOC), which deals with organizational or process change synchronization. An effective utilization of MOC is very crucial in a highly complex environment as in a drilling rig. The events on Piper Alpha platform put forward a big learning lesson as to the importance of such processes. Another critical component of the operations is the permit system. Permits are control measures designed to ensure multiple operations do not intersect or counter act to cause incidents. Permits are system requirements on any typical process that if over sighted can lead to significant events.

Drilling process typically involves highly pressurized environment, extremely combustible components, usage of radioactive sources, explosives meant for controlled explosion & high-speed moving components. These consolidated in space crunched offshore rigs mandate strict access control and barricade mechanism that forms the lifeline of system control to ensure safety. In a statistical survey conducted in US, about 70% of all fatalities in mining industry have occurred in Oil and Gas. The risks involved in drilling operation are multi-dimensional and highly cost implicated. The need for timely review and preventive intervention to mitigate such risks is largely adopted in drilling industry.

5 Well Barrier

As per API RP 96, a minimum of 2 tested & independent barriers are required for drilling a safe well. The primary barrier during a drilling operation is the drilling fluid that maintains well integrity in terms of well control. Advanced technology like MPD/UBD takes additional support of RCD, Auto Choke & NRVs. It is the first barrier against any well bore intrusion. Secondary well barrier come into picture when the primary barrier fails to maintain well integrity. Hence, primary well barrier is typically enclosed by secondary barrier. The most important form of secondary well barrier is well control equipment like BOP stack, Chokes, etc. The drilling process follows a set procedure of Safe Operating Practices that ensures that well barrier integrity is not compromised. SOPs and MOCs play a very important role in ensuring that complexity of operations does not fail this primary requirement. However, history has shown us instances in form of the Macondo Incident that primary barriers are under risk of compromise due to unforeseen circumstances followed by a sequence of process that materializes to failure of secondary barrier that leads to an Accident Event.

6 Risk Assessment

Drilling is a hazardous operation, and assessment of safety in processes associated is measured as a factor of risk. Risk is defined as probability of accident likelihood and the magnitude of loss to health, property or environment. Risk management system involves several major steps. The first step would be Hazard identification, which includes physical, mental, chemical or biological identifications. The second step is to decide who may be harmed and how. The third step would be assessing the risk and implementing control measures. Final step includes recording the findings and reviewing the control measures. Several qualitative and quantitative methods have been developed to assess risk management system in harsh environments. Qualitative risk assessment is intended to identifying likelihood and impact of risk event occurring on the overall process. On the other hand quantitative methods specify probability density of overall consequence of risk in the process. Among the various risk evaluation/analysis methods used in industry, the most widely used are Physical Inspections at various levels of assessment, Operational charts & flow charts for processes, safety review check list analysis, Relative ranking or "What if" analysis. Some of the focused methodology that involve case by case approach are Hazard and

Operability Study (HAZOP), Failure Modes and Effect Analysis (FMEA), Fault Tree Analysis (FTA), Event Tree Analysis (ETA), Layer of Protection Analysis (LOPA), Cause Consequence Analysis (CCA) and Human Reliability Analysis (HRA) [4].

Bow-Tie diagrams have been considered as one among the best tools to demonstrate the behavior of a risk that can cause an accident to its further consequences. It describes the cause effect relationship of an accident by systematic work through hazard and its management. A typical Bow-Tie diagram is shown in Fig. 1. The left bow corresponds to all possible causes of a hazard or a top event that defines fault tree consisting of events and logic gates. These events can be primary or basic events, initiating events, external events, conditional events, undeveloped events and undesired or critical events. The fault tree is represented by Boolean logical AND and OR gates that interconnect events and specific conditions as shown in Fig. 2. Exclusive OR gates, Priority AND gate and Inhibit gates are also used in FTA. The top event will happen if output of AND gate goes high which means that all inputs to this particular gate should go high. However, the output of OR gate goes high if

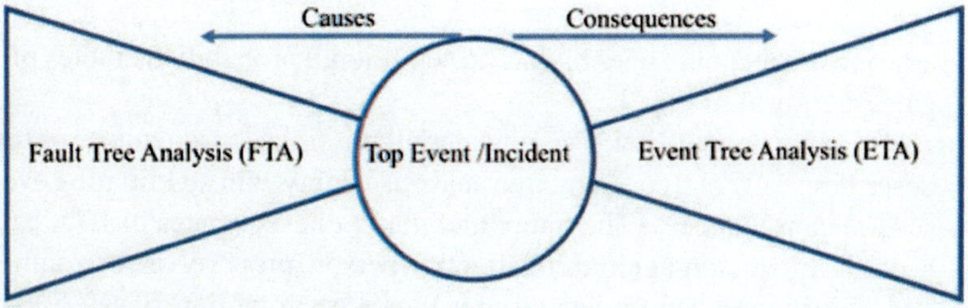

Fig. 1 Typical bow-tie model

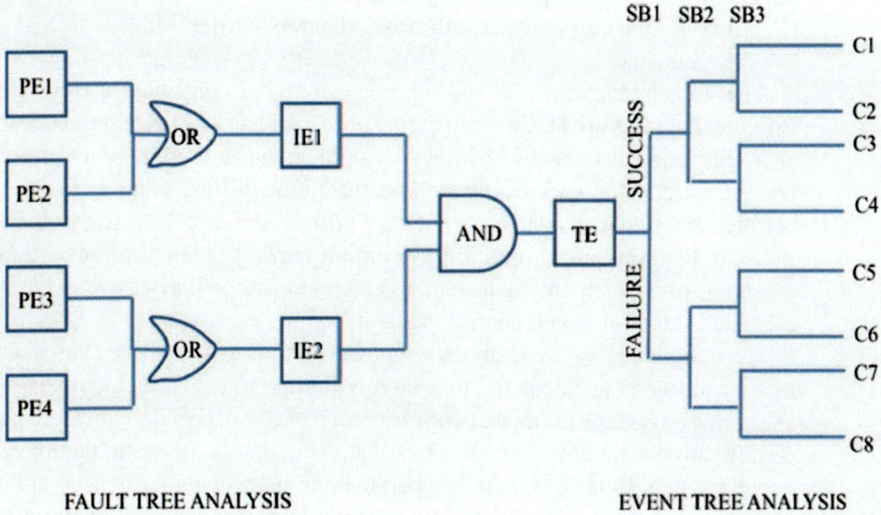

Fig. 2 Bow-tie model with events and logic gates

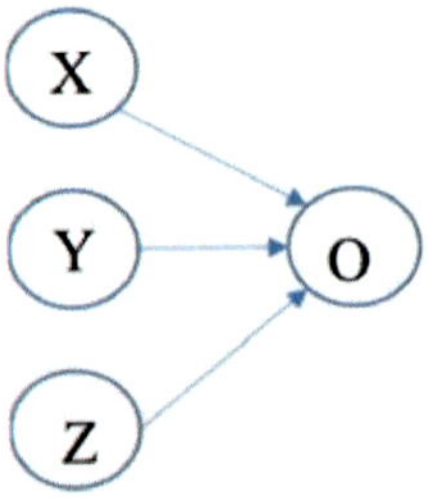

X	Y	Z	O (OR)	O (AND)
0	0	0	0	0
0	0	1	1	0
0	1	0	1	0
0	1	1	1	0
1	0	0	1	0
1	0	1	1	0
1	1	0	1	0
1	1	1	1	1

Fig. 3 Conditional probability tables for OR and AND gates

any one of the inputs goes high. The conditional probabilistic tables of AND and OR gates are given in Fig. 3.

The right bow of Bow-Tie indicates all possible consequences of top event which describes an event tree. Event tree analysis begins with an initiating event and ends in several consequences. The minor and major consequences of ETA can be classified according to their magnitude of affecting person, property or environment. The minor and major consequences depend upon the success or failure of safety barriers installed in the system of event tree. For example, during course of drilling operations, the consequences can be safe operation, kick, blowout, kill well, shutdown, catastrophic effects etc. [5]. As per Fig. 2, turning on safety barrier 1, 2 and 3 results in consequence C1. Similarly the consequence will be C2 if safety barrier 3 fails with SB1 and SB2 in successful state.

The fault tree, event tree and top event together contribute a Bow-Tie model. Bow-tie diagrams are also used to define proactive barriers to limit the occurrence of an accident and also reactive barriers to diminish the magnitude of its consequences. Oil & Gas Industry, specifically well control in the drilling industry leans on robust, certified & tested well control barriers as a primary defense. Reactive barriers include influx mitigation barrier, ignition prevention barrier, escalation prevention barrier, emergency management, management and organization barrier that assist is various escalated stages of a well control incident.

Even though Bow-Tie analysis is a powerful tool to determine causes and effects of a hazard or an accident and thus largely limited to a post incident analysis or for scholastic advantage rather that assisting in a reactive or dynamic incident prevention or mitigation. The key characteristics that drive this is the static nature of its fault and event tree. Hence it is handicapped in defining probability density and updating conditional dependency of risks of a process. Bayesian network or Belief networks overcame these limitations [6].

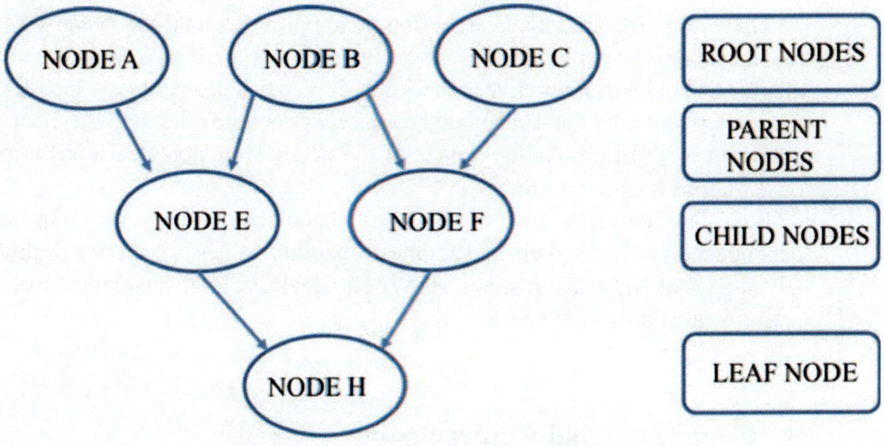

Fig. 4 A Bayesian network model (*Source* 3: LPPI 50, p. 143)

BN represents joint probability distribution of variables in the system. It is a graphical system having both qualitative and quantitative parts with nodes/variables and links connecting them. The qualitative part is that it has a Directed Acyclic Graph (DAG) in which each node represent system variable and arrows show the relationship between nodes. The directed arcs connect pairs of nodes that illustrate a cause effect relationship as shown in Fig. 3. A node is called a parent node if there is a directed arc connecting it to another node called the child node. Nodes, which have no parent, are known as root nodes as shown in Fig. 4. The quantitative part is the conditional probabilistic table (CPT) of each variable that defines the probability density [7]. The probability of child nodes is calculated by considering the probabilities of the parent nodes and conditional probability tables (CPT) by obeying the fundamental well-known "Bayes theorem" of conditional probability.

In a typical Bayesian Network, the JPD (Joint Probability Distribution) of $X = \{A1, \ldots., An\}$, is defined as shown by the equation:

$$P(X) = \prod_{i=1}^{n} P(A_i | Pa(A_i)) \tag{1}$$

where $Pa(A_i)$ is the parent set of A_i.

BN helps to update the prior probability of events such that we have new information called evidence E thus giving an updated probability called posterior probability [8]. This is the conditional probability value that is available after taking into account the evidence of a process. The equation below represents posterior probability of variables

$$P(X|E) = \frac{P(X, E)}{P(E)} = \frac{P(X, E)}{\sum_X P(X, E)} \tag{2}$$

These networks find application in human reliability analysis because of the fact that the model can perform both forward or predictive analysis as well as backward or diagnostic analysis. Predictive modeling is done when probability values are defined a priori for root nodes and calculated by inference for the other nodes. The diagnostic approach is when probability values of the nodes are calculated a posteriori when observations become available.

Bayesian Networks have the ability to handle probability variables, represent dependencies between them and update probability values over period of time. These advantages of BN make it superior over classical risk analysis methods such as BT analysis due their static nature.

7 Conclusion and Future Scope

This paper discusses an overview of aspects of drilling and its associated risks. Major risk assessment management systems and methods or strategies implemented to assess risks are also discussed. A Bow-Tie model system illustrating the cause effect relationship of a top event can be mapped into a graphical network with a set of variables called Bayesian Network. The occurrence probability of an event in harsh environment (drilling operation) can be predicted by monitoring the condition of data parameters in a process and their dependent relationships from process specific data. Data-driven risk assessment methodology provides an efficient approach for predicting a dynamic system risk profile of a drilling operation or its associated processes. The time dependent risk profile generated from this methodology can be used to activate safety measures at any time the risk predicted exceeds the acceptable level.

Acknowledgements I express my sincere gratitude towards my Supervising Guide who constantly guided me for identifying a research area and supported me attend various technical lectures and webinars to help me build knowledge for the purpose of my research.

References

1. Abimbola, M., Khan, F., & Khakzad, N. (2014). Dynamic safety risk analysis of offshore drilling. *Journal Loss Prevention Processing Industrial, 30*, 74–85.
2. Bhandari, J., Abbassi, R., Garaniya, V., & Khan, F. (2015). Risk analysis of deepwater drilling operations using Bayesian network. *Journal Loss Prevention Processing Industrial, 38*, 11–23.
3. Abimbola, M., Khan, F., Khakzad, N., & Stephen, B. (2015). Safety and risk analysis of managed pressure drilling operation using Bayesian network. *Safety and Science, 76*, 133–144.
4. Kanes, R., Ramirez Marengo, C., Abdel-Moati, H., Cranefield, J., & Vechot, L. (2017). Developing a framework for dynamic risk assessment using Bayesian networks and reliability data. *Journal Loss Prevention Processing Industrial, 50*, 142–153.
5. Khakzad, N., Khan, F., & Amyotte, P. (2012). Dynamic risk analysis using bow-tie approach. *Reliability Engineering and System Safety, 104*, 36–44.

6. Khakzad, N., Khan, F., & Amyotte, P. (2013). Dynamic safety analysis of process systems by mapping bow-tie into Bayesian network. *Process Safety Environment, 91*(1–2), 46–53.
7. Adedigba, S. A., Oloruntobi, O., Khan, F., & Butt, S. (2018). Data-driven dynamic risk analysis of offshore drilling operations. *Journal of Petroleum Science and Engineering, 165*, 444–452.
8. Khakzad, N., Khan, F., & Amyotte, P. (2013). Quantitative risk analysis of offshore drilling operations: A Bayesian approach. *Safety and Science, 57*, 108–117.

Hazards and Risk with Heavy Machineries Operation at Construction Site: Preventive Approach

Bikarama Prasad Yadav, Saara Vashishtha, and Deep Mehta

1 Introduction

Machines are used everywhere to bring ease to our lives. On a construction site, a number of large machines are used to ease the difficult tasks such as lifting heavy material, earthwork operations or hoisting while drilling. Such machinery is known as heavy machinery. The use of heavy machinery reduces the time of operation and makes the task less intensive as it completes the task that would manually take hundreds of people and weeks of labour in a few hours [1]. When it comes to operating these heavy machineries such as cranes, it becomes a very complicated process as it requires skill and trained operators. According to Vietnamese standard, there are various standards and regulations, one must know about the insecure areas and what kind of safety precautions must be taken while lifting the device. Apparently, there are no such rules attached or made for both insecure areas and for lifting devices which tends to create a lot of problems for the operator, hence, identification of insecure areas plays an important role depending on how small or large it is. Neither too small is good nor too big. Both have their flaws, small results in danger and too much of a large insecure area increases the cost of fencing and other materials [2, 3]. Massive construction project may involve more than a hundred cranes at a time. The design and use of these cranes vary from project to project. They are further classified into mobile and tower cranes depending on their availability. Mobile cranes are usually smaller in size with aloft capacity of around 20–80 tons. They consist of boom angles which can be moved in any direction, making them accessible to use. These boom angles are basically driven by the motor drum on which cables are wounded on them. Tower cranes are the ones which can be found inside or outside to a building structure [4]. In a case study by graphic product staff with OSHA presented the boom angle failure due to the lack of knowledge of the operator about the length of the

B. P. Yadav (✉) · S. Vashishtha · D. Mehta
University of Petroleum and Energy Studies, Dehradun, India
e-mail: bikarama@gmail.com

N. A. Siddiqui et al. (eds.), *Advances in Construction Safety*,
https://doi.org/10.1007/978-981-19-4001-9_11

boom which lead to the collapse [5]. Similar collapse of crane was observed due to lack of maintenance of machinery in a study presented by marine insite [6]. The heavy machineries they use such as cranes should be properly inspected, checked and should be in proper working condition. Most importantly, these all machines should be guarded and well barricaded to avoid fatalities. We know accidents are inevitable but by using various safety measures and reducing the risk to an acceptable level, we can achieve safety and also prevent these uncontrollable accidents. The main objective of this paper is to identify the existing hazards and their control measures to prevent the number of uncontrollable accidents [7].

2 Causes for Heavy Machinery Accident and Safety Precautions

There could be many occupational hazards related to heavy machinery. Some of these are listed such as electrocution, struck or hit by a moving vehicle, fall from moving vehicle, getting caught or crushed by some heavy machinery construction equipment.

As discussed above, dealing with heavy machinery can be quite hazardous, hence, safety needs to be assured at every step of the deal. Right from the purchase to the maintenance and repairing of the equipment, safety needs to be assured all the time. For achieving this, there must be proper division of responsibility every time the heavy machinery is under operation and every employee must take their responsibility with great seriousness. Communication plays a vital role at the time of the use of heavy duty equipment, appointing a worker for the job of spotting who can directly communicate with the driver provides better understanding of the environment to the driver. Although the spotter appointed might be at risk of being struck so they must stay out of the path of the machine. Providing proper signage increases the communication amongst other employees. The driver must be sure that there is nobody behind them while moving. The spotter helps the driver to communicate through the blind spots of the driver and so their coordination and communication are crucial. The knowledge of the working environment is another aspect that ensures safety at the time of usage of heavy machinery, each and every employee at the workplace must be aware about the various blind spots of the machinery that are used on their sites be it bulldozer, cranes, scrapers, excavators or trenchers. The working land at the time of loading or unloading operations must be free from people and must be properly levelled [8]. The presence of power lines under or above the ground must be checked prior to starting the operations of the equipment. The working environment should be free from all kinds of obstacles as far as possible and the existing obstacles must be well identified. Use of PPE must be mandatory always with heavy machinery. Helmets, high-visibility vest and safety shoes are the most important and must be worn at all points of time. The complete knowledge of the equipment and the hazards associated with it must be provided to the driver. The machine should be fit for working and only then it must be employed for operation and the inspection checklist must be prepared following the proper codes. It is very obvious that the maintenance and repair of machinery is directly related to safety and hence maintenance records must be inspected from time to time [9].

2.1 *Safety Prevention for Crane*

Hazards Related to Crane Safety

Various equipment consists of uncovered moving parts which is a potential hazard if working near it. The points of the location of such moving parts are called crunch or squeeze points. As the name suggests, the risk of hand or feet getting trapped (squeezed or crunched) in them leading to unwanted injuries. Usage of lock out system while working on maintenance and repairing on squeeze points can be helpful in preventing accidents. PPE has hand gloves and safety shoes mandatorily worn while involving work near squeeze points. The sound levels when they surpass 85 db are injurious, hence to tackle it, use of hearing protection devices (HDP) earmuffs, and earplugs must be done. When working around cranes, the falling hazard is the most prominent. The falling object, material or parts of object are observed which may lead to severe head damage, hard hats and helmets can protect against this hazard. The use of hard hats or helmets must be mandatory at work at heights as well and other places, where the danger of collision of head with objects is present. Foot injury due to striking or trapping of the foot with the machinery is commonly observed at sites, and in order to prevent it, the use of appropriate safety footwear must be mandatory. These also provide protection against the cuts perils, electrical shocks and high temperatures that might also be present [10].

2.2 *Safety equipment Required for Cranes*

There are a bunch of safety devices which can be used for crane safety purposes and to minimize the rate of accidents and mishap. Anti-current devices: when crane components come in contact with overhead power lines, there is flow of energized current. In order to prevent its transmission from coming in contact with the personnel, these anti-current devices can be used to prevent the transmission. These include insulating barriers usually made up of rubber in order to prevent the contact. However, they cannot be used as a primary solution as it can be sensitive oriented and cannot be effective in case of multiple power heads. They can be used as a secondary solution only when it is feasible [4]. Anti-upset devices: sometimes, due to the hardware failure, the crane can be overturned. In order to prevent its tipping and fall out, these anti-upset devices can be used. These devices include boom and load angle indicators providing brake mechanism and various sensors for safe operation of the crane. While

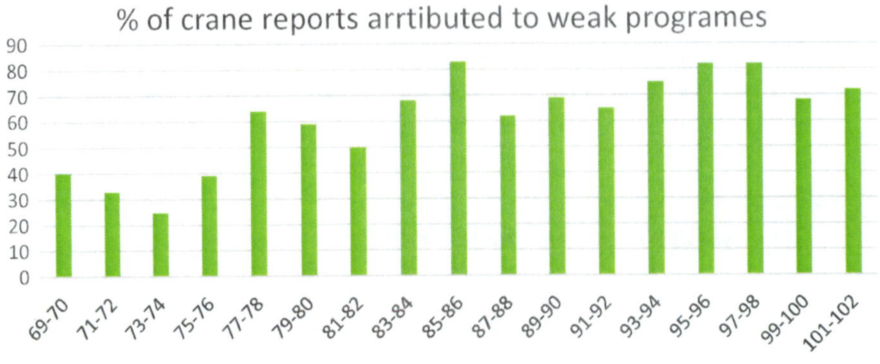

Fig. 1 % of crane reports attributed to weak programmes [11]

lifting heavy amount of load, there could be case of overloading of material, hence, monitors are installed which are connected to the sensors displaying the percentage of the weight carried and also gives out an alarm signal which reduces the mental pressure of the crane operator and also prevents the overloading capacity. Protection mechanism for operators: in case of rigging or electrical failure, there are chances of operator and personnel getting injured. They could get hit by the load destroying the crane's cab and causing injuries. However, reinforcement of the crane's cab can be essential for protection. The cab must be provided with hard resistant glass with windshield in case of bad weather conditions. Anti-collision devices: due to the chassis, electrical failure or human error, these could result in distortion of crane's movement resulting in collision of crane and airplane. Hence, these anti-collision devices help in preventing the collision. It uses some kind of emergency flashlight, usually red colour which might help in anti-collision of these devices. Along with the light, there should be an alarm system also. The tracks on which cranes are mounted should be provided with limit switches thus, providing some restriction and the track end must installed with some barriers so that their motion can be restricted up to some extent [4] (Fig. 1).

2.3 *Safety Procedure for Crane Operation*

There are various safety procedures while lifting the load, electrical procedure or failure of components. These procedures help to provide proper training, identify the hazards and reduce the likelihood of getting injured [12]. Training is essential when it comes to operating a crane because it helps in reducing or eliminating the failures or errors induced by human factors. There are various requirements such as minimum age criteria, local language depending on the location, where the crane is being operated. Additional factors such as emotional tolerance, mental stability, freedom from drugs and other abusive substances are also considered. During the crane operation, riggers are at great risk of getting injured. Hazards such as lifting of load can result in failure of boom. As per OSHA guidelines, the rigger must be able to identify the rope criteria and specifications. Also, the maintenance personnel must be properly trained, the maintenance must include mechanical as well as hydraulic machines [12].

Certification is mandatory when it comes to certain regions. The rigger, operator and personnel must go through a training course in order to gain a certificate. The exam could be written as well as practical. There is a medical test and also a core test for better competency purposes. The test can be for different cranes such as lattice-boom or telescopic depending on the capacities of the crane. Some states may have made it compulsory to take the certification course in order to get employment. Installation and removal of the crane must be done under the supervision of a competent person. Inspection of soil quality and crane foundation must be certified by a qualified engineer [13]. Evaluation of building and foundation must be done prior to any new construction project. The inspection of the crane must be done such that it must have the capacity to rotate 360° without colliding with any object. Pre, during, and post inspection of the crane must be done in order to avoid any risk. All records of upkeep and fixes performed on cranes must be kept in the regulatory office of exchange for a time of in any event 7 years. While operating load, the operator has to rely on a rigger and signalman which cannot be a better option, as it gets risky, hence, electrical and radio communication must be provided. On site must have an illustration of hand signal poster for better understanding. Many times, the crane operator is not able to observe the power lines which result in electrocution of personnel. OSHA recommends that personnel must be kept in charge when the crane is approaching towards the power lines and give better communication signals in order to avoid the collision. When lines are rated 50 kV or below, the clearance between the crane and line must be 10 ft, and for lines above 50 kV, the clearance must be 10 ft with 0.4 inch additionally [13].

2.4 Role and Responsibilities

When working with a crane, there are certain responsibilities which need to be taken care of such as proper training requirements as well as documentation. The role of a site supervisor to the crane operator, each of them plays a prominent role in crane safety. Site supervisor: under his supremacy, he looks over the whole site, where the crane is supposed to be operated. He is responsible for making sure that the crane is inspected properly and is it in good working condition or not. He makes sure that the crane operators meet the adequate requirement and in which circumstances the operator should be allowed to work near the power lines. Crane user: he is the one who entirely controls how the crane must operate. The crane user makes sure that the supervisor and operator must be qualified enough for the crane operation. All the rules and regulation must be in compliance within the requirement. While performing any work, he ensures the crane meets the required lifting capacity. He is the one who appoints personnel who can take care of maintenance, repairing and testing of the crane. Crane owner: they can also appoint the personnel for crane maintenance and repairing work. They are usually available as a technical resource in case of any technicality issue arises in the crane. The whole crane responsibility depends on them, they ensure that the crane and its components are properly manufactured and meets the specific standard requirements. They conduct various inspection and testing programmes. They are the one who maintains all the information related to the rope installation as well [14].

3 Hazard and Risk Assessment

For any industrial area to be safe and secure, it is necessary to identify the risk and hazard associated with it and then either eliminate the risk or reduce it to an acceptable level. To avoid the unwanted event which can lead to undesirable consequences, it is necessary to carry out the hazard and risk assessment for that industry. Its main objective is to identify and analyse the hazards associated within the industry. It also includes various tools such as FMEA, What if, HAZOP, and checklist analysis. The usage of these qualitative tools can vary from company to company [15].

3.1 HIRA for Crane Safety

Prior to doing any activity with a crane, a danger evaluation should be attempted. The evaluation will be recorded as a hard copy and will consider the accompanying. The undertaking to be completed. The scope of techniques by which the assignment should be possible and the fittingness of utilizing the crane as opposed to a more secure strategy. The perils engaged with the lift and the related dangers. The hardware to be utilized for the lift-like slings, spreader bars and snares with fitting evaluations. The proposed course of movement while a heap is suspended. The unmistakable space and area of putting down the heap. The potential consequences of a disappointment of the crane or stuff. The chance of people entering the lifting zone. The area and kind of obstructions and notice signs. Crisis methods: any activity that might be happening in the at workplace could represent a danger. Any hazard/ dangers related that the lift is evaluated that doesn't relates straightforwardly to the crane's use [13]. Advantages: the hazard and risk assessment helps in making the crane an asset to the construction industry. It helps to identify the potential hazards and risks such as overloading and powerhead lines. It helps the employees to maintain a working distance so that they do not get stuck by the crane. Identification of uncovered electrical wires would help in avoiding electrical hazards. However hard we may try, there can still be equipment failure leading to collapse of the entire crane, but the engineering and administrative controls could help them to reduce to an acceptable level. The daily and pre-inspection

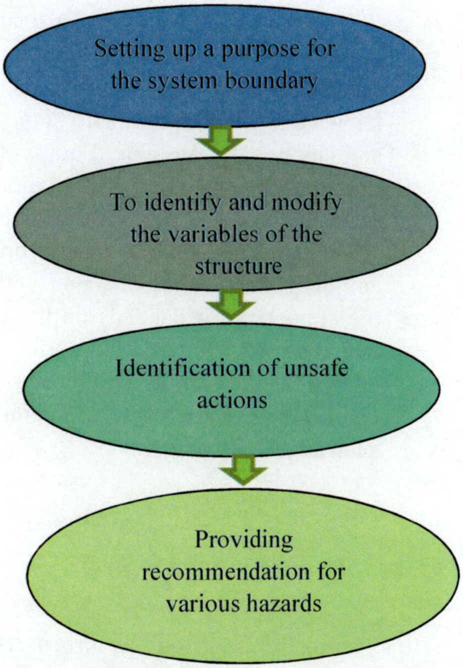

Fig. 2 Steps for hazard assessment of crane [13]

routine of the crane also help in identifying the damage components prior to that of crane operation [16] (Fig. 2).

3.2 Fault Tree Analysis

This method is used to determine the root cause of any accident. It is mostly used in cases where a combination of events causes the accidents. After the FTA, the preventive and mitigated measures can be suggested for reducing the like hood of occurrence of the event. The advantage of using this method of analysis is that it can be used qualitatively as well as quantitatively as per the requirement, other than that another benefit of using this method is that this provides a clear presentation of analysis even to a non-specialist user. This method is mostly the only used technique when it comes to complex system to generate likelihoods. This technique is time-consuming and can analyse one top event at a time. The use of this method may prevent the analyst to overlook the causes and modes of failure but this factor can be prevented by consulting expertise personnel. A conclusion of FTA done on incident involving crane is presented. It has been observed that there is an increased number of cases for failure of crane boom, derrick and rope sling leading to collapse and toppling of cranes. It is not only applicable to one type but also different types of cranes such as mobile, tower, overhead and crawler cranes [17].

The fault tree analysis is a qualitative tool which can be used to analyse the failure of cranes. The possible reasons for failure of crane can be: human error which can include lack of experience, inadequate skill, error in decision-making, lack of knowledge about equipment, some other factors can be physical and mental state and emotionally unstable. Equipment-related errors can be manufacturing defects, improper design and poor installation. Some other medium of factors depends on type of weather condition, type of soil and time of day. Not a single man, but management-related issues which include improper safe working conditions, lack of supervision, communication, poor planning with improper training can also lead to failure of the crane. It also depends on the type of complexity of work it is authorized or unauthorized. Human-based error has led to overloading of crane and machine-related issues such as failure of structure, alarm and brake. These are the major contributing factors for crane failure and improvement must be implemented for such areas [18].

3.3 Checklist Method

This method is an indirect form of risk assessment that can be used by many organizations and a large number of people. The checklists are used to identify the presence of risk in various methods. This further leads to identifying preventive and mitigation methods. This method is very fast and easy, hence, it is widely used. The reliability of a checklist is the degree to which it can be repeated and used to gain the same result. The reliability can be found out by using Kappa statics or correlation or analysis of variance (ANOVAs). The advantage of reliability of checklist is that they provide directional points or prompts to look for safety analysis to both experts and non-expert observers. Training is required for carrying out this process, and the quality of training provided may vary the outcome of the assessment that is carried out. Scientific knowledge is required to carry out the analysis [19]. Inspection checklists can be very useful for pre, during, and post crane operation. It includes maintaining record of work permit issued, maintenance and inspection of crane and its components such as control, brakes and other equipment which helps in identifying the existing hazards and implementing the necessary safety measures which can be done. Post operation includes proper inspection of crane safety and providing necessary recommendations at the end of each operation. A survey has been carried out for electric overhead travelling cranes based on the checklist provided. Basically, a checklist is taken into account for finding the physical interaction with the crane which helps to identify the hazards during operation, maintenance and other general lifting work. Based on this feedback for different elements, we get to know if the particular equipment is satisfactory or not, is it in proper working condition or not. The checklist can include hook blocks with following content such as its identification mark, capacity and condition of the hook. In case of hoist, the check points can be the condition of the wire rope, its diameter and type of construction. Types of limit switches. Floors and walkways followed by first aid, fire and emergency preparedness and other housekeeping conditions such as safe working load must also be included in the checklist [20].

4 Conclusion

Number of accidents are happening in construction industry involving heavy machineries and due to a few lagging factors, proactive communication and necessary technical training are proven best for minimizing accidents. Usage of technological means and methods is to be incorporated which must prevent these mishaps and accidents at large construction sites. Based on the analysis done in past, few other aspects are also highlighted which are insecure burden, absence of correspondence, absence of proper PPE, lack of seriousness towards responsibilities and improper site management which has led to majority of accidents. Risk minimization strategies are also very helpful to control over the unwanted incidents due to the usage of heavy machineries. Based on the critical analysis, it is concluded that checklist method is most efficient in crane operation. It is thorough and easiest method to identify the cause and to take immediate action. Since human error is also one of the reasons leading to incident in construction sector, FTA is combinedly used for better results. It has also helped in further planning and mitigating probable incidence due to the usage of heavy machineries.

References

1. Edwards, D. J., & Holt, G. D. (2009). Construction plant and equipment management research: Thematic review. *Journal of Engineering Design and Technology, 7*(2), 186–206. https://doi.org/10.1108/17260530910974989
2. Chu, X. N., & Thi, H. D. (2018). Determination of the Hazard area of crane and hurdle-using method for accident prevention. *International Journal of Applied Engineering Research, 13*(9), 6717–6722. [Online]. Available: https://www.ripublication.com/ijaer18/ijaerv13n9_29.pdf
3. Tower crane accident statistics. http://www.towercranesupport.com/09article.php (Accessed September 15, 2021).
4. Neitzel, R. L., Seixas, N. S., & Ren, K. K. (2001). A review of crane safety in the construction industry. *Applied Occupational and Environmental Hygiene, 16*(12), 1106–1117. https://doi.org/10.1080/10473220127411
5. Crane accidents|Graphic products. https://www.graphicproducts.com/articles/crane-accidents/ (Accessed September 15, 2021).
6. Case study: Deck crane failure sheds light on lack of maintenance. https://www.marineinsight.com/case-studies/case-study-deck-crane-failure-sheds-light-on-lack-of-maintenance/ (accessed September 16, 2021).
7. (PDF) Analyses of heavy construction equipment accidents and safety prevention. https://www.researchgate.net/publication/329988455_Analyses_of_Heavy_Construction_Equipment_Accidents_and_Safety_Prevention (Accessed September 6, 2021).

8. Pethaperumal, H., & Sivakumar, N. (2017). Effectiveness of mechanical material handling equipment safety in construction sites for operation safety and environmental health. *International Journal of Applied Environment Science, 12*(3), 541–552. Accessed September 6, 2021. [Online]. Available: http://www.ripublication.com
9. Machinery, plant and equipment (Labour administration and inspection). https://www.ilo.org/global/topics/labour-administration-inspection/resources-library/publications/guide-for-labour-inspectors/machinery-plant-equipment/lang--en/index.htm (Accessed September 6, 2021).
10. Lingard, H., Cooke, T., Zelic, G., & Harley, J. (2021). A qualitative analysis of crane safety incident causation in the Australian construction industry. *Safety Science, 133*, 105028. https://doi.org/10.1016/J.SSCI.2020.105028
11. Lloyd, R. L. (2003). *A survey of crane operating experience at US nuclear power plants from 1968 through 2002 (Report No. NUREG-1774).*
12. Im, S., & Park, D. (2020). Crane safety standards: Problem analysis and safety assurance planning. *Safety Science, 127*, 104686. https://doi.org/10.1016/J.SSCI.2020.104686
13. Jiang, L., Zhao, T., Zhang, W., & Hu, J. (2021). System hazard analysis of tower crane in different phases on construction site. *Advances in Civil Engineering, 2021*. https://doi.org/10.1155/2021/7026789
14. Responsibilities of supervisors, users and owners. https://www.nccco.org/nccco/news-center/archived-press-releases/news/2021/05/07/responsibilities-of-on-site-personnel-for-cranes2 (Accessed September 15, 2021).
15. Journal, I. (2017). Hazard identification and risk assessment in automotive industry hazard identification and risk assessment in automotive industry. *13*(5), 7639–7667.
16. Sunaryo. & Hamka, M. A. (January 2017). Safety risks assessment on container terminal using hazard identification and risk assessment and fault tree analysis methods. *Procedia Engineering, 194*, 307–314. https://doi.org/10.1016/J.PROENG.2017.08.150
17. Workplace Safety and Health Council. (September 2009). Crane safety analysis and recommendation report. *National Crane Safety Taskforce.*
18. Wang, Q., & Xie, L. (2012). Safety analysis of tower crane based on fault tree. *Applied Mechanics and Materials, 163*, 66–69. https://doi.org/10.4028/WWW.SCIENTIFIC.NET/AMM.163.66
19. Clift, L., Lawton, C., & Maguire, M. (2021). Checking the checklist: The effect of training on the application and effectiveness of checklist-based risk management report submitted to the IOSH research committee, 2013. Accessed: September 16. [Online]. Available: www.iosh.co.uk
20. Patil, R., & Shukla, V. Hazards identification of EOT cranes and their control measures, 2018. Accessed: September 16, 2021. [Online]. Available: www.ijariie.com

Advancement in Recent Trends, Challenges, and Precautionary Health Measures in Diabetes Management for COVID-19 Pandemic

Ajay Vasishth, S. Nagaveni, Gagan Anand, and Kanchana Latha Chitturi

1 Introduction

Today whole world is struggling with one of the dreadful communicable diseases, COVID-19, which is caused by a coronavirus (SARS-CoV-2) as shown in Fig. 1. This communicable disease is predominately spread by respiratory droplets, though other possible transmission routes cannot be ignored, as the virus can be present in the patients' body fluid, urine, and stool [1]. The disease can be started as acute minor self-limiting flu-like symptoms to aggressive respiratory failure, pneumonia followed by death. According to the 5 March 2020 report presented by WHO, the mortality rate associated with COVID-19 was approximately 3.4% (situation report e45 2019). However, in China, the mortality rate was way lower than the rest of the world, i.e. 1.4% [2].

Furthermore, in China, the mortality rate was lower than 1% compared to 1099 patients; surprisingly, the mortality rate was comparable to the severe periodic influenza [3]. Till 10 March 2020, India had approximately 39 confirmed cases of COVID-19. In the present scenario, the cases are increasing rapidly due to the non-availability of drugs and vaccines.

G. Anand
Department of Physics, University of Petroleum Energy Studies, Dehradun, India

K. L. Chitturi
Indirapriyadarshini Government Degree College (W), Nampally, Hyderabad, India

A. Vasishth (✉)
Department of Physics, Chandigarh University, Gharuan, Mohali 140413, Punjab, India
e-mail: dr.ajay2244@gmail.com

S. Nagaveni
Guru Nanak Institute of Technology (Autonomous) Campus, Ibrahimpatnam, R.R District, Telangana 501506, India

N. A. Siddiqui et al. (eds.), *Advances in Construction Safety*,
https://doi.org/10.1007/978-981-19-4001-9_12

Fig. 1 Coronavirus 2019

2 Previous Study

2.1 Diabetes Mellitus

Diabetes mellitus (DM) is a chronic metabolic multifactorial disease and a significant health issue due to the increased incidence rate globally. As per the recent WHO report, by the end of 2030, almost 366 million individuals will be affected by DM that accounts for 5% of the world's total population (Devehat et al. 2005), [4–6]. DM reflected as to be a dreadful disease because of its significant effect on various body organs. Additionally, DM is the primary causative factor of blindness in grown-ups. Furthermore, coronary heart disease is two to four times prominently associated with DM patients. Lastly, diabetic patients are more prone to long-term organ damage and dysfunction, including kidneys, heart, blood vessels, nerves, and eyes due to chronic hyperglycaemic condition.

2.2 Respiratory Infections Due to COVID-19

The individual above the age of 2 years suffering from DM is recommended to follow annual influenza and pneumococcal vaccinations since diabetic patients are prone to respiratory diseases. Therefore, such patients are a highly vulnerable group and

Fig. 2 MERS-CoV-19

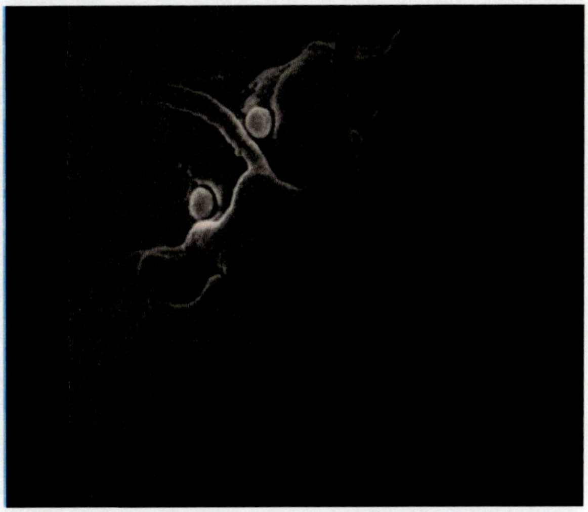

can be easily getting infected, Fig. 2 with Middle East respiratory syndrome-related coronavirus (MERS-CoV) and acute respiratory syndrome (SARS) coronavirus [7–9].

2.3 Symptoms of COVID-19 in Diabetic Patients

The 2020 year is considered as the COVID-19 year as countless individuals are suffering from the coronavirus. After coronavirus exposure, the virus can spread approximately in 2–14 days. Additionally, innumerable people deal with flu-like symptoms throughout the globe, including fever, sore throat, breathlessness, cough, body ache, loss of taste sensation, and smell sense. However, DM patients are more likely to develop symptoms, as mentioned earlier, and respiratory-associated complications, such as pneumonia and breathe shortness. Diabetic patients are more prone to viral infection due to less immunity, high blood glucose levels, and ineffective tissue healing process [10].

3 Infections in Patients with Type 1 DM, Type 2 DM, and Gestational Diabetes

Globally, approximately 425 million people are suffering from DM currently. DM can be categorized into three categories: type 1, type 2, and gestational diabetes mellitus. As per the statistical report on DM, 10% of diabetic patients are suffering from type1 DM. Type 1 usually associated with adolescents and young children, and type 2 DM is associated with middle age and adult patients. In type 1, due to autoimmune condition, the body indulges in the self-destruction of pancreatic beta cells and lead to insufficient insulin production.

In contrast, type 2 DM patients become less responsive to insulin. Lastly, gestational DM is associated with pregnancy. In all three conditions, patients are highly vulnerable to coronavirus infection due to compromised immunity [11, 12].

4 Materials and Methods

In the current study, only patients suffering from the diabetic condition were included. The viscosity coefficient values were higher in diabetic patients' blood compared to normal healthy individuals. The higher viscosity coefficient values are linked to insulin resistance and a low insulin level in body cells. Therefore, in diabetic condition, hemorheological parameters get disturbed. Increased blood viscosity levels in DM patients can lead to altered microcirculation and inadequate nutrition distribution to tissues.

Ultrasonic interferometer was deployed for estimation of ultrasonic parameters, including elastic and acoustic of blood constituents both in healthy and diabetic patients as shown in Fig. 3. After data analyses, the values of ultrasonic velocity were formulated in tabular format. As per the results, it is evident that the volume flow rate and viscosity coefficient rates are directly related to glucose levels; therefore, the values were higher in DM patients. During the pandemic condition, DM patients are facing more challenges due to the shortage of anti-diabetic drugs in the market. Additionally, The Centres for Disease Control and Prevention (CDC) recommended that diabetic patients follow their regular tests during pandemic conditions.

Fig. 3 Ultrasonic interferometer

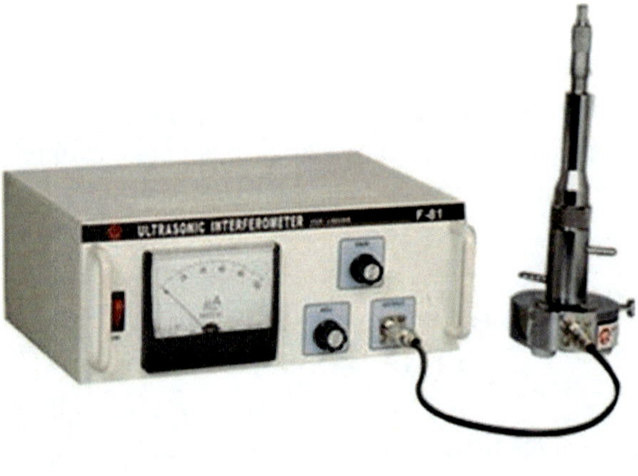

Table 1 Diabetic blood: data on average values of elastic, acoustic, and ultrasonic parameters

Sample code	Frequency v (MHz)	Ultrasonic velocity m/sec	Absorption coefficient α (cm^{-1})	Modulus of elasticity (× 10^{11}) dyne/cm^2	Loss modulus (× 10^8)
DB1	1	1586	0.014	0.420	2.111
	2	1740	0.022	0.512	5.000
	3	1855	0.0224	0.966	28.700
	5	1971	0.025	1.000	157.275
	10	1850	0.011	6.440	541.300
DB2	1	1520	0.024	0.210	0.640
	2	1622	0.026	0.550	6.538
	3	1705	0.032	0.580	10.200
	5	1781	0.033	1.100	64.500
	10	1705	0.018	3.000	600.400
COVID	1	1555	0.018	0.21	0.640
	2	1685	0.022	0.55	6.538
	3	1797	0.028	0.58	10.20
	5	1965	0.031	1.10	64.50
	10	1846	0.016	3.00	600.40

5 Experimental Setup

5.1 Ultrasonic Interferometer

From Table 1, average values on the diabetic human blood and its constituents of elastic, acoustic, and ultrasonic parameters can be given. The absorption coefficient, ultrasonic velocity, loss modulus, and modulus of elasticity were evaluated at the various frequency by using ultrasonic interferometer.

6 Measures to Minimize the Risks Due to COVID-19

6.1 Specific Measurements in Diabetic Patients

This is mandatory for diabetic individuals to control the glycaemic level to minimize infection risk as well as the severity of the infection. It is imperative to monitor blood glucose levels frequently. Good glycemic control [13, 14] may reduce the chances of spreading bacterial pneumonia. The specimens were taken from the family members of Dr. S. Nagaveni, who were in the age group of 41–55 years.

Now, we can focus on the specificity of COVID-19 as well as its spreading in diabetic people in detail. It can be concluded that diabetic people are more prone to infection from SARS-CoV-2. Prevailing to kidney disease, frailty, and heart disease further contributes to this disease's high risk.

7 Preventive Measures for Diabetic Patients in COVID-19

1. People should maintain a minimum of 6 feet or 2 m as social distance as this virus is usually transmitted through the diffusion of tiny droplets into the air during sneezes or coughs by the infected people.
2. Through touching, this virus can also transmit from person to person.
3. Diabetic people can focus more on preventing measures to minimize the infection, such as:

   - Frequent washing of the hands with alcohol-based sanitizers or soap is the best preventive measure for diabetic people. In diabetic persons, running nose, cough, and fever after appropriate health precautions need further testing in respective healthy centres. For the people with type-1 diabetes, frequent close monitoring of urinary ketones and blood glucose is necessary if hyperglycemia with fever occurs.
   - The normoglycemia can be maintained by correctional bolus when frequent changes occur in dosage. Anti-hyperglycemic agents can result in hypoglycemia or volume depletion or [15] which must be avoided.

8 Results and Discussions

It is very noticeable that has to be observed from diabetic mellitus [16], the earlier study with diabetic people concluded the high coefficient of viscosity in diabetic blood in comparison with normal blood due to either insulin resistance or low insulin level that can disturb the hemorheological parameters in diabetes mellitus. Using ultrasonic interferometer determined acoustic, elastic parameters of different constituents in normal and diabetic persons of blood [16] and represents the increase in volume flow rates and the coefficient of viscosity dependent on glucose levels and increase of radius correspondingly due to high density of the blood. In COVID-19 pandemic, diabetic people are facing a higher risk of lousy diagnosis across the world. A large population segment represents the worldwide prevalence of diabetes during the COVID-19 pandemic [17]. The poorer prognosis of diabetic individuals in the context of cardiovascular disease, comorbidities, age, hyperglycemia, obesity, and hypertension [18] results in a higher risk of infection in diabetic individuals. However, the scenario is more complicated as it requires societal factoring such as ethnicity and deprivation in relevance during the COVID-19 pandemic to survive

Fig. 4 Ulrasonic velocity versus frequency for all the samples

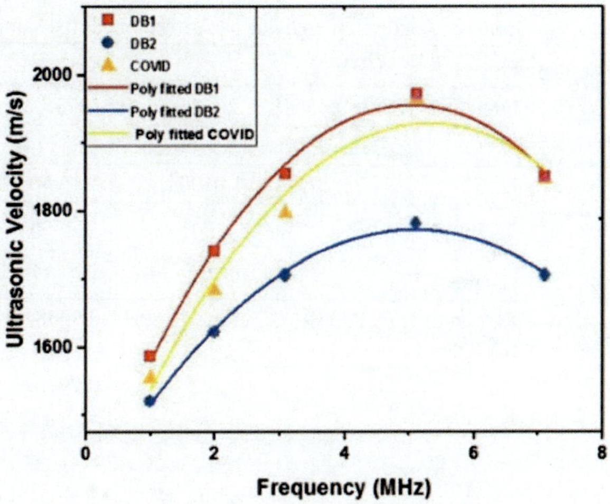

for dietetic peoples. A physician can focus more on balancing the health status of the diabetic person along with glucose-lowering treatments (Nagaveni et al. 2013, 2014). Vasishth [19] associating with COVID-19 infection treatments. It can also be concluded that diabetes management in diabetic patients with infection due to COVID-19 resulted in clinical challenges and risks.

The ultrasonic velocity of diabetic blood and COVID-19 blood is shown in Fig. 4.

From the Fig. 4, it is clear that the ultrasonic velocity linearly increases with frequency up to a threshold frequency in all three cases. As observed for COVID-19 blood, the threshold velocity is same as diabetic 2 sample. This arises due to similar viscosity of the blood in both the cases. Beyond the threshold frequency, the velocity decreases as observed. The results infer that the ultrasonic interferometer is just not sufficient to identify the COVID-19 infected blood but can be used to analyse the ultrasonic velocity of COVID-19 infected blood.

The statistical analysis of polynomial fit of second order was performed on the data values, as shown in Fig. 4. We find that the actual values and the interpolated values agree well. We can conclude that the polyfit Eq. (1) as shown below can be used as a tool to predict the futuristic values for ultrasonic velocity based the frequency (Table 2).

$$y = y_0 + ax^2 + bx \tag{1}$$

9 Conclusions

Diabetic management helps identify and evaluation of diabetes mellitus; it has many possible applications in the medical field and studies biophysical aspects of normal

Table 2 Regression analysis performed on all the samples

Plot	DB1	DB2	COVID
Intercept	1369.65548 ± 26.19321	1373.7398 ± 14.9907	1333.9204 ± 62.185
A	233.19307 ± 15.9160	157.17423 ± 9.1089	224.69459 ± 37.786
B	−23.20729 ± 1.91004	−15.50262 ± 1.09314	−21.22747 ± 4.5346
Residual sum of squares	505.87746	165.69637	2851.2762
R-square (COD)	0.99399	0.99581	0.9709
Adj. R-square	0.98798	0.99161	0.9418

and diabetes mellitus of plasma. It is essential in different drug delivery systems in the medical area provided by physicists and developed in biological, biochemistry, and pathological divisions. However, virus neutralization potential with mAbs needs time during this COVID-19 pandemic season, which should be more specifically virus targeted. The advancements and research need to be formulated to explore the clinical trials for therapeutic and prophylactic approaches against COVID-19.

References

1. Del Rio, C., & Malani, P. N. (28 February 2020) COVID-19-new insights on a rapidly changing epidemic. *Journal of American Medicine Assocation.*
2. Guan, W. J., Ni, Z. Y., Hu, Y., Liang, W. H., Ou, C. Q., He, J. X., Liu, L., Shan, H., Lei, C. L., Hui, D. S. C., Du, B., Li, L. J., Zeng, G., Yuen, K. Y., Chen, R. C., Tang, C. L., Wang, T., Chen, P. Y., Xiang, J., Li, S. Y., Wang, J. L., Liang, Z. J., Peng, Y. X., Wei, L., Liu, Y., Hu, Y. H., Peng, P., Wang, J. M., Liu, J. Y., Chen, Z., Li, G., Zheng, Z. J., Qiu, S. Q., Luo, J., Ye, C. J., Zhu, S. Y., & Zhong, N. S. (28 February 2020). China medical treatment expert group for covid-19. Clinical characteristics of coronavirus disease 2019 in China. *National English Journal of Medicine.*
3. Fauci, A. S., Lane, H. C. (2020 February 28). Redfield RR. Covid-19—Navigating the uncharted. *National English Journal of Medicine.*
4. Le Devehat, C., Vimeux, M., & Khoclabandehlou, T. (2004). Clinique Hemorheol Microcirculation, *30*, 297.
5. Diem, P., Kalt, L., Haveter, U., Fajfr, R., Reihl, B., & Bever, U. (2004). *Diabetes Technology Therapeutics*, *6*(6), 790.
6. Agardh, C. D., Agardh, E., & Torffuit, O. (1997). *Diabetes Research and Clinical Practice, 35*, 113.
7. Yang, J. K., Feng, Y., Yuan, M. Y., Yuan, S. Y., Fu, H. J., Wu, B. Y., Sun, G. Z., Yang, G. R., Zhang, X. L., Wang, L., Xu, X., Xu, X. P., & Chan, J. C. (June 2006). Plasma glucose levels and diabetes are independent predictors for mortality and morbidity in patients with SARS. *Diabet Medicine*, *23*(6), 623e8.
8. Yang, K., Horvat, N., Guerreiro, N. F. C., de Castro, I., & de Giassi, K. S. (12 November 2019). Spectrum of clinicaland radiographic findings in patients with diagnosis of H1N1 and correlation with clinical severity. *BMC Infectious Diseases*, *19*(1), 964.

9. Somasundaram, N. P., Ranathunga, I., Ratnasamy, V., et al. (2020). The impact of SARS-CoV-2 virus infection on the endocrine system. *Journal of the Endocrine Society, 8*(4), 1–22.
10. International Diabetes Federation. (2019). *IDF diabetes atlas* (9th ed.). International Diabetes Federation.
11. Garg, S. K., Rewers, A. H., & Akturk, H. K. (2018). Ever-increasing insulin-requiring patients globally. *Diabetes Technology and Therapeutics, 20*, S21–S24. World
12. Rodbard, D. (2019). State of type 1 diabetes care in the United States in 2016–2018 fromT1D exchange registry data. *Diabetes Technology and Therapeutics, 21*, 62–65.
13. Charalampopoulos, D., Hermann, J. M., Svensson, J., et al. (2018). Exploring variation in glycemic control across and within eight high-income countries: A cross-sectional analysis of 64,666 children and adolescents with type 1 diabetes. *Diabetes Care, 41*, 1180–1187.
14. Pettus, J. H., Zhou, F. L., Shepherd, L., et al. (2019). Incidences of severe hypoglycemia and diabetic ketoacidosis and prevalence of microvascular complications stratified by age and glycemic control in U.S. adult patients with type 1 diabetes: A real-world study. *Diabetes Care, 42*, 2220–2227.
15. Beck, T. W., Bergenstal, R. M., Cheng, P., et al. (2019). The relationships between time in range, hyperglycemia metrics, and HbA1c. *Journal of Diabetes Science Technology, 13*, 614–626.
16. Nagaveni, S., Vasishth, A., & Radha Krishna, T. (2014). Studies on rheological behavior on human erythrocytes of normal and diabetes mellitus. *International Journal of Pharmacology and Pharmaceutical Sciences, 2*(3), 18–21.
17. March, C. A., Flint, A., DeArment, D., Gilliland, A., Kelly, K., Rizzitano, E., Chrisman, A., Muzumdar, R. H., & Libman, I. M. (2020). Paediatric diabetes care during the COVID-19 pandemic: Lessons learned in scaling up telemedicine services. *Endocrinology, Diabetes and Metabolism*, 67.
18. Nagaveni, S., Vasishth, A., Radha Krishna, T. (2015) Effect of elastic and acoustic properties of human blood and its plasma. *International Journal of Engineering Research and Management Technology, 2*(6), 1–7.
19. Vasishth, A., Singh, B., Nagaveni, S., & Singh, S. (November 2016). In vitro red blood cells photohemolytic studies by antidiabetics, antihypertensives and their combinations. *International Journal of Pharmaceutical Science and Health Care*, 6(6).
20. Coronavirus disease 2019 (COVID-19) situation report e45.
21. Turezynski, B., Michalska-Malecka, L. S., & Szezesny, S., Romaniuk, J. (2005). *Diabetes Complication, 19*(2), 128.
22. Song, Z., Xu, Y., Bao, L., Zhang, L., Yu, P., Qu, Y., Zhu, H., Zhao, W., Han, Y., & Qin, C. (14 January 2019). From SARS to MERS, thrusting coronaviruses into the spotlight. Viruses (1), 11.E59.
23. Nagaveni, S., Vasishth, A., & Radha Krishna, T. Comparative study of excess permittivity in normal and diabetes mellitus of human erythrocytes, published. In *Second international conference on advances in electronics, electrical and computer engineering EEC 2013*, (pp 402–406). https://doi.org/10.3850/978-981-07-6935-2_82
24. Nagaveni, S., Vasishth, A. Applications of nano technology in diabetes, presented in national conference on green nano technology. GRNATE-2014 held at Chandigarh university, Punjab, India on 27 May, 2016, (pp. 64–75).

An Overview on Construction and Demolition Waste Regulation and Strategies

Vishal Kumar Singh, Ashwin Malviya, Yendamuri Sivaji Raghav, and Suvendu Manna

1 Introduction

The construction sector is one of the major sector that is directly related to the economic growth of a country [1]. There are two primary categories of companies in the construction industry: large consulting firms and construction-focused organizations which work on various infrastructure and development projects [2] and construction organizations which focus on building works, tunnels, road construction, bridges, and airfields. This requires large amount of foreign and local investments which increased the opportunity for employment and contributes to the national economy. Despite being a profit-driven sector, construction neglects the impact of its activities on the environment and is therefore also referred to as an oblivious industry. With increase in the opportunity, construction activity has led to urbanization which is a major reason for increase in construction waste [3].

Globally, for management of construction waste, people have selected landfill deposition. More than 820 million tonnes of building waste in Europe in 2017 was deposited in landfills. Likewise, 14.9 million tonnes of CDW is being dumped into landfills yearly in Hong Kong [4]. USA has started recording the construction and demolition waste generation data from 2015 which was 13.2 million tons, and there is gradual increase in the waste generation, and it was recorded 14.3 million tons in 2018 [5]. However, the construction waste generation varies according to the country where Spain (70%) is the major contributor followed by UK (50%), Hong Kong (38%), China (30–40%), European Union (30%), and USA (29%) [6].

As per the official statistics, about 12–15 million metric tonnes of building and demolition waste is generated every year in India [7]. India has a growing economy which has brought significant growth in construction activities. With increasing the

V. K. Singh (✉) · A. Malviya · Y. S. Raghav · S. Manna
Sustainability Cluster, Department of Health Safety and Environment, University of Petroleum and Energy Studies, Dehradun 248007, India
e-mail: vishalksingh3011@gmail.com

N. A. Siddiqui et al. (eds.), *Advances in Construction Safety*,
https://doi.org/10.1007/978-981-19-4001-9_13

number of constructions activities, it is important to monitor the amount of this generated and to analyse the practice for proper handling and disposal of the waste [8]. The report by building material and technology promotion council states that India's metropolitan regions create 750 million tonnes of C&DW each year [9]. In general, the construction and demolition waste includes materials including bricks, concrete, dirt, rock, wood, wall coverings, plaster, drywall, plumbing fixtures, and roofing shingles. Most of those construction and demolition waste might has a detrimental influence on waste management and strategy implementation. Thus, during the contract process, design, acquisition, transportation, management, and all construction waste generation possibilities should be calculated. Inadequate infrastructure and financing, lack of definite responsibilities, roles of the authorities, insufficient rules, legal framework, and poor enforcement have made the problem alarming, more complex, and challenging in developing countries. Due to the absence of information on building site, trash generation and data gathering on construction waste management in India is difficult. There are many small, medium, and big building projects in India. For smaller building sites, waste management is not implemented onsite. To run a successful construction company, an inventive and smart system is essential [10].

With the awareness of the environmental effects of building waste, both national and international governments have enacted various legislation and regulations to encourage more sustainable construction practices [11]. In this chapter, we have described the existing legal framework on construction waste management in India alongside with other countries and waste reduction strategy to promote sustainable practices in the construction sector.

2 Global Waste Management Regulations

2.1 Australia

There have been many programmes developed by multiple Australian states that focus on reducing construction waste by employing many and novel recycling and reuse methods. To fix the problem of such waste management and disposal is a priority for the government of the Australian Capital Territory to recycle/compost its construction and demolition waste onsite [12]. Members of the Master Builders Association Western Australia in Western Australia can get information and resources from the Waste Authority to decrease waste and enhance reuse and recycling [13]. The Australian Building Codes Board (ABCB) enforces the Building Code to implement sustainable building practices. Even though these approaches for construction and demolition waste management are well developed, but in a real scenario, these practices are not effective. This encourages contractors to use waste-reducing measures, like prefabricated components and skilled workers. The primary challenge to waste management promotion in construction is public disinterest due to increased processing costs, as well as their worry over the quality of reclaimed materials. Lack of information also contributes to a major cause of construction and demolition waste management's limited effectiveness. The diverse viewpoints and often conflicting interests of the major stakeholders that are involved in construction and demolition waste management make it difficult to understand and achieve consensus on construction and demolition waste management issues [14].

In 2010, a new framework to promote sustainability in Australia, called the National Waste Policy "Less Waste, More Resources," was developed [15]. These are two approaches to managing construction and demolition waste. One option priority reducing the amount of garbage generated while the other involves getting all state governments to promote best practices for recycling materials used in construction and demolition projects. Enforcement of building rules places substantial obstacles in the way of waste management for several stakeholders. The building code of Australia limits the reuse of construction and demolition waste and prevents the builders to develop sustainable materials, the BCA, which is a consortium of builders and industry groups, fails to give a quantifiable way to evaluate the success of sustainable building techniques [16].

2.2 Asia

In the construction sector, many nations in Asia have a unique definition of C and D trash. Although several nations have started to use the 3R principles for all types of construction and demolition waste management, the most common approach has been in metropolitan areas. Countries like Japan, Hong Kong SAR, Sri Lanka, Singapore, and Malaysia make up the list of other such places. The awareness raising on the problems associated with C and D garbage management is implemented in several nations as well. Most countries in Asia have no rules or regulations in place that control the reduction and reuse of C and D waste. Some of the policies are already in place, while others are currently being developed. In 1991, the Japanese government issued the first law governing the use and recycling of materials due to non compliance of the standard due to minimal recycling of concrete and nominated other ministries for recycling of their materials where the other ministries will be responsible for their material. So, the Japanese ministry of land, infrastructure and transport, called for by-products of concrete and soil, asphalt concrete, and wood to be included into building [17].

In the last decade, rising urbanization and the renovation of both older cities and newer towns have led to a rise in CDW usage. There should be a standard practice that the construction and demolition waste should be deposited in a centralized location and processed as per the statutory requirements. In China, the first construction waste management policy was issued in 1995 which defines the management of urban construction materials and equipment. But the amount of waste generated in this sector is a substantial amount of waste which is dumped directly into the landfills [18].

Reducing waste is the best method among the three basic waste management method that is reduce, reuse, and recycle. Most architects and designers in Sri Lanka are new in management of construction and demolition waste in construction sector. Lack of allocation of funds and limited research for waste management has created a gap in the construction waste management. The existing regulation in Sri Lanka is not sufficient for the management of construction waste because it does not include the regulation for C and D waste management [19].

The Environment Protection Department (EPD) and Environment and Food Bureau are responsible for creation and implementation of environmental policies in Hong Kong. The policy making in Hong Kong is decentralized. All statutory, as well as non-governmental, agencies, including the legislative councils, the housing authority, and task-oriented groups can all assist in waste policy development and implementation. The Hong Kong government run has sixteen landfill sites out of which only three are operational. The government has relied more on the landfills which created concern for the society. The Environment Bureau has set goals to reduce the waste by 40% in next ten years [20].

2.3 Europe

Many European nations currently have recyclable waste streams for CDW to meet the target of 70 per cent recycling. As per the European statistics, 80% of total waste creation occurs because of mass movement in Member States like the Netherlands, Germany, and Denmark [21]. For instance, high collection rates of well-segregated CDW are achieved in Spain, but the market uptake of recycled materials is really low; large storage areas at treatment plants have essentially become temporary landfills.

While CDW management plans and strategies are prevalent in Europe, waste management plans must be outlined specifically, as per EU regulations [22]. The results that are achieved, while similar in many ways, are affected by the way the project is implemented and how this affects the people and locations it impacts. Regardless of how good management plans are, CDW regulations and policing must also be applied in order for the construction and demolition waste to be successful.

New building is rapidly increasing in Russia. Construction has become 3.4 times larger in price-to-volume terms in the period of five years, from 2000 to 2005, and amounted to around $101.6 billion [23]. In Russia, the total amount of construction trash produced in 2019 was 62.57 million tonnes, which ranks construction waste at 0.83% of all industrial garbage created in the nation. While it may seem negligible, 62.57 million tonnes is a large amount of garbage, placing Russia ahead of the majority of European nations when it comes to building and demolition waste [24].

In keeping with the rules and frameworks set by the European Parliament and the European Council, the EU strives to regulate waste management in the EU Member States. All rules and frameworks included in the Circular Economy Package are present in the Organic Production Regulation (EU Regulation 2018/851) as well as additional publications, including the Construction and Demolition Waste Management Protocol [25].

At least 95% of European case studies on CDW recycling revealed recycling efforts where any recovery operation is performed in order to recycle waste into goods or chemicals [26]. While Denmark and the Netherlands have had a lot of success with the CDW recycling programme, this has mostly been done utilizing a number of different instruments. Along with other factors, environmental credit systems, such as Building Research Establishment Environmental Assessment Method (BREEAM) or Leadership in Energy and Environmental Design (LEED) as well as taxes or levies imposed by the public administration, help contribute to improved outcomes.

2.4 United States

The Resource Conservation and Recovery Act (RCRA) controls the USA federal solid waste management system. As with RCRA, construction and demolition debris is classified as a solid waste under construction and demolition laws and hence may be a hazardous waste. Although it has not included as a specified hazardous waste. Even though some building materials have the potential to be hazardous waste generators. Construction and demolition waste as per RCRA hazardous waste rules can affect many aspects of its management, depending on the volume [27]. For generating systems that produce more than 100 kg of hazardous waste per month, an operating hazardous waste management facility must decontaminate and transport the waste. As per rule 257 of jurisdiction which exclude hazardous waste including conditionally exempted small quantity generator (CESQG) from landfill disposal which restricts CDW into landfill disposal. According to the USA EPA, new clarifications state that home waste debris including lead paint does qualify for the RCRA domestic waste exclusion, and as such is not a hazardous waste. Under the Comprehensive Environmental Response, Compensation, and Liability Act (CERCLA) commonly known as Superfund, the person who created the CDW is responsible for any hazardous materials that they release. To ensure that, waste management companies are held responsible for the pollution from the management of construction and demolition debris.

Construction and demolition debris management standards have been issued by every state. These includes identifying which types of waste items constitute construction and demolition debris and what components must be eliminated. These parameters vary greatly from state to state, depending on each state's water-monitoring regulations, liner construction, site limitations, financial assurance, training needs, and recycling procedures. Construction and demolition debris poses a unique threat to both human health and the environment, as each state has different, unique characteristics, and variables such as annual rainfall, annual temperature range, land availability, and geologic stability, as well as regulators' perceptions of construction and demolition waste risk. With the ability to design laws based on their circumstances, state governments are allowed to impose rules, but because there is no guidance from the federal government, regulations are inconsistent throughout the country.

3 Legal Framework in India

"The Municipal Solid Wastes (Management and Handling) Rules, 2000", published notification number S.O. 908(E), dated the 25th of September 2000 by the Government of India provides a framework for managing Municipal Solid Wastes generated in the urban areas of the country.

The published rules, namely "The Solid Waste Management Rules, 2015" has a separate chapter on the construction and demolition wastes which was published by the Central Government in the Ministry of Environment. The rules shall be called Construction and Demolition Waste Management Rules.

3.1 Duties of the Waste Generator

As per the notified construction and waste management rules 2016, the responsibilities have been defined.

1. Every waste generator shall firstly be responsible for collecting, segregating concrete and soil, and storage of the construction and demolition waste generated at the site as directed by the local authority.
2. The generator shall also ensure that other types of solid wastes should not get mixed up with the construction wastes.
3. In a project, if the waste generation is more than 300 tons in a month, they need to segregate waste among categories like concrete, soil, steel, wood and plastics, and bricks and mortar. They should prepare and submit a waste management plan to the concerned authorities and should get approval before any construction and demolition activity and keeping the concerned authority informed about the activities which would be carried out from planning phase till the implementation phase.
4. One of the duties of the waste generator is also to keep the construction and demolition waste in the identified location or within their premises and send the waste to the collection centres authorized by the government or hand over the waste to the authorized processing facility of construction and demolition waste, and there should not be any obstruction to the traffic, public, or drains due to deposition of this waste
5. As per the notification of the concerned authority waste generators who are producing more than 20 tons in a day or 300 tons per project in a month shall also pay relevant charges for collection, transportation, processing, and disposal of wastes. The rates are fixed by the concerned local authority or by other authorities designated by the State Government [28].

3.2 Duties of Service Provider and Their Contractors

1. A clear waste management strategy within six months should be prepared by the service provider for the inclusion of all the necessary parts of waste management, such as segregation, storage, collection, reuse, recycling, transportation, and disposal.
2. These individuals should clean the area every day while they are there, regardless of how long their project lasts, how much garbage they create, and whether it is construction or demolition waste.
3. If the service provider does not have the logistics services/support to carry out the work that was given to them then they should take permission from the local authorities and tie up with the waste removal service providers. The costs that are decided by the local authority or the government should be charged to the service provider [28].

3.3 Duties of Local Authority

1. To make sure that waste management regulations are followed, local authorities first must set guidelines for appropriate waste management and obtain comprehensive plans from the generators on waste their waste management.
2. The local authorities should identify the phases and procedures and finally the clean-up in the completion of project in construction waste management.
3. If required to take assistance from concerned authorities for safe disposal of building and demolition waste mixed with industrial hazardous or toxic materials.
4. The local authority is also required to make provisions for installing containers at acceptable places for waste collection and disposal, and make sure that waste is removed from the container when it is full.
5. The collected waste should be transported to the authorized sited for treatment and disposal using own resources or by assistance from service providers.
6. The generators should be incentivized to salvage, treat, and recycle waste.
7. The local authority will review and authorize the waste management plan of the generators no later than one month after the building plan has been approved, provided that the waste management plan has been submitted by the date of approval of the building plan.
8. The generation of waste should be tracked, and database should be prepared on yearly basis.
9. Expert institutions should be consulted for the management of construction and demolition waste for recycling into products in most effective way.
10. For information, education and education of construction and demolition waste the local authorities should collaborate with the expert institutions.

11. There should be provision of incentive for making material out of construction and demolition waste and use in paver blocks, road pavements, and rural roads [28].

3.4 *Criteria for Storage, Processing, or Recycling Facilities for Construction and Demolition Waste and Application of Construction and Demolition Waste and Its Products*

Selection of a construction and demolition waste storage and processing or recycling facility should follow the criteria as per schedule I of construction and demolition waste management rules 2016. The operators must apply for authorisation in Form-I according to the requirements of the State Pollution Control Board. The report should be submitted in Form II to state pollution control. The materials prepared using construction and demolition waste as used in sanitary landfill should be provided as per schedule II [29].

4 Method of C and D Waste Reduction

4.1 *Field Study for Waste Management*

In order to observe the site management practices that are capable of minimizing waste generated by construction activities, a total of six construction sites were studied over a period of 30 months. Waste mitigating management practices were observed and documented over the period, and clarification of intent was also made by engaging the project team over the period of field study [29].

4.2 *Statistical Analysis of Waste Generated*

The statistical analysis of the waste generated at the site is done by gathering the data and information about different category and types of waste generated at the site. Samples of the wastes should be collected and should be send to Government Labs in order to see the environmental impacts of the waste. Then it should be seen the number of wastes generated from different locations that at which place more wastes are generated and at what place lesser wastes are generated so that that area can be well maintained, and all the measures can be taken to limit the waste disposals from that area, moreover the average consumption of the wastes must also be done from the sites for the analysis.

4.3 *Quantitative Study by Questionnaire Survey*

One of the ways to manage construction waste at the site is to prepare a questionnaire involving various contractors who have knowledge of the waste generated at the construction site and how to manage and handle those wastes. This questionnaire survey will help in reaching to a large amount of audience in the way that the questionnaire can be send to all the employees and to other persons so that they can know about different types of wastes generated at the site and what probable measures they can employ for controlling that waste. The questionnaire can include questions on different waste generating at the site, waste handling methods at the site, storing areas for wastes, are they generating any wastes, where are they disposing wastes, etc. Questions should also be asked from onsite personnel about the waste management system at their site.

4.4 *Detection of the Construction Activities in Generation of Reusable Waste*

Integration of different components is required for reuse of construction material. Maximum inputs are required from onsite reuse of material and specifically require for excavation, demolition materials, and soil remain. Some previous studies have suggested that various activities have potential for diverting the construction waste from landfill, and this will reduce the need of transportation and recycling. The construction waste coming out from the industry can be reused in other forms like making of bricks, in aerosol industry, and in other industries.

4.5 *Designing of Project for Waste Collection at Site*

Drawings and design are made for the storage of the construction wastes at the site. These drawings need to be followed and adhered so that proper storage of the construction wastes can be done at the site. Working in compliance with the drawings and design is one of the key elements for the waste management at the site and if proper storage areas are planned in accordance with the layout, then it would be very easy for waste management. Drawing is made in such a manner that a designated and a bigger area is given for the storage of the construction waste because a large quantity of construction waste is generated from the site. The design of the project says that the wastes to be kept and stored at the locations which are designed as per the designs and not at any location which they find comfortable.

4.6 Waste Segregation

There is large number of different wastes generated at the construction site, and these wastes need to be properly identified so that they can be stored at their designated locations and can be easily disposed off from the site. The waste should be divided in the categories such as hazardous wastes, non-hazardous wastes, bio-medical wastes, and proper location of these three categories of wastes should be marked at the construction site so that they can be kept at their location. Proper segregation of wastes places an important role in waste management at the site because waste can be easily identified and can be disposed of accordingly. If proper segregation is not done for construction waste at the site, then all the waste will be placed at one site, and there will be difficulty in identifying to which category the wastes belong.

4.7 Safe Material Storage Facilities for Storing of Waste at the Site

This is an important part that all the materials at the construction site should be stored in a proper manner and at safe location and at safe places. This plays an important role because if all the materials will be safely stored then generation of the wastes will also be very less and their stacking too. It should be noted that if the waste materials are stored in a safer manner at safer locations, then they can easily be disposed off from the site as and when required. All construction wastes shall be sorted onsite into inert and non-inert components. Non-inert materials (wood, glass and plastic) shall be recycled or reused and disposed to landfill.

4.8 Preventing Waste Mixture with the Soil

Construction waste is created in large amounts at the building site, which means it is possible that these construction waste products should not be combined with the soil at the construction site. Waste building materials might impair the soil's ability to sustain weight, which is not suitable for landscaping.

4.9 Disposal of the Wastes Generated at the Construction Site to the Registered Vendors

Construction wastes which are generated at the site needs to be disposed of at regular intervals so that they are not accumulated at the site and waste management can be done properly. Disposal of wastes plays a crucial role in the waste management system at the site because if the wastes are not getting disposed of from the site, and in that case, the waste management is not useful. For proper waste disposal, registered vendors need to be hired. Proper waste disposal form and check sheet to be filled prior to sending any wastes from the site, and there should be regular monitoring and check on the contractor to whom contract has been given for waste disposal and ensure that the waste is disposed in a designated place as per CPCB guidelines as shown in (Fig. 1) [28].

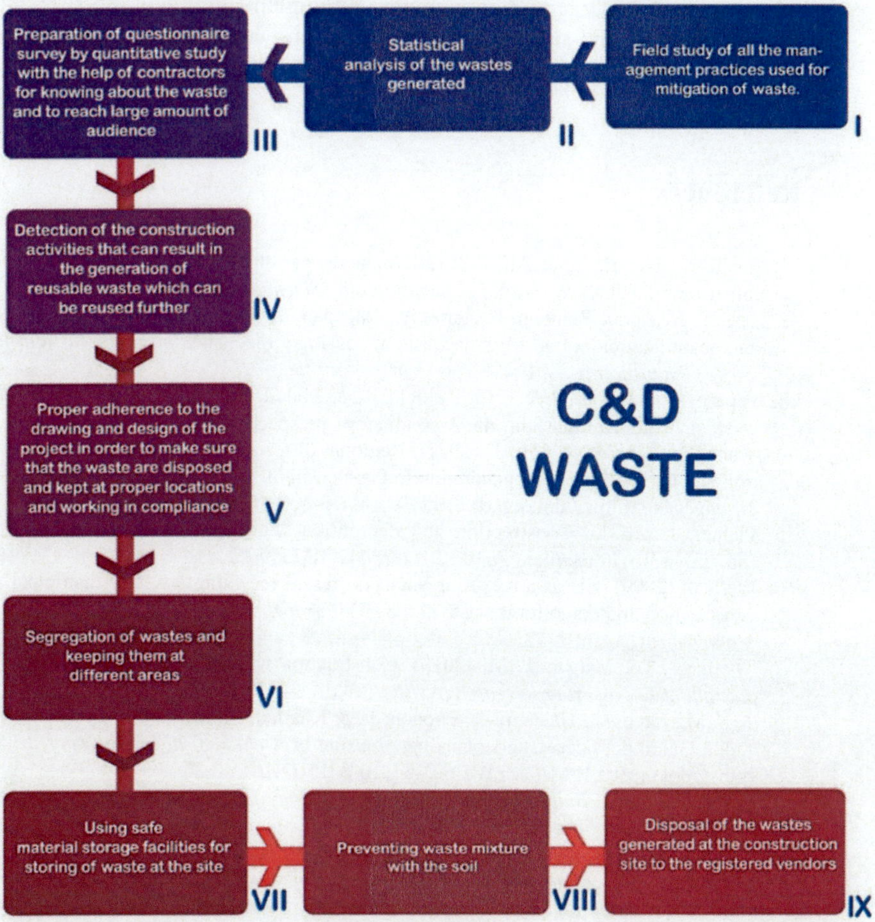

Fig. 1 Methods for construction and demolition waste reduction [28]

5 Conclusion

The construction and demolition waste contributes a major part in environment degradation globally. Various countries have implemented the C and D waste policies to reduce the environment load although their policies are not that effective. Their current policies are based on the 3R principle. In India, the government has formed the construction waste management policy for regulating the construction waste in 2016, but the implementation of policy is not that effective. Therefore, we have suggested new waste reduction methods. By implementation of our strategies in the current policy might strengthen the construction waste management and will bring more focused direction towards sustainability goal.

References

1. Saadi, N., Ismail, Z., & Alias, Z. (2016). A review of construction waste management and initiatives in malaysia. *Journal of Sustainability Science and Management, 11*(2), 101–114.
2. Lee, C. F., Abdul Rahman, I., Asmi, A., Nagapan, S., & Khalid, K. I. (2013). Classification and quantification of construction waste at housing project site. *International Journal of Zero Waste Generation, 1*(1), 1–4.
3. Papargyropoulou, E., Preece, C., Padfield, R., & Abdullah, A. (2011). Sustainable construction waste management in Malaysia: A contractor's perspective. *Citeseer*.
4. Wang, Z., Xie, W., & Liu, J. (2021). Regional differences and driving factors of construction and demolition waste generation in China. *Engineering, Construction and Architectural Management*. https://doi.org/10.1108/ECAM-10-2020-0887
5. Pichtel, J. (2021). Construction and demolition debris. In *Waste management practices* (pp. 637–650). https://doi.org/10.1201/9781420037517-25
6. Etc/Scp. (2009). EU as a recycling society. Present recycling levels of municipal waste and construction and demolition waste in the EU. *European Topic Centre on Resource and Waste Management, April*, 1–73.
7. Thomas, J., & Wilson, P. M. (2013). Construction waste management in India. *American Journal of Engineering Research (AJER)*, *2*(n.d.), 9–12.
8. Arif, M., Egbu, C., Haleem, A., Ebohon, J., & Khalfan, M. M. (2009). Green construction in India: Gaining a deeper understanding. *Journal of Architectural Engineering, 15*(1), 10–13. https://doi.org/10.1061/(asce)1076-0431(2009)15:1(10)
9. BMTPC. (2015). *Building materials and technology promotion council, Creating enabling environment for affordable housing for all. A newsletter of BMTPC fuekZ .k lkfjdk special issue municipal solid waste management soli*, (vol. 7).
10. Olawumi, T. O., & Chan, D. W. M. (2019). Critical success factors for implementing building information modeling and sustainability practices in construction projects: A Delphi survey. *Sustainable Development, 27*(4), 587–602. https://doi.org/10.1002/sd.1925
11. Wahi, N., Joseph, C., Tawie, R., & Ikau, R. (2016). Critical review on construction waste control practices: legislative and waste management perspective. *Procedia—Social and Behavioral Sciences, 224*, 276–283. https://doi.org/10.1016/j.sbspro.2016.05.460
12. Australian Capital Territory Government. (2011). *ACT waste management strategy*, (p. 56).
13. Li, R. Y. M. (2015). Construction safety and waste management: An economic analysis. *Springer International Publishing*. https://doi.org/10.1007/978-3-319-12430-8
14. Yuan, H., & Shen, L. (2011). Trend of the research on construction and demolition waste management. *Waste Management, 31*(4), 670–679. https://doi.org/10.1016/j.wasman.2010.10.030

15. Department of Agriculture, W., & E. (2009). *Australia's national waste policy—Less waste, more resources—Home page*. https://www.environment.gov.au/protection/waste/how-we-manage-waste/national-waste-policy
16. Stewart, P., Robinson, J., & Low, S. (2012). The business case for sustainable design—The city of Melbourne CH2 project. *Construction Economics and Building, 5*(2), 58–70. https://doi.org/10.5130/ajceb.v5i2.2961
17. Rao, A., Jha, K. N., & Misra, S. (2007). Use of aggregates from recycled construction and demolition waste in concrete. *Resources, Conservation and Recycling, 50*(1), 71–81. https://doi.org/10.1016/j.resconrec.2006.05.010
18. Huang, B., Wang, X., Kua, H., Geng, Y., Bleischwitz, R., & Ren, J. (2018). Construction and demolition waste management in China through the 3R principle. *Resources, Conservation and Recycling, 129*, 36–44. https://doi.org/10.1016/j.resconrec.2017.09.029
19. Ranjan, H., Karunasena, G., & Rathnayake, U. (2014). Construction and demolition waste management gaps in construction industry. In *Proceedings of the 7th FARU international research symposium*. https://www.researchgate.net/publication/324496561_Construction_and_demolition_waste_management_gaps_in_construction_industry
20. Fabian, N., & Lou, L. I. T. (2019). The struggle for sustainable waste management in Hong Kong: 1950s–2010s. *Worldwide Waste: Journal of Interdisciplinary Studies, 2*(1). https://doi.org/10.5334/WWWJ.40
21. Eurostat. (2017). *Eurostat—Data explorer*. http://appsso.eurostat.ec.europa.eu/nui/show.do?lang=en&dataset=env_wasgen
22. European Union Waste Framework Directive. (2008). Directive 2008/98/EC of the European parliament and of the council of 19 November 2008 on waste and repealing certain directives (Text with EEA relevance). ***Official Journal of the European Union***. Queen's Printer of Acts of Parliament.
23. Oleynik, S. (2016). Amounts and sources of construction and demolition waste. *Russian Journal of Resources, Conservation and Recycling, 3*(1). https://doi.org/10.15862/02rro116
24. Law, F. (2000). *Federal law no. 89-FZ of June 24, 1998 On production and consumption waste. Consultant plus*. http://www.consultant.ru/document/cons_doc_LAW_19109/ (Access 06 June 2020).
25. European Commission. (2018). EU construction and demolition waste protocol and guidelines, internal market, industry, entrepreneurship and SMEs. In *European commission, policies, information and services*.
26. Tam, V. W. Y., & Tam, C. M. (2012). Construction and demolition waste. In *International encyclopedia of housing and home* (pp. 215–218). https://doi.org/10.1016/B978-0-08-047163-1.00596-8
27. Clark, C., Jambeck, J., & Townsend, T. (2006). A review of construction and demolition debris regulations in the United States. *Critical Reviews in Environmental Science and Technology*, 36(2), 141–186. https://doi.org/10.1080/10643380500531197
28. Central Pollution Control Board (CPCB). (2017). Guidelines on environmental management of C&D waste management in India. *Prepared in Compliance of Rule 10 Sub-Rule 1(a) of C & D Waste Management Rules, 2016, 1*(2), 1–39.
29. CPCB. (2020). *CPCB|Central pollution control board*. https://cpcb.nic.in/

Mitigation of COVID-19 Through BBS Approach in Construction Sector

Kumar Ashish, Bikarama Prasad Yadav, M. B. Sharma, Abhishek Nandan, and N. A. Siddiqui

1 Introduction

The COVID-19 was first detected in the year of 2019 at Yuwhan China Hubei provision. Corona virus is severe disease of middle respiratory syndrome and acute respiratory syndrome. Recently, imparting of virus was seen from animal to human, which is known as zoonotic, SARC-CoV transmitted from cats to humans, MERS-CoV is transmitted from camel to humans are few examples of such transmission of virus also [1]. The common signs of infection which are seen usually between 2 and 14 days after the catching of infections which are cough, fever, shortness of breath, pain, severe acute respiratory syndrome, loss of taste or smell, kidney failure and death [2]. COVID-19 has already taken so many lives not only in India but all over the world which had lead a tremendous throwback to world economy. On dated 20 december 2021 approximately 34.7 million confirmed cases and 1783, 818 fatal cases has been recorded [3]. At present scenario, world is trying to combat the COVID-19 with a systematic approach, presently world is using the hierarchy of control of ERICPD [4], i.e. elimination, reduction, isolation, control, PPEs and the most important discipline. GDP is the sum of gross value added by all resident producers in the economy plus any product taxes and minus any subsidies not included in the value of the products. Indian construction industry employs over 30 million people and creates assets worth over ₹ 200 billion. It contributes more than 5% to the nation's GDP and 78% to the gross capital formation. Over 122 million people in India lost their jobs in April, according to estimates from centre for monitoring indian economy. Around 75% of

K. Ashish · A. Nandan
Cargill India, Pune, India

B. P. Yadav (✉) · M. B. Sharma
University of Petroleum and Energy Studies, Dehradun, India
e-mail: bikarama@gmail.com

N. A. Siddiqui
GD Goenka University, Gurgaon, India

N. A. Siddiqui et al. (eds.), *Advances in Construction Safety*,
https://doi.org/10.1007/978-981-19-4001-9_14

them were small traders and wage labourers. Tamil Nadu was amongst the worst-hit states which estimated unemployment rate in April was the highest amongst states and its labour participation rate amongst the lowest. Hence, to boost up the GDP construction industry need to boosted up the work as well [5]. Now after months long lockdown, we cannot simply rely on the lock down as economic activities need to be carried forward. We must incorporate some mandatory changes in the system and focus on changing the "At risk" behaviour of the employees for better results. These huge number of catastrophic event and transmission of infection need to be stopped and by changing the "At Risk" behaviour of the people we can overcome the same [6].

2 Review of Literature

In the past, researchers has found that COVID-19 has been of various types in which the most are not deadly and injurious to human like HcoV-OC43, HcoV-2293E, HcoV-NL63 and HcoV-HKU1 which causes mild respiratory problem whereas two type of these viruses are very much dangerous for human are SARS-CV, MERE-CoV which led to many catastrophic events. Out of these, **SARS-CoV-2** recently developed lead to a lot of catastrophic event in world [7]. To combat the N-CV, front line workers are exposed to these viruses on daily basis it has been also discussed the various ways by which viruses can be transmitted from one person to another, mainly by sneezing, airborne transmission, coughing, touch [8]. As it is mentioned by F. Pasco and his team, they found that the number of reported in construction industry COVID-19 cases being increased from 0.38 per 1000 to 9.3 per 1000 in august 2020. Construction industries have contribution of 13% of GDP, study reveals that, due to the COVID-19 there is decrease in construction activities 30% to 90%. This has given a significant setback for the person working in these sectors. COVID-19 has impacted a lot and has delayed the project a lot and it has been covered in the force majeure event which refer to a contractual clause regarding an extraordinary event beyond the party's controls. As per WHO under the pandemic situation automatically activates this clause. After this also many organizations suffered a loss and lead to loss of employment. It has been seen that whole world was under lockdown during this COVID-19 situation, the spike of the virus was spreading though out the world. During the period of 15 Jan 2019—1st April 2019, it has been observed COVID-19 first case and later world was under lockdown which sack the employment of a lot of worker [8]. After unlocking in the world still the three is rise in the COVID-19 cases, still it has been found the case are increasing day by day still as per WHO report. According to WHO, on 8th January 2022, there have been 302,999,389 confirmed cases of COVID, including 2,690,731 deaths and a total of 9,371,326,391 vaccine doses have been administered [9]. Today, world is using the technique of ERICPD—Elimination, reduction, isolation, controls—Administrative control and Engineering Controls, PPEs and Discipline [10]. For understanding the transmission of the COVID-19, we need to find the main reason, which is expected to be the ***Behaviour*** for the transmission of the COVID-19. For understanding the behaviour of the health care workers, employee, a behavioural analysis is needed to identify the person with "**SAFE BEHAVIOUR**" & "**AT RISK BEHAVIOR**". Industrial Psychology Research Centre, University of Aberdeen, Scotland, United Kingdom, has developed model for evaluating various behaviour of the nurses and doctor at the workplace. They have expressed that the safety culture depend upon the product of individual values and peer group values, attitudes, perceptions competencies and pattern of the behaviour [11]. Williamson and his team has described that employee behaviour is the mirror of the workplace ethics and standard [12]. Health and Family welfare, Govt. of Chhattisgarh have also developed a basic guideline for preventing transmission of infection amongst the people and workers by developing protocol for cleaning of blood/fluid spills, for deep burial, do and don't s while handling bio medical waste, zoning in OT, distance between two bed in the ward and disposal of the waste using colour bags. For preventing the un-authorized entry of the infected person inside the hospital various measures shall be followed which is described under Guideline of COVID-19 temperature screening given by U.S Department health and family welfare, which explain the various measure such as DIDO and FIFO process for the screening of the person by FLIR infrared camera [13]. Centre for diseases control and prevention has recommended for the various way we could protect the worker from the getting infected during the working like sanitizing the area, limiting the tool sharing, decreasing the number of shift, washing hand at different time during the shift, informing supervisor when fell sneeze, cough, fever and providing training to worker for hand washing, contaminated water [14].

3 Material and Methods

While developing a plan of action towards improving the behaviour of the employee at the workplace and also to prevent the transmission of this infection, the use of behaviour based safety in the system is a crucial aspect. Firstly, Identification of various activities that are been carried out over the construction site, then segregation of the work based on interview amongst the various person working in construction industry and based on the exposure, working hour and areas. Further preparing SOP for understanding the change in the process due to COVID-19 scenario. Preparation of checklist will be done based on the SOP and the set of input data for analysing the behaviour of workers in the construction industries. Lastly, analyses of the data for determining UNSAFE and SAFE ACT.

After the brief description of the order of work hereby explaining the mode in which the work is being carried out further (Fig. 1).

For the progress of the work, the SOPs are created which are shown below (Table 1). With the reference to it question is developed for the various area of concern, the questionnaires are based on the behaviour of the person, which is shown below.

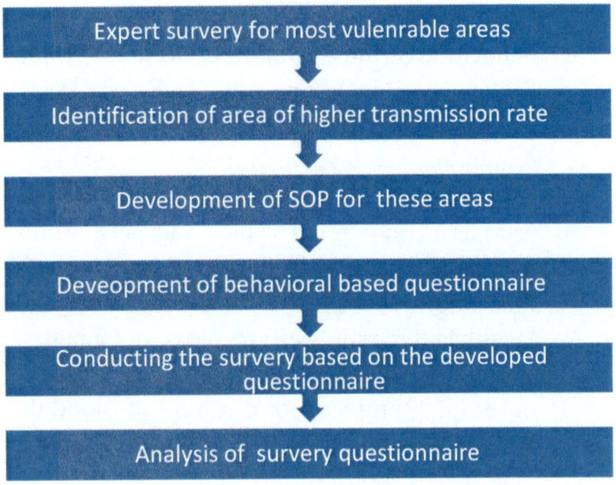

Fig. 1 Order of work for improving workers behaviours

Table 1 SOP and questionnaire

Major areas	SOP	Questionnaire
Security personal at gate	√	√
Housekeeping staff	√	√
Store	√	√
Welding cutting and fabrication	√	√
Canteen	√	√
OHC	√	√
Infectious fabrics	√	√
Covid suspected fluid and blood	√	√
Use and disposal of the PPE's	√	√
Labour camp	√	√

4 Results and Discussion

Based on the data gathered from the SOPs procedural points. Further it has been categorized as a pie chart of specific activity which shows the SAFE and UNSAFE BEHAVIOUR (Fig. 2) of the people working at the specific area as shown below.

These results are shown below (Table 2) is the combination of all the safe and unsafe behaviour. It shows that 30% of the people are leading to unsafe behaviour at workplace although 70% of them are adopting safe behaviour and practices as aligned to the SOP. As per the DU-POINT Bradly curve, it is known that the industry where the number of unsafe behaviours are more, it could lead to pull back of the culture.

The same results are represented (Figs. 3 and 4) in bar charts and pie charts which clearly explain the safe and unsafe behaviour according to the works/jobs.

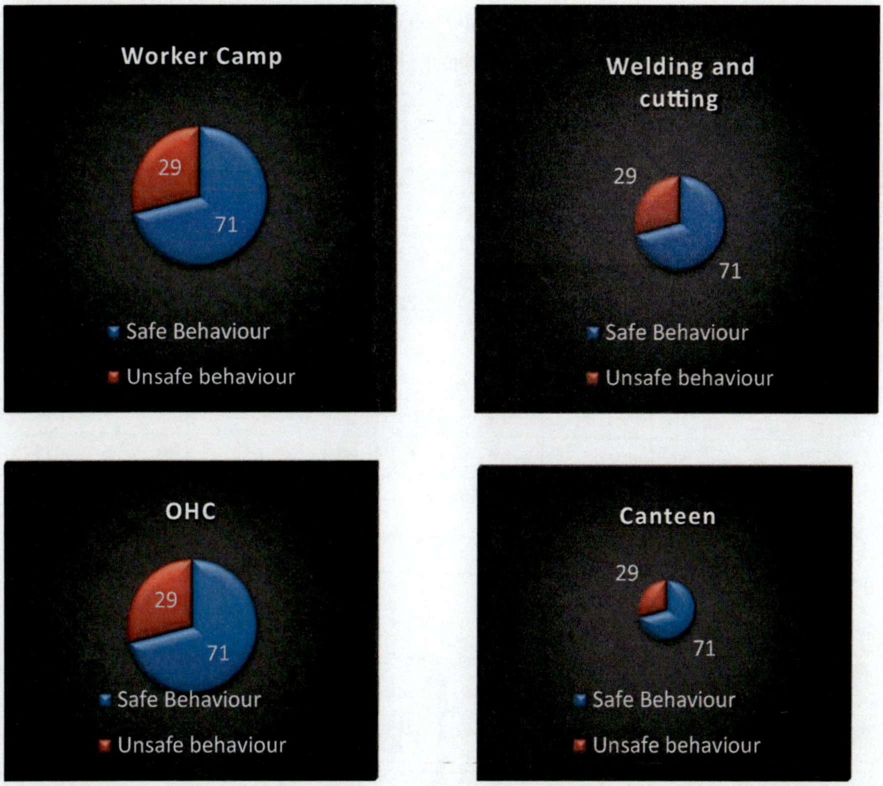

Fig. 2 Safe and unsafe behaviour based on the activity

5 Conclusion

From this work, it is found that majority of the health care units are lacking in implementation of these important aspects which came as a major barrier in prevention from COVID-19. It is been observed that most of the staffs does not not follow the basic simple rule of 20 s-hand washing, touch free type door and sanitizer dispenser are not yet installed at in most of the area of the facilities. In most of the construction projects, basic facilities regarding workers covid managements are not available. In many cases, it is found that the covid coordinators remains present at workstations are not aware of their responsibilities in some of the cases. Isolation barriers has important aspect to create the gaps but due to the lack of such measure at work site, unauthorized entry or screening mechanisms were not much effective. Tools and equipment's which has direct cause of being touched amongst the workers are also one of the sources identified as a probable cause. Majority of the reasons observed during this study lacks behind the training, awareness and lack of behavioural approaches amongst each other. Behavioural approaches suggested in this work can be major

Table 2 Comparing of safe and unsafe behaviour for the specific SOP

S. No.	Name of SOP	Safe behavior	Unsafe behavior	% of safe behavior	% of unsafe behavior
1	PPEs use and disposal	143	57	71	29
2	Infectious fabric	117	43	73	27
3	Donning and doffing	143	57	71	29
4	Security personal	125	75	62	38
5	Canteen	222	88	72	28
6	OHC	73	27	73	27
7	Welding and cutting	69	31	69	31
8	Vehicle entry	93	37	72	28
9	Labour camp	86	54	61	39
10	Total	1071	469	70	30

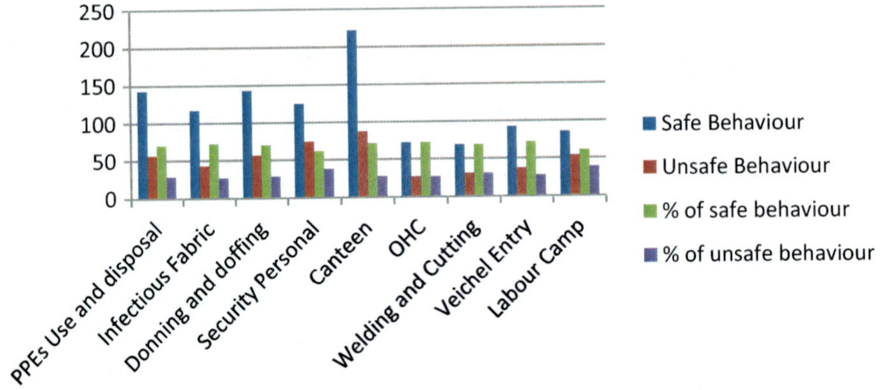

Fig. 3 Comparison of safe and unsafe behaviour specific to SOP

Fig. 4 Comparison of safe and unsafe behaviour

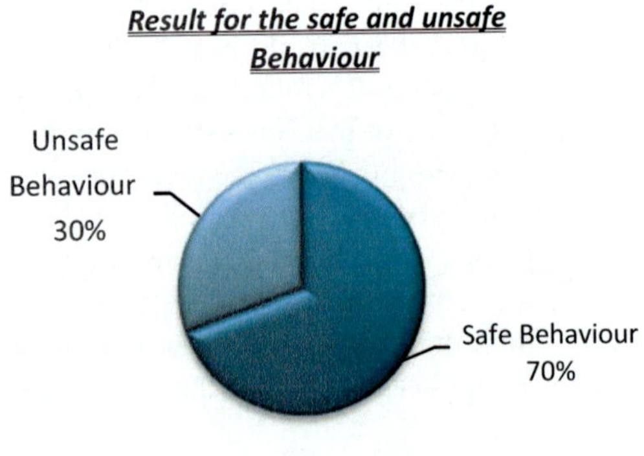

changes in the current scenario for which workers needs various planned training and inclusion of technical and behavioural contents along with onsite examples.

References

1. Kumar, S., Maurya, V. K., Prasad, A. K., Bhatt, M. L. B., & Saxena, S. K. (2020). Structural, glycosylation and antigenic variation between 2019 novel coronavirus (2019-nCoV) and SARS coronavirus (SARS-CoV). *VirusDisease, 31*(1), 13–21. https://doi.org/10.1007/s13337-020-00571-5
2. Li, Q., et al. (2020). Early transmission dynamics in Wuhan, China, of novel coronavirus-infected pneumonia. *New England Journal of Medicine, 382*(13), 1199–1207. https://doi.org/10.1056/nejmoa2001316
3. Wang, W., Tang, J., & Wei, F. (2020). Updated understanding of the outbreak of 2019 novel coronavirus (2019-nCoV) in Wuhan, China. *Journal of Medical Virology, 92*(4), 441–447. https://doi.org/10.1002/jmv.25689
4. "pmed".
5. Selvik, J. T., Elvik, R., & Abrahamsen, E. B. (2019, April). Can the use of road safety measures on national roads in Norway be interpreted as an informal application of the ALARP principle? *Accident Analysis & Prevention, 135*, 105363. https://doi.org/10.1016/j.aap.2019.105363
6. van den Broucke, S. (2021). Why health promotion matters to the COVID-19 pandemic, and vice versa. *Health Promotion International, 35*(2), 181–186. https://doi.org/10.1093/HEAPRO/DAAA042
7. Asrani, P., Hasan, G. M., Sohal, S. S., & Hassan, M. I. (2020). Molecular basis of pathogenesis of coronaviruses: A comparative genomics approach to planetary health to prevent zoonotic outbreaks in the 21st century. *OMICS: A Journal of Integrative Biology, 24*(11), 634–644. https://doi.org/10.1089/omi.2020.0131
8. Cheng, V. C. C., Lau, S. K. P., Woo, P. C. Y., & Kwok, Y. Y. (2007). Severe acute respiratory syndrome coronavirus as an agent of emerging and reemerging infection. *Clinical Microbiology Reviews, 20*(4), 660–694. https://doi.org/10.1128/CMR.00023-07
9. Chan, J. F. W., Lau, S. K. P., To, K. K. W., Cheng, V. C. C., Woo, P. C. Y., & Yue, K. Y. (2015). Middle East respiratory syndrome coronavirus: Another zoonotic betacoronavirus causing SARS-like disease. *Clinical Microbiology Reviews, 28*(2), 465–522. https://doi.org/10.1128/CMR.00102-14
10. Gralinski, L. E., & Baric, R. S. (2015). Molecular pathology of emerging coronavirus infections. *The Journal of Pathology, 235*(2), 185–195. https://doi.org/10.1002/path.4454
11. Drosten, C., et al. (2003). Identification of a novel coronavirus in patients with severe acute respiratory syndrome. *New England Journal of Medicine, 348*(20), 1967–1976. https://doi.org/10.1056/nejmoa030747
12. Medical, C., & Officer, H. (2020). Chief medical & health officer Raipur selection of agency for "providing services for management of.
13. Kaur, S., Bherwani, H., Gulia, S., Vijay, R., & Kumar, R. (2021). Understanding COVID-19 transmission, health impacts and mitigation: Timely social distancing is the key. *Environment, Development and Sustainability, 23*(5), 6681–6697. https://doi.org/10.1007/s10668-020-00884-x
14. Bobby, S., & Singh, S. (2016). Impact of constructive safety climate on worker's safety. *IOSR Journal of Mechanical and Civil Engineering, 02*(02), 59–64. https://doi.org/10.9790/1684-15010020259-64

A Study on Effective Implementation of BBS Programme Through Behaviour Maturity in FMCG Company

Prashanth Kumar Reddy, Surendar Varadharajan, and Bikarama Prasad Yadav

1 Introduction

Behaviour of the workforce contributes maximum in building a Benchmark Environment Health and Safety system of any organization. In particular, it directly measures the effectiveness of the safety culture in the organization [1]. Behaviour of the active workforce, i.e. taking from Behaviour of the Manager towards enforcing Safety and to a Supervisor, who is responsible for supervision and to worker, who is actually doing the job shall come in a category of common Behaviour trait that helps in integrating, implementing and enforcing safety in the organization [2].

Behaviour-based safety is still a challenge in FMCG companies due to its kind of moving machinery, unskilled/contract workers, production pressure, unavailability of proper monitoring system and weak EHS culture. Previous data and records of an FMCG company reflect that there is significant increase in the incident rate of the company during 2015–2017 in which 70% are due to unsafe behaviour, after that company launched its 1st cycle of behaviour-based safety programme to improve the safety by providing trainings. Though the incident rate has fallen down but it has not touched the zero-incident target, which further motivates the company to launch 2nd and 3rd cycle to find out the gaps in the programmes and to develop action plan to tackle the existing challenges. Now the company has launched its 3rd cycle of behaviour maturity assessment to know the current status of the Behaviour-based safety and to find out the gaps in the existing system and to further improve the EHS performance of the sites.

P. K. Reddy
Larsen and Toubro- Geostructure, Chennai, India

S. Varadharajan (✉) · B. P. Yadav
Sustainability Cluster, University of Petroleum & Energy Studies, Dehradun, India
e-mail: vsurendar1990@gmail.com

N. A. Siddiqui et al. (eds.), *Advances in Construction Safety*,
https://doi.org/10.1007/978-981-19-4001-9_15

2 Literature Review

Behaviour-based safety is not a new term in Health and Safety management, initial contribution to this field is made by H. W. Heinrich (Assistant Superintendent in a Traveller Insurance Company) in early 1930s and 1940s based on several accidents claims from clients of his employer, Heinrich classified the accidents into two categories based on the root cause of the accident, i.e. unsafe act and unsafe condition, and he concluded his work by saying 88% of all the accidents happened are due to the unsafe acts of the employees [3]. Later in 1986, DuPont ten year study on safety training observation programme published that almost 96% of the industrial accidents occur due to unsafe act or human negligence and thereafter till date many scientists, organizations and several surveying agencies has their own studies on the contribution of the behaviour-related errors in overall industrial injuries and illnesses but a very few stated that unsafe acts contribution is less than (<) 88% for total industrial injuries and illnesses. Considering this, even today majority of the industries include BBS programmes in their Health and Safety Management System.

In an organization, Behaviour-based safety refers to a set of safety programmes to change the behaviour of the workforce in order to mitigate or prevent the work-related injuries and illness [4]. Earlier studies suggest that the majority of the accidents, i.e. 88% do happen because of the unsafe acts and carelessness of the employees. Organizations with effective safety programmes particularly involving Behaviour-based safety are always having a good control on the workplace injuries. Generally a typical Behaviour-based safety programme is meant to impact the attitude of the workforce towards safety [5], which is a main problem in any industry.

Why BBS is important? In an organization of any kind, be it from high risk oil and gas operations to a considerable low risk office atmosphere, attitude of the individual towards workplace safety is a key, and it directly reflects the incident frequencies and injuries associated with it. Behaviour is just a reaction to the person's attitude, which is again a built up individual trait that cannot be changed but it can be controlled through behavioural change and/or modifications. So, upon implementing a successful BBS programme, which will eventually control the attitude of the workforce in following safety standards. This reduces the direct and indirect costs associated with the workplace injuries as well as time spent on incident and accident investigations [6].

A successful BBS programme highlights significant areas to work and improve as safety is concerned and also [7], and it helps employees in understanding what critical and repetitive behaviours are generally been a root cause of the injuries and act accordingly by having an action plan to prevent the re-occurrence. It enhances the safety culture by making workforce actively involving in the system [8].

Behaviour-based safety studies are theoretically good but coming to application, it is very critical to have a right BBS model or strategy to be in place for desired result. ABC model, Antecedents (A), Behaviour (B) and Consequences (C), often considered as the core element of the Behaviour modification in which the behaviour is said to be triggered by a set of Antecedents which actuates the behaviour followed by the Consequences which are post-behavioural outcomes for the individual behaviour [9]. This Antecedents give us the triggering situations and favourable conditions for a particular behaviour to happen [10] and the Consequences part of the ABC model explains why that particular behaviour is adapted and this can be corrected through Consequence Management, i.e. Positive reinforcement, Negative reinforcement, Punishment and Extinction.

According to BBS experts, there is no specific and exact method to follow for capturing and improving the Behaviour of the workforce but all the BBS strategies built on the core ABC model [11]. BBS programmes and their implementation within the organization can also be integrated with other strategies for continuous and effective monitoring.

In majority of the industries, particularly in manufacturing industries, it is believed that occupational hazards can be effectively controlled through detailed regulation [12] but in practical, it is not. Just having a traditional approach doesn't help in achieving 'zero injury' at workplace. It requires additional efforts through BBS programmes like safe workplace behaviour coaching, implementation, etc., also it requires a competent workforce, who understands and involves in the BBS programmes well. Though industries investing their time and efforts to successfully implement and sustain BBS programmes, many such programmes are not effective in terms of the time and cost involved [3] and this is considered to be mainly because of the lack of awareness [13] on the BBS programmes among the workforce.

According to (Kaila 2008) study, in India, the awareness of the safety programmes among the Managers is limited to only 75–85% and in non-Managers, it is in the range of 70–80% only, leaving a scope of around 20–25% to improve [13] in awareness part of the BBS programmes.

Many studies on the effective implementation of BBS programmes have not considered or partially considered the significance of the workforce awareness and maturity regarding BBS programmes [14]. Success of the any Behaviour-based safety programme depends mostly on the awareness of the workers about its complete flow and outcomes.

A case study by (Harbans Lal Kail 2014) about implementation of BBS programme at a multinational organization [15] gives an idea about setting up an effective BBS programme plan for a typical industry of any kind; Study and data of (M. Fleming and R. Lardner, Chartered Occupational Psychologists) show the significance and impact of awareness among employees on the implementation and success of BBS programmes.

This study involves the assessment of a BBS programme awareness and/or maturity in an Indian food manufacturing company. The inference from this assessment is used to improve and sustain the existing BBS programme.

3 Methodology

The methodology of this project is based on DMAIC approach, i.e. Define, Measure, Analyse, Improve and Control (DMAIC) is a data-driven strategy used to improve processes. This approach can be used as a strategy to improve any existing programme or process [16] and here in this study, it is applied on EHS Behaviour Standard of Britannia to improve the Behaviour Maturity and thus enhancing the safety culture [17].

3.1 Define

In this study, the problem statement is to know the EHS Behaviour Standard Maturity and Implementation gaps at chosen site and the reason for the defining this problem statement is to enhance the safety culture by actively involving workforce in EHS Behaviour Standard programmes and to start with this Understanding of Organizational BBS programme is must.

3.2 Measure

What are the available sources to measure the EHS Behaviour Standard Maturity? And how to measure them? These are two things to consider under this section and in this study, the EHS Behaviour Standard Maturity is measured through the data available like Behaviour Observation Register (BOR), CCTV observations, etc., which records the following data: (i) number of behaviour observations reported and (ii) number of people violated the behaviour standard of the organization. Based on the available data further, study is carried out.

3.3 Analysis

In this phase, all the available data is analysed by taking out a variety of parameters and attributes into account like:

(i) Type of violations observed
(ii) Critical and repetitive unsafe observation/violations
(iii) Detailed analysis, i.e. area wise, equipment wise and department wise observations
(iv) Finding out the 'target population' critical people for further study
(v) Conducting the EHS behaviour maturity assessment and
(vi) Analysis and gap finding in EHS behaviour standard.

3.4 Improve

Further to the analysis part, based on the gaps identified in EHS Behaviour standard, an action plan consisting of Corrective action and Preventive action (CAPA) is introduced and implemented to improve the existing EHS Behaviour Standard Programme.

3.5 Control

A continuous monitoring has to be done after successful improvement of the programme and organization need to ensure sustain the programme throughout.

DMAIC framework can be applied for an existing system in order to improve or in the initial phases of any new project proposal as per the requirement of the problem statement (Table 1).

4 Understanding BBS in the Organization

EHS Behaviours in the organization depend on following factors [18]:

- **Individual factors**—Individual characteristics play a significant influence in establishing a better EHS culture at work locations. Workers', Supervisors', Managers' and contractors' behaviour is influenced by a variety of individual characteristics.
- **Management system**—Training and competency, assurance, personnel, job design, tools and equipment, procedures and so on all play critical roles in improving the company's EHS performance.
- **Organizational culture and values**—Employees are motivated to perform safely all of the time by management commitment, employee involvement and lessons learned from previous incidents on the job site. A good business constantly pays attention to improving perception by soliciting their advice, establishing a robust EHS management system by offering, including employees in safety conversations and motivating them to conduct their job in a safe manner.

Table 1 Behaviour theme

Theme	Everyone	Supervisor	Manager
Standard	Follow rules	Ensure compliance	Set high standards
Communication	Speak up	Encourage the team	Communicate openly
Risk management	Be mindful	Promote risk awareness	Confront risk
Involvement	Get involved	Involve the team	Proactively involve

After considering all inputs and influencing factors, the organization has arrived to a 4 × 3 matrix model to access the actual Behaviour of the workforce at sites.

5 Behaviour Theme

Considering all critical incidents, model performers and practical expertise in human factors at different roles in the organization, it has come to a consensus and framed a Behaviour theme constituting four (4) fundamental parameters that forms the basis of Behaviour safety.

This theme is applied to a different level of workforce, i.e.

- Everyone
- Supervisors
- Managers.

The four parameters in theme are:

- Standards
- Communication
- Risk Management
- Involvement.

Integrating these parameters with the different level of workforce will form a 4 × 3 matrix of Behaviour theme as given in below Table.

6 Measurement and Analysis

Measuring the working Behaviour of the site can be extracted from Behaviour observation book, which is maintained at designated area across all sections of the plant. Once the data is extracted, then same has to be analysed. These behaviours are to be categorized further area wise to know the critical points in the plant where incidents due to certain behaviours are high.

7 Data Sources

As a part of measuring and analysing the Behaviour of the site data from various sources has to be referred.

Documents that are referred for this includes:

(i) Behaviour observation register
(ii) Proactive identification initiative—unsafe acts

(iii) CCTV observations
(iv) EHS data—(Worker Interaction and General Site Observations).

8 Selection of Target Population

Based on the Behaviour data available at site, target population of workers have been identified based on a criteria which includes:

- Repeatability of the Behaviour violation (2 or more than 2)
- Severity of the Behaviour violation (Intermediate to high is taken)
- Area wise violations (critical areas with previous accident history)
- Worker category (Contractual/Company Trainee).

9 Survey Questionnaire/Survey Sheet

After reviewing the past incidents and the nature of the employee's engagement, survey questions are made to check the effectiveness and maturity of ongoing BBS programme.

The sheet consists of three sections:

1. Everyone—In this section, questions related to task of workers will be asked.
2. Supervisors—In this section, questions related to Supervisor's theme will be asked to Supervisors.
3. Managers—In this section, questions related to the themes of the Managers will be asked.

10 Improve and Control

Inference is taken once after the completion of Behaviour maturity survey at site level and using the gaps found during the survey, an improvement action plan is made to increase the effectiveness of the programme. These inferences are provided with particular Corrective action and Preventive action by the surveyor to Control and Sustain the EHS Behaviour Standard Programme at site.

11 Results and Discussions

An analysis of a total 2249 Behaviour Observations were reported and documented in first three quarters of financial year 2019–20 is done and found the most critical and repetitive Behaviour at the site level.

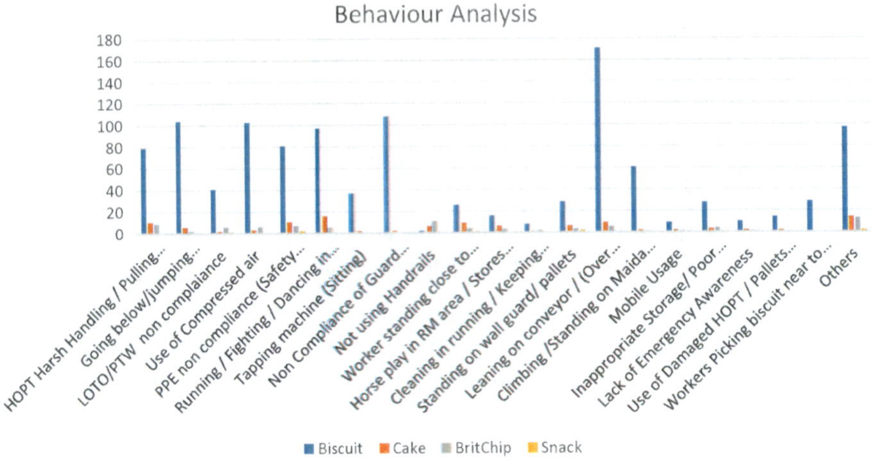

Fig. 1 No. of behaviour observations and section wise

The detailed bar graph in Fig. 1 shows the number of observations reported to the given type of behaviour across all the sections at site.

Analysis of individual sections is also done along with an equipment specific and zone specific behaviour analysis.

11.1 Analysis Based on Behaviour Theme

After these, violations are categorized under different employee section, they are grouped again into four (4) based on the type of violation according the EHS Behaviour Standard theme with respect to the particular employee category. In this case, out of 1108 violations under Everyone category, 707 are due to Follow Rules violation, 58 are due to Speak Up violations, 299 are due to not being Mindful while doing work and 44 are because of not involving in the safety system (Fig. 2).

11.2 Target Population Selection

Target people were identified from the data taken from Behaviour Observation Register and unsafe acts reported at site. Their priority (High and Intermediate) for interaction is taken based on their Repeatability of the Behaviour/Criticality of the Behaviour (Fig. 3; Table 2).

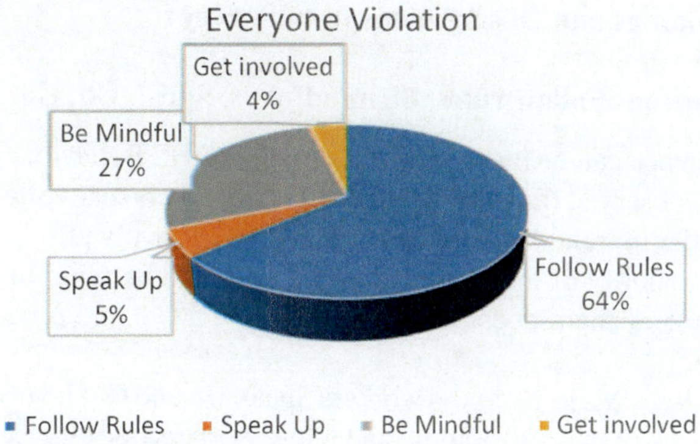

Fig. 2 Analysis of behaviour

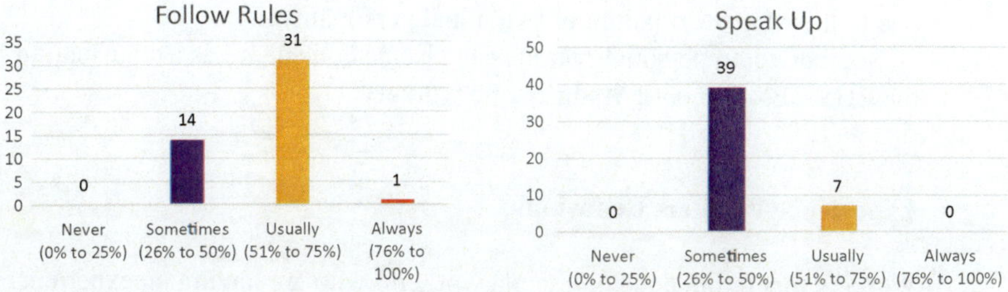

Fig. 3 Survey results

Table 2 Target population

Population	Identified	Target
Total people identified	224	
People available at site for survey/interaction (Till survey start date)	113	100 People
% People available	50.45%	Achieved

11.3 Results of the Survey (Response from the Survey)

After the successful survey of the identified and available Target Population across all the divisions and sections, the following results are obtained.

Key Findings and Observations from Survey

Comparison (Follow rules, Be mindful vs. Speak Up, Get involved)

An inference can be made after the survey of EHS Behaviour standard maturity, and it can be observed from the responses of the workers that, follow rules and be mindful are similar in trend likewise Speak Up and Get involved.

The majority of the workers Follow rules and be mindful Usually (50% < 75%) and Some of them does Sometimes (25% < 50%) and a very few does is Always (>75%).

Similarly, Majority of the workers Speak Up and Get involve in the EHS activities Sometimes (25% < 50%) only and a few workers does it Usually (50% < 75%).

This similarity is responses which are seen because of the difference in the Workers individual personality trait, i.e. if a worker always follow rules and be mindful while working, but he/she is not active to Speak Up and Get involved in the safety system due to their Introvert nature of individual personality.

So, individual personality traits can also be considered as a contributing factor to the EHS BBS Standard Maturity of a worker.

Experience Workers Behaviour

Workers (operators and contractual workers) who are having an experience prior to joining this company, particularly in food industries are comparatively good in understanding the company's EHS Behaviour standard and its outcomes then a fresher. They are able to explain the clear outcomes of the BBS programmes. Overall maturity of the experienced workers is seen to be good across all the sections of the plant.

Contractual Workers Behaviour

Contractual workers feel that they are always treated as second priority. They lack maturity in all BBS themes compared to Company employees and especially all contractors lagging in Speak Up and Get involved sections, i.e. they are not made part of the active EHS system. Though a proper communication and a system are made ready available at the shop floor to report the unsafe acts, majority of them feel they do not have authority to report (Fig. 4).

This is the area where EHS department should look into and integrate these EHS Behaviour maturity assessment points with an action plan to make contractual workers feel responsible and involved in the safety system.

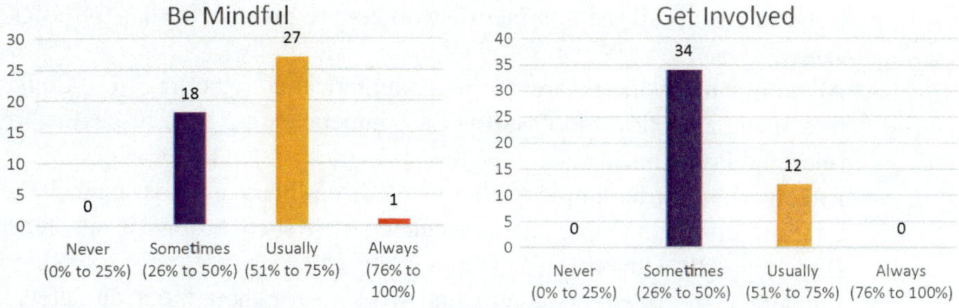

Fig. 4 Output of the survey

Follow Rules

Inferences from Survey

- Enthusiasm towards learning safety at work is itself very low in company workers category
- Majority of the company employees and officers are NOT fully aware of Operational Control programme (OCP) and Alternate LOTO Procedures (ALP). They know some procedure is in place such as ALP, but not clear in terms of understanding
- Contractual workers understanding is limited to certain sections only (Their Area of work only). Majority of the company workers take job seriously and uses right tool for the job (Within their Understanding*) and miss use or false use of equipment is rarely seen.
- People have no clue on consequence policy of the company (They just have a fear and doesn't know what are the consequent actions for certain behaviour Violations (Ex: Not applying LOTO Properly, Not using walkways and working in running machine etc.)
- Contractual workers who are having experience more than a Year are observed to be good in following the Britannia Safety System compared to workers with experience <1 Year"
- Only a few Company workers give effective suggestions and sometimes they are demotivated to suggest because of the ignorance behaviour of Supervisors
- Majority of the company employees having a peer pressure to suggest anything to Supervisor.

Recommendations

- Need to ensure the operator's understanding of safety at workplace by increasing interactive sessions (Ex: Through Planned Personnel Contact)
- All OCP's and ALP procedures are to be translated into local language.

- Need to ensure all the contractual workers are given refresher trainings on site safety.
- All company workers are to be communicated regarding the Consequence Management Policy and ensuring their understanding with respect to following rules and Procedures
- Effective and strict implementation of disciplinary actions against the violators and communicating the consequences of such actions in all departments (Involving HR actively)
- Encourage the operators/contractual workers for their input on safety aspects and make them feel involved in the system by implementing their suggestions immediately (If applicable)
- Line Safety Monitor (LSM) shall have a frequent meeting with operators to know and understand their views on improvements (Documentation of such minutes of meeting also helps to keep track)

Speak up

Inferences from Survey

- Some of the company workers feel fear to ask and get clarification. It is due to production pressure and thinking that they might irritate Supervisor
- Workers feel they completely understand their work area and coming to new joiners and contractual workers, they completely relay on Supervisor input in Toolbox Talk.
- Understanding of unsafe acts, near miss, unsafe condition and Proactive Identification Initiative (PII), an initiative of company to report unsafe conditions is NOT very satisfactory with the company employees
- Many people just inform PII/UA to Supervisor and ignore due to their work pressure (They don't go and write themselves) and a few workers feel 'What is the need of Writing? when we tell to Supervisor ' i.e. lack of knowledge on importance of the reporting)
- Lack of quick response from management also demotivating workers to report in some cases
- Contractual workers feel they do not have authority to report or wright in PII/Behaviour Register
- Workers are feeling that they are NOT given ownership and once the issue is raised they tend to ignore the follow-up part (Supervisor inability to involve the people)
- Generally company operators resists other operator/worker doing unsafe behaviour but fail to question him/her due to production delay pressure and/or ignorance.
- Majority of the company workers feel they do not have authority to stop the work that may stop or delay the production.

- They warn people whom they know personally and people under them but failed to do so in case of other department operators and Supervisors
- They try to stop if any such dangers only when safety officer is around so that he/she supports them.
- They do feel that 'Stop the work in danger' culture is not in the company, i.e. (They rarely seen an operator taking such stand)

Recommendations

- Making the Tool box meeting more interactive by answering their questions and allowing them to express themselves.
- Making operators and workers are clear with understanding of UA, UC and NM and importance of reporting by regular interaction with them.
- Building a habit of reporting in writing among operators and workers immediately after the observation (Ex: By making sure this Reporting will contribute in their appraisal/growth from HR end)
- Responding quick and taking action to close the incident will boost the reporting
- Making sure contractual workers are also involved in the system (Ex: Ask Every Contracting Supervisor to report and involve him/her in the loop)
- Interacting and involving the operator in guard implementation and other safety KAIZEN implementation
- Appreciating and rewarding such challenging behaviour help in setting an example (Appreciating with Thank you card during Tool box meeting or before senior management)
- Building the culture of 'Stop the job if in Danger' among the operators and contractual workers

Be Mindful

Inferences from Survey

- Experience employees are observed to be more vigilant compared to fresher who are poor or weak in understanding the safety at machinery.
- Majority of the workers blindly relay on senior operator or Supervisor (i.e. they do what they were told to do without questioning their Safety in it)
- Workers are aware of only particular hazards and they are not knowing the dynamic nature of hazard at his/her workplace (ex: when a maintenance person working at height involving electrical work near the Utility pipes, he tend to lose the continual awareness of hazard)
- Poor anticipation of possible risks due to routine nature of work
- Lack of Emergency Evacuation Awareness among the contractual workers (Ex; Not clear on the evacuation during fire scenarios)

- Only a few people does a proper planning before going to take up any job and rest of the operators simply start the task based on their previous experience with similar task or equipment
- Some people directly relays on cooperator and assumes things during LOTO (Ex: thinking that other operator is at the electric panel and he takes care of LOTO their)
- New joiners assume things more compared to experienced officers due to their hurriedness and/or lack of understanding

Recommendations

- Company operators shall be given refresher training on hazard awareness and its identification
- Introducing and involving operators in Job Safety Analysis (JSA) or existing HIRA to make them aware of continual hazards with the activity/machine
- Display of virtual images showing dynamic nature of hazard during different activities
- Mass training at the shop floor on emergency evacuation (contractual workers)
- Risk awareness among new joiner is to be evaluated by planning interactive risk awareness sessions at certain frequency
- Strengthening the work permit system implementation along with a checklist for activities (ex: some regular cleaning activities involving work at height, etc.) to make sure operator plans and organize his job.
- Ensuring quality supervision from Line Safety Monitor
- Identifying and demarking of every possible electric panel, machinery along with display of LOTO procedure (Ex: Where to apply? and How to apply for particular machine) and making it easily understandable for new joiner.

Get Involved

Inferences from Survey

- Though they look out during danger activities, they do ignore or leave attention during the normal activity though it has a considerable risk
- Though LSM concept introduced, their contribution in building EHS awareness is very Low
- Workers contribution is poor in EHS-related discussions
- Department Safety Committee meetings are not held frequently (If held they are not scheduled and informed well in advance)
- They attend only trainings and they are less spoken in safety committee meeting due to peer pressure
- No information of such initiatives to workers (Ex: Not all workers are aware of safety programmes conducted by EHS department)

- Workers lack of interest to attend (Ex: Feels waste of time)
- Contractual workers are always given second preference in all aspects from communication to participation.
- Lack of interest and ignorance from senior operators when they are assigned with fresher operators (No effective sharing of EHS knowledge; Ex: They just train him how to operate without giving them full picture of hazards posed with the machine)
- They share EHS knowledge, only during their leisure time

Recommendations

- Strengthening the LSM responsibilities and giving them lead to ensure all operators within his line are responsible for the contractual workers safety
- Ensuring the effectiveness of the safety committee meeting and sharing all the points discussed in the meeting positively with all the workforce
- Need to conduct an area/department wise EHS meetings headed by area Supervisor (Ex: If larger workforce involved)
- Effective Communication of Training and Local programmes through a dedicated Safety Sign Board in every Section
- Management Motivation and Inspiring actions to make operators workers to participate in the programmes
- Involving contractual workers (Ex: Planning such programmes specific to contractor also helps in reaching out more effectively)
- Assigning a senior operator and Supervisor as a 'BUDDY' to a new joiner will help in effective knowledge sharing

12 Conclusion

The survey conducted among workers gives the status, effectiveness and maturity of ongoing EHS Behaviour programme at site level. According to the average responses obtained in Everyone survey (BIL operators and contractual workers), a majority of the workers behave in line with ideal EHS Behaviour Standard at Sometimes (25% < 50%) only and a very few workers (2) among target population are found to be doing the same, Always (>75%). So, there is a scope of improvement up to 50–75% in terms of making workforce, Supervisors and Managers aware of significance and outcomes of BBS programmes. By this study, it is conclusive that the Behaviour Maturity of the workforce involving workers, Supervisors and Managers also a considerable factor in determining the effectiveness of the BBS Programme implementation.

References

1. Vinodkumar, M. N., & Bhasi, M. (2010). Safety management practices and safety behaviour: Assessing the mediating role of safety knowledge and motivation. *Accident Analysis and Prevention, 42*(6), 2082–2093.
2. Neal, A., Griffin, M. A., Neal, A., & Griffin, M. A. (2002). Safety climate and safety behaviour. *Australian Journal of Management*.
3. Can, B. S., Hazardous, B., & Your, T., Warning! Behavior-based safety can be hazardous to your health and safety program! A union critique of behavior-based safety.
4. Ocon, R., & Mcfarlane, O. (2007, June). Reducing employee injuries through behavior based safety, pp. 1–9.
5. Ismail, F., & Hashim, A. E., Steps for the behavioural based safety: A case study approach. *International Journal of Engineering and Technology, 4*(5).
6. Jasiulewicz-kaczmarek, M., Szwedzka, K., & Szczuka, M. (2015). Behaviour based intervention for occupational safety—Case study. *Procedia Manufacturing, 3*, 4876–4883.
7. Salem, O. M. (2006, May). Effectiveness of behaviour based safety interventions to reduce accidents and injuries in workplaces: Critical appraisal and reduce accidents and injuries in workplaces.
8. Van Loggerenberg, N. J. F. (2014, June). Achieving a total safety culture through behavior based safety, establishing and maintaining an injury.
9. Fleming, M., & Lardner, R. (2002). Strategic plans to implement Safety programs in health and safety management.
10. Kumar, S. (2001). ABC of PRA: Attitude and behaviour change *27*, 70–73.
11. Behavior-based safety behavior-based safety (2005, July).
12. Persekutuan, W. (2015). Level of awareness on behaviour-based safety (BBS) in manufacturing industry towards reducing workplace incidents. *International Journal of Education and Research, 3*(1), 77–88.
13. Kaila, H. L. (2015). Behaviour-based safety in Indian organisations. *Journal of Health Management, 3*(2009), 489–500.
14. Staff, E. H. S. T. (2000). The six biggest mistakes in implementing a behavior-based safety process.
15. Kaila, H. L. (2013). A case of behaviour based safety (BBS) implementation at a multinational organisation. *Journal of Organisation & Human Behaviour, 3*.
16. Al-tarawneh, R. M. (2019). The effect of using six sigma approach in improving the quality of health services in the Jordanian Ministry of Health. *International Business Research, 12*(12), 11–28.
17. Sokovic, M. (2010). Quality improvement methodologies—PDCA cycle, RADAR matrix, DMAIC and DFSS. *Journal of Achievements in Materials and Manufacturing Engineering, 43*(1), 476–483.
18. EHS behaviour standards, Britannia BBS modules & behaviour themes

Material Handling Hazards and Control Measures in Construction Industry

Bikarama Prasad Yadav, Sanket Shitole, Pradeep Kumar, Vishal Kumar Singh, and Robin V. John Fernandes

1 Introduction

The term material handling refers to the delivery, movement, storage of products, or material from one location to another with the help of equipment's such as forklift and crane manually [1]. Work that requires manual handling is prevalent in areas such as construction [2], manufacturing [3], mining [4], retailing [5], and agriculture. In an industrial context, manual handling of objects has long been a source of concern for occupational health experts trying to avoid injuries. Heavy-lifting tasks increase the risk of low back pain and other MSDs [5]. The material handling system should be well known to workers so that everyone on site is aware of the hazards associated with material handling throughout the whole process like material inspection, storage, movement, and use. The material handling activities carried out in construction sectors are with the help of equipment and by moving material manually. The utilization of mechanical equipment to handling materials can improve efficiency but it can increase the chances for worker's injuries [6]. Moving material manually from one location to another is the minimal effective method and considered as riskiest system as it may expose to work-related injuries like musculoskeletal disorders (MSDs), its injuries or pain in the joints, ligaments, muscles, and structures that support limbs and back [7]. However, moving material manually is the only choice in many times [8]. The likelihood of manual material handling in construction sites is very high and because of the nature of the job, it is unavoidable sometimes. The man–machine interfaces in construction sites often result in poor fitness and reduces the work capacity of workers [8]. The category of handling materials accounts for a large number of fatal and non-fatal accidents due to improper manual lifting or dangerous loading of materials [9]. In construction sector, 85% accidents happened due to the fall of a person, the fall of an object, and material handling and responsible

B. P. Yadav (✉) · S. Shitole · P. Kumar · V. K. Singh · R. V. John Fernandes
University of Petroleum and Energy Studies, Dehradun, India
e-mail: bikarama@gmail.com

N. A. Siddiqui et al. (eds.), *Advances in Construction Safety*,
https://doi.org/10.1007/978-981-19-4001-9_16

for the majority of fatality [6]. Improper handling of material handling equipment accounts for up to one-third of all fatality in the construction sector and about 6% of total building cost are incurred due to these accidents [6]. These incidents can be prevented by implementing proper safety management systems and procedures at construction sites. The workers should focus on lifting devices as and when required. As well as lifting techniques and use of protective equipment can help to reduce injuries. To protect themselves from injuries and ease of productivity, the workers need to know that new or repaired equipment is in proper working condition as well as they must know about safe operating procedures [1]. In this chapter, we have discussed about the hazards which are associated with the manual handling and the accidents which can take place by negligence or without following the standard operating procedures. In order to reduce the chances of accidents cause due to manual handling, the control measures which are required to be taken by the workers to prevent any musculoskeletal disorder (MSD).

2 Risk Factor of Manual Material Handling

To analyse the risk factors in construction industry, it is necessary to identify the parameters such as preliminary survey of the construction site, identification of issues from the survey, questionnaire for workers and employees, data collection, and analysis of the responses received [10].

Injuries to employees can be caused by a variety of circumstances, such as the working surface, the workplace, and working in harsh environmental conditions [11]. Only 12% of workers' performance are impacted by working in severe circumstances, whereas 82% are comfortable. Since night shift employees also suffer from eye irritation and tiredness, headaches, and discomfort because the workplace is not lighted to the necessary standards (150 lux) as per workplace lighting rules and principles [12].

Ground-level workers' elbow discomfort is caused by a variety of occupational risks including climbing, weather conditions, rapid changes in position, repetitive labour, workload, and uncomfortable posture. When it comes to ankles, there are a number of risk factors that might lead to injury. Due to the fact that ground-level workers perform MMH exercises in an open environment, stress plays a significant role in influencing the specialists in view of kneeling, stooping, bending, bringing down, and dull work, as well as ascending, grasping, holding, extending, and a combination of all danger factors [10]. In particular, business-related risk factors are connected with techniques of labour, off-kilter postures, continuous bowing, and contorting and lifting [13].

3 Ergonomic Risk Factor in Manual Handling

3.1 *Awkward Posture*

Posture involves a pattern of the distribution mass of the body parts and angular relation [12]. Postures are affected by the factors such as excessive force and holding more time [14]. Performing static and dynamic manual material handling tasks requires the use of muscles. While a person is doing static work, they maintain their posture, but when they are doing dynamic work, their skeletal muscles contract and release in synchronization [15].

Awkward working posture happened when the specialist is needed to achieve an undertaking in unnatural stance like curving, twisting, extending, and coming to because of helpless workstation plan and occupation measure. Due to inadequate workstation design and job procedure, the worker is compelled to do a task in an uncomfortable position (such as twisting and bending), stretching, or reaching. When a worker's back is forced into flexion due to an obstruction at their workstation, the item moves away from the spine, increasing pressure on the spine [16]. A person's shoulder and arm will hurt if they work in a posture where the hands are over 100% of the shoulder's height [17]. Performing manual material handling in an unstable position, such as on an uneven walking surface and increases the risk of injury. Moving weights up slopes might cause back pain. Apart from that, posture is also connected to how long a position can be held for [17].

3.2 *Excessive Load*

Injuries were more likely to occur while carrying large and heavy loads instead of lifting them [18]. This can lead to overexertion of contacting muscles [19]. Pressure imparted to the muscle when carrying and holding heavy loads. It has been demonstrated that holding the heaps on the shoulder coming about in expanded of pressing factor and torment on the stacked muscle [20].

3.3 *Extreme Temperature*

At the point when labourers are performing manual materials taking care of related with conveying task in a hot work space, they feel fatigue and discomfort. Working in outrageous cold or hot temperature may expand low back symptoms [8]. Performing work in seriously cool climate for delayed period prompts decreases in manual expertise and increases accidents [21]. Moreover, it additionally causes the adaptability of muscle and joints decline and may improve the probability of musculoskeletal wounds [21].

4 Safety Measure to Prevent Accidents or Injuries

It is possible that unexpected and unplanned actions might lead to accidents that cause significant or minor injuries, and in some circumstances, even fatalities. According to the laws and regulations, standards must be followed. Regular inspections of machine safety and employees with the appropriate educational credentials can help prevent accidents [22]. Most of the construction sites follow and implement safety procedures still, unexpected activities can lead to accidents which may result in injuries and in certain cases casualties too. Despite the fact that it is difficult to limit the accident to zero, it can be prevented by implementing the rules and regulations, frequent inspection of machines, abiding by the standards, and employees with required qualifications and training [22].

Implementation of a safety management system for material handling equipment is challenging specifically in the construction sector where material handling equipment is intensively used so it requires continuous studies to enhance operational safety [6]. In future, execution of the instructions and suggestions of safe working procedures for the material handling activities will reduce accidents in the construction sector [23].

Implementation of further suggestions can prevent accidents arising from material handling equipment's in the construction industry:

(a) **Proper housekeeping**: The various factors contribute to housekeeping, such as clean passageways with marking, proper storage of materials or tools, etc. If things are placed in proper order, a significant number of accidents are likely to fall.
(b) **Mechanical equipment safety**: Mechanical material handling equipment like forklift, cranes, and dock leveller should be maintained in sound condition with the proper inspection and testing as per government norms.
(c) **Lighting, ventilation, and noise**: These are the factors that contribute physical environment of a worker. Provision of adequate lighting and ventilation could result into an increase in work efficiency as well as a decrease in the accidents
(d) **Training and instructions**: To enhance the work site's safety system, both the training and proper instructions are important factors. Continuous efforts are required in order to prevent accidents and improve safety awareness at worksite.
(e) **Personal protective Equipment (PPE)**: Appropriate PPE's like hard hats, shoes, gloves should to be used and replaced whenever required [22].

5 Conclusion

In the construction sector, most of the accidents or injuries occur due to material handling activity. These are due to lack of material handling training, improper lifting techniques which results in musculoskeletal disorders, improper material handling equipment, etc. Ergonomics hazards should be taken into consideration while handling material manually as these accounts most of the injuries. Proper risk assessment of material handling activity may help employer to reduce associated risk. This review shows various aspects of material handling hazards and if management implement safety measures on a regular basis for the construction industry. In this work, various means and methods are suggested due to which material handling work injuries can be controlled. It has also covered the negligent attitude, behaviour, and activities where largely MSD's and other injuries happened.

References

1. Sunderam, P., & Viswanathan, E. (2014). Development of material handling system in a manufacturing company. *International Journal of Innovative Research in Science, Engineering and Technology (An ISO 3297:2007 Certified Organization), 3*(1), 8909–8913. www.ijirset.com
2. Holmstrom, E. B., Lindell, J., & Moritz, U. (1992). Low back and neck/shoulder pain in construction workers: Occupational workload and psychosocial risk factors-part 2: Relationship to neck and shoulder pain. *Spine, 17*(6), 672–677. https://doi.org/10.1097/00007632-199206000-00006
3. Latza, U., Pfahberg, A., & Gefeller, O. (2002). Impact of repetitive manual materials handling and psychosocial work factors on the future prevalence of chronic low-back pain among construction workers. *Scandinavian Journal of Work, Environment and Health, 28*(5), 314–323. https://doi.org/10.5271/sjweh.680
4. Latza, U., Karmaus, W., Stürmer, T., Steiner, M., Neth, A., & Rehder, U. (2000). Cohort study of occupational risk factors of low back pain in construction workers. *Occupational and Environmental Medicine, 57*(1), 28–34. https://doi.org/10.1136/oem.57.1.28
5. Plamondon, A., Delisle, A., Trimble, K., Desjardins, P., & Rickwood, T. (2006). Manual materials handling in mining: The effect of rod heights and foot positions when lifting "in-the-hole" drill rods. *Applied Ergonomics, 37*(6), 709–718. https://doi.org/10.1016/j.apergo.2005.12.003
6. Anil Kumar, C. N., Sakthivel, M., Elangovan, R. K., & Arularasu, M. (2015). Analysis of material handling safety in construction sites and countermeasures for effective enhancement. *Scientific World Journal, 2015*.https://doi.org/10.1155/2015/742084
7. Nurmianto, E., Ciptomulyono, U., & Kromodihardjo, S. (2015). Manual handling problem identification in mining industry: An ergonomic perspective. *Procedia Manufacturing, 4*, 89–97.https://doi.org/10.1016/j.promfg.2015.11.018
8. Reiser, R. F., Wickel, E. E., & Menzer, H. M. (2008). Lumbar mechanics of floor to knuckle height lifting on sloped surfaces. *International Journal of Industrial Ergonomics, 38*(1), 47–55. https://doi.org/10.1016/j.ergon.2007.08.002
9. Niskanen, T., & Lauttalammi, J. (1989). Accidents in materials handling at building construction sites. *Journal of Occupational Accidents, 11*(1), 1–17. https://doi.org/10.1016/0376-6349(89)90002-3
10. Ray, P. K., Parida, R., & Saha, E. (2015). Status survey of occupational risk factors of manual material handling tasks at a construction site in India. *Procedia Manufacturing, 3*, 6579–6586. https://doi.org/10.1016/j.promfg.2015.07.279
11. Albers, J. T., & Cheryl, F., E. (2007). Ergonomics for construction workers. *CDC NIOSH, 92.* www.cdc.gov/niosh/eNews
12. Reilly, T. (2020). Introduction to ergonomics. In *Ergonomics in sport and physical activity*.https://doi.org/10.5040/9781492595458.0004
13. Parida, R., & Ray, P. K. (2012). Study and analysis of occupational risk factors for ergonomic design of construction worksystems. *Work, 41*(SUPPL.1), 3788–3794. https://doi.org/10.3233/WOR-2012-0679-3788

14. Kage, V., & Putti, B. B. (2015). Effectiveness of stretching and strengthening exercises (Janda's Approach) in subjects with postural backache: A randomized controlled trial. *International Journal of Physiotherapy and Research, 3*(6), 1301–1306. https://doi.org/10.16965/ijpr.2015.195
15. Konz, S. (1994). Manual materials handling. *International Journal of Industrial Ergonomics, 14*(3), 263–264. https://doi.org/10.1016/0169-8141(94)90102-3
16. Randall, S., & Jeter, G. (1997). *A guide to manual materials handling and back safety*. https://www.navfac.navy.mil/content/dam/navfac/Safety/PDFs/ergo_page/tools/GuideMaterialHandlingBackSafety.pdf
17. Miedema, M. C., Douwes, M., & Dul, J. (1997). Recommended maximum holding times for prevention of discomfort of static standing postures. *International Journal of Industrial Ergonomics, 19*(1), 9–18. https://doi.org/10.1016/0169-8141(95)00037-2
18. Schultz, K., & Galante, J. J. (2005). Ergonomic guidelines for manual material handling. In *8th Annual Applied Ergonomics Conference Proceedings* (pp. 1021–1060). https://ci.nii.ac.jp/naid/10028063587/
19. Marras, W. S. (2003). Occupational biomechanics. In *Occupational ergonomics: principles of work design* (pp. 10-1–10-39). https://doi.org/10.1016/0376-6349(85)90024-0
20. Persson, A. L., Hansson, G. Å., Kalliomäki, J., Moritz, U., & Sjölund, B. H. (2000). Pressure pain thresholds and electromyographically defined muscular fatigue induced by a muscular endurance test in normal women. *Clinical Journal of Pain, 16*(2), 155–163. https://doi.org/10.1097/00002508-200006000-00009
21. Al Amin, M. S., Nuradilah, Z., Isa, H., Nor Akramin, M., Febrian, I., & Taufik. (2013). A review on ergonomics risk factors and health effects associated with manual materials handling. *Advanced Engineering Forum, 10*, 251–256.https://doi.org/10.4028/www.scientific.net/aef.10.251
22. Pethaperumal, H., & Sivakumar, N. (2017). Effectiveness of mechanical material handling equipment safety in construction sites for operation safety and environmental health. *International Journal of Applied Environmental Sciences, 12*(3), 541–552. https://www.ripublication.com/ijaes17/ijaesv12n3_12.pdf
23. Muralitharan, T., & Elangovan, T. (2015). Safety and analysis in material handling of construction industry. *Middle-East Journal of Scientific Research, 23*(3), 523–528. https://doi.org/10.5829/idosi.mejsr.2015.23.03.22119

Review on Construction Waste Management: India Versus Malaysia

Bikarama Prasad Yadav, Hemand Chandran, Sarath Ajithkumar, P. Mondal, Vishal Kumar Singh, and Vishal Sharma

1 Introduction

Demolition and construction waste is one of the main waste sources in today's national waste streams portfolios. Construction and demolition waste has gotten a lot of attention in the last few decades from both practitioners and researchers all over the world [1, 2]. It is characterized by waste generated during the demolition, construction or deconstruction and repair of any form of building or civil work, as well as during natural calamities [3]. As compared to other forms of industrial wastes that are typically produced at a single manufacturing facility have dispersed nature which poses one of the most significant challenges for proper management and logistics. The cost of transporting and disposing of site waste, as well as the cost of material procurement, has been discovered to be two cost factors for builders as a result of waste produced on construction sites. The building and demolition waste accounts for a significant portion of global waste output, with production reaching nearly 3 billion tons across 40 countries as of on 2012 [4]. That is a lot higher than the 0.80 tons per person per year estimated for municipal solid garbage generation [5].

According to Hong Kong data, total waste received at landfills reached 13,844 tons per day in 2012 or 5.05 million tons per year, with C and D waste accounting for nearly 25% of total waste. In 2008–2009, Australia created 19.0 million tons of waste; of this total waste stream, 8.5 million tons were disposed to landfill, while 10.5 million tons or 55%, were collected and recycled [6]. China's yearly construction waste generation is estimated to be around 2 billion tons, accounting for 80–90% of total municipal waste. In 2016, around 9.6 million tons of solid waste were generated,

B. P. Yadav (✉) · H. Chandran · S. Ajithkumar · P. Mondal · V. K. Singh
University of Petroleum and Energy Studies, Dehradun, India
e-mail: bikarama@gmail.com

V. Sharma
Amazon India, Gurgaon, India

N. A. Siddiqui et al. (eds.), *Advances in Construction Safety*,
https://doi.org/10.1007/978-981-19-4001-9_18

with about 47% of that waste coming from numerous building and demolition sites located in and around the lively Abu Dhabi City region [7]. India produced 530 million tons of construction waste, making it the world's second-largest generator [4].

Building and demolition waste management is frequently associated with high costs due to logistics and the comparatively low added value from its recovery and recycling. Some of the world's most critical developing economies also lack proper construction and demolition waste recovery systems, resulting in a massive build-up of construction and demolition waste in landfills. Reducing construction site waste will lower both the cost of purchasing raw materials and the cost of removing the waste produced on the job site. It can also reduce waste due to inefficiency on the job such as source separation which can reduce the amount of waste generated by non-segregated disposal.

The construction waste management organization faces a number of significant issues that are unreliable waste source for recycling, no incentives for recycling activities, expensive land use, insufficient attention to waste minimization design, there are no rules governing on-site sorting, non-regulated landfill activities, lack of management between departments and government administration [8]. In this chapter, we discuss what are the challenges faced during waste management in construction industry in India and waste management strategies are used in construction projects. We will also discuss the future trends in construction waste management.

2 What All Are Considered to Be Waste?

Various construction activities such as excavation or formation, civil building construction, site clearance, dismantling activities, roadwork and building restoration produce construction waste such as concrete, lumber, steel, earth rubble and site clearing materials. Prevalent cultural values and behaviours have led many people to assume that they should discard items until they are no longer needed [9]. Although these wastes exist which are thrown away and not being reused or properly disposed. Some of the wastes are listed which are generated in the construction site.

1. Timber: The majority of the waste wood came from the formwork process. Waste was produced primarily as a result of work done on the materials to make them fit, the necessary shape and size of the moulded concrete as well as from rough stripping methods. Better preparation by the subcontractor to make the formwork "fit" with minimal change and better care during the formwork stripping would have reduced waste. Since there was easy crane access to position bins into the working platform, waste timber products produced by formwork were deposited into bins at the work area. As a result, a significant amount of material may be separated for recycling [10]. At the ground floor level, problems included the reckless contamination of timber with foreign substances such as masonry or other waste. The entire load of wood became non-recoverable,

forcing the waste contractor to dispose of significant amounts of wood waste as general waste.

2. Masonry and plasterboard: The bulk of these were used for partitioning purposes during building. Since the masonry blocks purchased and used were standard sizes available from the manufacturer, waste was reduced significantly [11]. During building, waste occurred as blocks was left over from different work areas and no attempt was made to collect and reuse them. Other minor waste was produced by broken blocks or unusable off-cuts. Plasterboard was prone to damage both during handling and after installation [12]. The Project Manager, in collaboration with the plasterboard subcontractor and plasterboard manufacturer, planned the sheet sizes needed at various stages during the project to reduce waste. Despite the presence of a bin for concrete and masonry waste, no such materials were successfully isolated. On this site, both concrete and masonry wastes were disposed of as general waste. Plasterboard waste was piled up near the work areas in stockpiles. The waste was deposited into suspended crane bins during the clean-up process, as defined for metal items [13]. Segregation was relatively easy due to the large quantities of this material. As a result, a large quantity of plasterboard waste was successfully separated. Since the manufacturer did not have the facilities to reprocess this product, there were no recycling options for plasterboard on this project.
3. Paper products: On-site, a considerable volume of paper packaging for products had to be disposed of properly. Packaging was usually meant for disposal and served as a protection to products during handling, so it was not essential during the product use process. Due to sorting issues, very few paper products made it into the designated bin and were instead discarded as general waste [14].
4. Metals: Reinforcement, steel partition framing and roofing off-cuts were the major sources of metal waste. Reinforcement waste was small, and it mostly came from miscellaneous spare parts left over after the project was completed [15]. Off-cuts and changes to sheet materials and flashings to accommodate roof penetrations and geometry resulted in metal roofing waste. The need for size changes to suit the application was blamed for metal stud waste. The metal scrap bin was centrally situated between the two apartment buildings on the ground floor. The crane gathered all waste into one suspended container at each floor level during garbage collection period [16].

3 Design Specifications for Construction and Demolition Site

1. The products must be identified accordingly which all will be used on the job site, as well as those that will produce waste that may be collected separately, reused, repurposed or disposed of properly.
2. Establishment of a precise set of attainable waste management objectives.
3. Identifying a coordinator for the waste management plan.

4. Create a well-defined procedure for collection of waste. These procedures should be clearly stated in working orders and disseminated to all employees and contractors on the job site, Proper training to workers in this regard and communication within the management and workers should be maintained properly for performing good with waste management, constant feedback from on-site personnel is also critical.
5. Estimation of overall waste generated based on the material and a waste generation timetable in order to determine adequate disposal options over the life of the construction site. The construction of a school generates far less waste than the construction of an office building due to the usage of pre-fabricated and precast parts.
6. Identification of different hazardous waste, as well as a procedure for isolation, storage and transportation of these waste should be in place
7. In case of non-hazardous materials, required pre-treatment should be done before sending it off-site.
8. Once a proper plan is developed for waste management, precise location for the collection of waste should be marked and labelled containers should be in place. Small bins for collection near the generation point should be kept.
9. Containers and bins should be labelled and colour coded properly. Sketches or pictures of what all items can be dropped in each container should also be in place.
10. Record of all documents associated with waste management such as receipts and training records should be stored electronically or physically. A copy of its waste management system should be made available in the plant all the time [17].

4 Construction Waste Management in India

Construction sector in India plays a major role in providing job and economic growth as any other country in the world. According to India's 11th 5-year plan, construction industry will be the largest economic activity surpassing agriculture [9]. As per studies, about 70 billion US dollars are being invested in India every year in construction activities and it is going to raise in the coming years. Additional 50 billion dollars will be needed to meet the annual growth of 15%. Indian construction industry is a generator of bulk waste as any other country, and it is estimated that about 10–12 billion tons of construction waste is being generated solely in India every year [4]. India is among the top 10 countries which depends on nature for raw materials. Studies states that if we recycle construction waste, 12 million tons of virgin materials can be saved every year [18]. But in India using recycled products is still a stigma among workers and contractors even though research prove that it does not make in difference in strength of material (Fig. 1).

Even though there are regulations for managing construction waste, contractors and workers are neglecting the laws and disposing waste as they please due to the

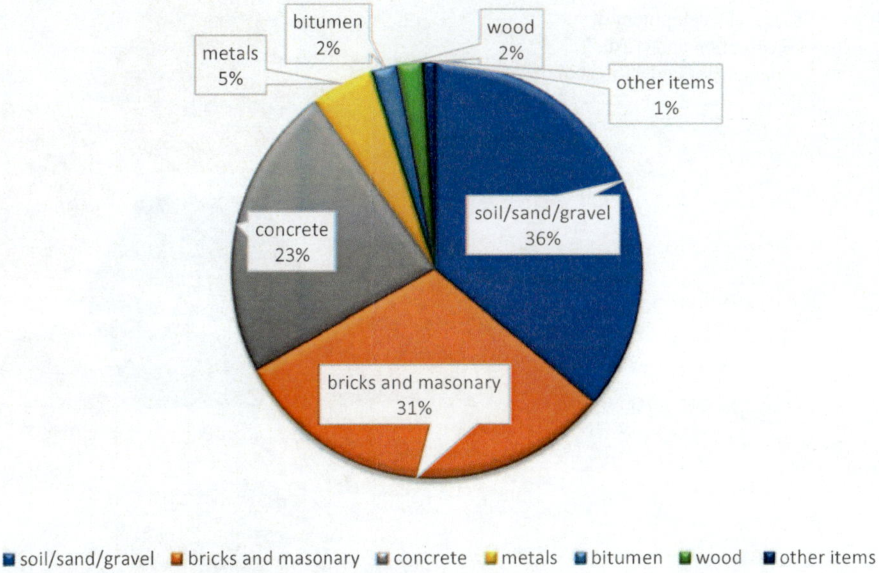

Fig. 1 Composition of waste generated in India

lack of supervision from the concerned authority. Because of this attitude, a many more problems like environmental degradation and social issues have created. As per reports, India is the country which utilizes its construction waste the least which only recycles about one per cent of its construction and demolition waste [19]. Countries like Germany, USA, Denmark and UK have developed methods that will allow them to recycle 80 to 90% of the construction demolition waste generated in an economically feasible manner [20] (Fig. 2).

Construction and demolition waste was considered as a municipal sewage waste in India until they realized the bigger problems that this waste created. Legislation was passed which allowed the construction process to begin only after completing the demolition waste management plan and to start facilities in every city to manage construction waste in accordance with Swachh Bharat Mission. As per the estimation, about 53 cities should have started recycling construction waste by 2017 but only about 13 cities have implemented it so far [21]. Even though India is lacking in recycling, many good initiatives and models for proper supervision are underway for proper management of construction and demolition waste.

5 Construction Waste Management in Malaysia

The construction sector is a key source of income and employment in every country, which has given rise to social and economic progress. The inadequate implementation and lack of waste management methods in construction projects have resulted

Fig. 2 Development of construction and waste management legislations in India

environmental problems and unlawful dumping the amount of construction waste drastically increased due to rapid development and urbanization [22].

Solid Waste and Public Cleansing Management Corporation of Malaysia states that about 8 million tons of waste are generated from different construction projects in the country every year. This issue gets worsen because of the tremendous amount of waste produced and its quality. Apart from this construction waste, Malaysia produces about 25,600 tons municipal waste daily. Recent studies state that construction waste possesses a serious threat to environment as well to public health [23].

Monitoring, collection, conveyance, processing and waste disposal are indeed objective of waste management. Every government has to make stringent rules regarding construction waste. Government should not only create policies and laws they should also make sure these are in practice. Malaysian government has allocated landfill areas for the disposal of construction waste but due the mismanagement of the contractor governments efforts is in vain. Due to the growing population and urbanization, standard of living is increasing and there are not enough places for dumping construction waste which has generated many illegal dumping sites around Malaysia. Out of the total waste produced only 76% of waste is only collected successfully by the Malaysian government out of its 5% is recycled and rest is disposed into as landfills [24] (Fig. 3).

Fig. 3 Timeline of solid waste management transition in Malaysia

5.1 Construction Waste Management Initiatives in Malaysia

Malaysia greatly depended on landfills for disposing construction waste but now they realized its harmful effect and they are running out place to dump waste, which force them to find alternatives and formed organization to enforce the disposal of waste management and to introduce sustainable building concept, which started to show changes in the concept of people, and they started using pre-fabricated materials from construction, therefore, the waste generated will be less. They also started using industrial building system (IBS) which will help them controlling waste generated during construction phase and its environmentally friendly [25].

The challenges which the government face in achieving a good waste management system are insufficiency in funds, regulations, technology, recycling market and awareness [26] (Table 1).

Table 1 Percentage of construction waste deposited in Malaysia in the following years [23]

Treatment	Percentage of waste disposed		
	2002	2006	2020
Recycling	5	5.5	12
Composting	0	1	8
Incineration	0	0	12.3
Inert landfill	0	3.2	13.9
Sanitary landfill	5	30.9	40
Other disposal site	90	59.4	13.8

6 Construction and Demolition Waste Management Strategy

6.1 Material Efficiency at Sites

In construction field, "waste prevention" through modern building design is one of the focus areas but it is critical to pinpoint the particular element that generates a significant amount of waste [27]. During construction the materials that get damage, any last-minute changes to the existing plan, packages of different materials, any material got contaminated due to unwanted materials or even the excavated soil if unused is considered as waste. According to studies, about 33% of waste can be reduced by proper site management and by smart design ideas [28]. Reusing materials from old building is a brilliant idea to reduce the waste generation and also helps in cutting cost. Main advantage of this technique is that it reduces the number of new materials needed as most of need can be fulfilled by the existing or old ones [27].

6.2 Construction and Demolition Waste Recycling (C&D Waste Recycling)

C&D waste recycling is described as the process of breaking down C&D trash into homogeneous material for a lower value use or introduced as the replacement feedstock for re-manufacturing. C&D waste recycling includes collection, sorting, recycling, storage, transportation and disposal [1, 13]. It is the process of recycling garbage into useful materials. Its magnitude is governed by the trash's ultimate use. With regard to solid waste recycling, such as curbside and drop-off recycling, C&D trash recycling may simply involve sorting, or crushing waste into smaller parts if it is for low demand, such as backfilling or land reclamation. If it is for high-level use, such as the manufacturing of new materials, C&D trash may undergo more difficult treatment, similar to organic waste recycling, both with the use of advanced technologies [29].

Off-site or centralized recycling and on-site or decentralized recycling are the two types of C&D waste recycling. When C&D waste is generated on-site, it is frequently sorted immediately to evaluate whether on-site reduction, reuse or recycling are possible [30]. The segregated trash would otherwise be landfilled, with the non-inert portion going to public fill banks and the inert portion going to public fill banks. In many places, mixed C&D waste can be delivered to off-site sorting or recycling facilities for sorting and further treatment [31]. Eco-parks with specialized recycling facilities can also be found. Recycling waste is sent to these off-site facilities using off-site separating stations and even public fill banks. After that, the imported recycled goods would be offered on the market. On-site recycling is the focus.

6.3 Economic Cooperation and Development Organization

The Organization for Economic Cooperation and Development (OECD) is a great structure where the governments of 30 democracies collaborate to confront the environmental, economic and social concerns of globalization. The organization provides a forum for countries to share policy experiences, explores solutions to shared challenges, identifies best practices and collaborates on domestic & global policy coordination [32]. Australia, Greece, Finland, Canada, Czech Republic, Austria, Italy, Mexico, Poland, Sweden, Iceland, Denmark, Ireland, Netherland, Portugal, Switzerland, Belgium, France, Slovakia, Turkey, Spain, Norway, Luxembourg, Hungary, Germany, Japan, New Zealand, Korea, UK and USA are all members of the OECD [33].

The first step in figuring out how to treat CDW is to figure out what it is made up of chemically. This is a tough undertaking since the composition of CDW is exceedingly variable because it is defined by the behaviour that generates it rather than its chemical composition (Table 2).

Table 2 Percentage waste category in construction waste [20]

Waste category	% min-max range
Concrete and masonry	40–84
Concrete	12–40
Masonry	8–54
Others (minerals)	2–9
Wood	2–4
Metal	0.2–4
Gypsum	0.2–0.4
Plastic	0.1–2
Miscellaneous	2–36

7 Future Trends in Waste Management

One of the main challenges with CDW's recycling is the lack of trust in the range of products available. According to European commission before demolition of anything a pre-demolition audit should be conducted, it is a vital component in planning how to carry out the demolition process and it should be build or restored [22]. The main aim of the audit should be to identify all the materials that undergo recycling or reusing of some sort. By analysing the pre-demolition audit report, the quantity and quality of the recovered material can be known and this will help in improving CDW portion without adding much cost. Pre-audits are very necessity in hazardous waste management. During 1960s and 1970s, many buildings were constructed using hazardous materials such as asbestos which are very dangerous to human beings and considered as carcinogens [10]. A large portion of those buildings are being removed or refurbished. In this case, in order to safely dispose hazardous waste, pre-demolition audit is handy. In this context, the presence of PCBs in CDW has been a major source of worry in several demolition projects. PCBs were widely employed in the making of sealants until they were forbidden in the 1970s as a persistent organic pollutant due to their toxicity to the environment. The connection between CDW and recycled aggregate quality is clear [34]. Many countries have introduced systems to analyse the quality of recycled materials. A system to track the recycled materials is under consideration to regulate the quality of recycled products from the start to finish of the construction

A new rule was introduced during 2006 which created two standards in recycling plants as high environmental risk and low environmental risk which have different cost for treatment. Quality assurance would be provided by a system like Tracimat, which would create a protocol for the pre-demolition audit, selected deconstruction, harvested material quality and treatment process, ensuring the quality of recycled commodities while minimizing management costs [35]. To summarize, the majority of the proposed new CDW management schemes tend to ensure the quality of the final product because current sorting, logistics and treatment practises are technically feasible, and the technologies are available and typically less expensive than treatments for other wastes [30]. On the other hand, the complex and changeable chemistry of CDW necessitates adequate management and evaluation of recovered materials in order for them to be utilized in new construction projects in a timely and cost-effective manner.

8 Conclusion

A huge amount of construction and demolition waste accompanied by poor management causes environmental degradation, social and economic problems. Rapid urbanization and population growth have become a huge problem in maintaining a long term CDW management in both developed and developing countries. Construction and demolition waste has skyrocketed, posing serious threats to urban sustainability and survival in terms of economic values and also in terms environmental safety. Building and demolition management system needs huge amount of money to function properly due to the huge amount that needed to be spent logistics and materials recovered will not incur significant value. Majority of the countries have the issue of managing construction and demolition waste, most of the time this end up in landfills due to improper collection facilities and regulations. In comparison with India, Malaysia has more effective construction waste management method which performs 80–90% of construction waste recycling which is not only reducing the waste quantity but also the cost in construction. India can implement the policy as implemented by Malaysia and follow the future trends by making amendment in their existing policy. This will definitely improve the scenario in the waste management and also reduce the waste dumped in the landfill.

References

1. Loosemore, M., Lingard, H., & Teo, M. (2007). Waste management in the construction industry. *Design and Construction, October 1997*, 256–275. https://doi.org/10.4324/9780080491080
2. Mahalakshmi, R., Aswin, K. A., & Kumar, A. Design of fuzzy logic based maximum power point tracking controller for solar array for cloudy weather conditions. (n.d.). Retrieved April 20, 2021, from https://scholar.google.com/scholar?hl=en&as_sdt=0%2C5&q=Mahalakshmi%2C+R.%2C+K.A.+Aswin%2C+and+A.+Kumar.+Design+of+Fuzzy+logic+based+maximum+power+point+tracking+controller+for+solar+array+for+cloudy+weather+conditions.+in+Power+and+Energy+Systems+Conference%3A+Towards+Sustainable+Energy%2C+2014.+2014.+IEEE.&btnG=
3. Macozoma, D. S. (2002). Construction site waste management and minimisation. *International Report for International Council for Research and Innovation in Building and Construction (CIB)*.
4. Jain, S., Singhal, S., & Jain, N. K. (2021). Construction and demolition waste (C&DW) in India: Generation rate and implications of C&DW recycling. *International Journal of Construction Management, 21*(3), 261–270. https://doi.org/10.1080/15623599.2018.1523300
5. Mihai, F. C. (2019). Construction and demolition waste in Romania: The route from illegal dumping to building materials. *Sustainability (Switzerland), 11*(11), 0–20.https://doi.org/10.3390/su11113179
6. Qiao, L., Liu, D., Yuan, X., Wang, Q., & Ma, Q. (2020). Generation and prediction of construction and demolitionwaste using exponential smoothing method: A case study of Shandong Province, China. *Sustainability (Switzerland), 12*(12). https://doi.org/10.3390/su12125094
7. Aleksanin, A. (2019). *Development of construction waste management. 06040.*
8. Ma, M., Tam, V. W. Y., Le, K. N., & Li, W. (2020). Challenges in current construction and demolition waste recycling: A China study. *Waste Management, 118*, 610–625. https://doi.org/10.1016/j.wasman.2020.09.030
9. Faruqi, M. H. Z., & Siddiqui, F. Z. (2020). A mini review of construction and demolition waste management in India. *Waste Management and Research, 38*(7), 708–716. https://doi.org/10.1177/0734242X20916828
10. Akhtar, A., & Sarmah, A. K. (2018). Construction and demolition waste generation and properties of recycled aggregate concrete: A global perspective. *Journal of Cleaner Production, 186*, 262–281. https://doi.org/10.1016/j.jclepro.2018.03.085
11. Thomas, J., & Wilson, P. M. (2013). Construction waste management in India. *American Journal of Engineering Research (AJER), 2*(n.d.), 9–12.

12. Saadi, N., Ismail, Z., & Alias, Z. (2016). A review of construction waste management and initiatives in Malaysia. *Journal of Sustainability Science and Management*, *11*(2), 101–114. https://www.researchgate.net/publication/316544681
13. Ponnada, M., & Kameswari, P. (2015). Construction and demolition waste management—A review. *International Journal of Advanced Science and Technology*, *84*, 19–46. https://doi.org/10.14257/ijast.2015.84.03
14. Nizam, M., & Yusoff, B. I. N. (2010). *Waste minimization by recycling of construction waste. November*. https://core.ac.uk/download/pdf/159177830.pdf
15. Tingley, D. D., & Allwood, J. (2014). Reuse of structural steel: The opportunities and challenges. *European Steel Environment & Energy Congress, 15–17 September 2014, Teeside University*, *June*, 15–17. https://www.researchgate.net/profile/Danielle-Densley-Tingley/publication/279441808_Reuse_of_structural_steel_the_opportunities_and_challenges/links/559271e308aed6ec4bf88415/Reuse-of-structural-steel-the-opportunities-and-challenges.pdf
16. Bowyer, J., Bratkovich, S., Fernholz, K., Frank, M., Groot, H., Howe, J., & Pepke, E. (2015). *Understanding steel recovery and recycling rates and limitations to recycling*. Dovetail Partners Inc., April 2018, 1–12. https://www.dovetailinc.org/report_pdfs/2015/dovetailsteelrecycling0315.pdf
17. Central Pollution Control Board (CPCB). (2017). Guidelines on environmental management of C&D waste management in India. *Prepared in Compliance of Rule 10 Sub-Rule 1(a) of C & D Waste Management Rules, 2016*, *1*(February), 1–39.
18. Priya, S. D., Ambika, D., Nandhini, V., Rubini, V. S., & Poovizhi, G. (2020). Assessment of construction and demolition waste management in India. *International Journal of Advanced Science and Technology*, *29*(1), 1715–1730. http://sersc.org/journals/index.php/IJAST/article/view/3740
19. Chellappa, V., Srivastava, V., & Salve, U. R. (2021). A systematic review of construction workers' health and safety research in India. *Journal of Engineering, Design and Technology*. https://doi.org/10.1108/JEDT-08-2020-0345
20. Poon, C. S. (2007). Management of construction and demolition waste. *Waste Management*, *27*(2), 159–160. https://doi.org/10.1016/J.WASMAN.2006.10.012
21. Praveen, S. (2019). Swachh Bharat: A step towards environmental protection. *Clean India for New India*, *July*, 87–92.
22. Papargyropoulou, E. F. F. I. E., Preece, C., Padfield, R., & Abdullah, A. A. (2011). Sustainable construction waste management in Malaysia: A contractor's perspective. *Citeseer*. http://citeseerx.ist.psu.edu/viewdoc/download?doi=10.1.1.842.6622&rep=rep1&type=pdf
23. Yusri Yusup, F., Widad Fadhlullah, A., Teng, T. T., & Universiti Sains Malaysia. Pusat Pengajian Teknologi Industri. (2015). Waste management in construction industry—A review on the issues and challenges. In *4th International Conference on Environmental Research and Technology (ICERT 2015)*, *May*, 100–106.
24. Noor, R. N. H. R. M., Endut, I. R., Ridzuan, A. R. M., Dahalan, N. H., Yunus, J. N., & Tutur, N. (2018). Major challenges to improve the performance of construction waste management process: A case study in Klang Valley. *AIP Conference Proceedings, 2020*, 30008. https://doi.org/10.1063/1.5062680
25. Agamuthu, P., & Victor, D. (2011). Policy trends of extended producer responsibility in Malaysia. *Waste Management and Research, 29*(9), 945–953. https://doi.org/10.1177/0734242X11413332
26. Ishak, M. B., & Samah, M. A. A. (2010). Strict liability versus policy and regulation for environmental protection and agricultural waste management in Malaysia. *EnvironmentAsia*, *3*(Special Issue), 11–19. http://dlc.dlib.indiana.edu/dlc/handle/10535/7057
27. Adams, K., & Hobbs, G. (2017). ***Materials resource efficiency in construction.***
28. **Ruuska, A., & Häkkinen, T. (2014). Material efficiency of building construction.** *Buildings*, *4*(3), 266–294. https://doi.org/10.3390/buildings4030266
29. **Tam, V. W. Y., & Lu, W. (2016). Construction waste management profiles, practices, and** performance: A cross-jurisdictional analysis in four countries. *Sustainability (Switzerland)*, *8*(2). https://doi.org/10.3390/su8020190

30. Mohd Nasir, S. R., Othman, N. H., Mat Isa, C. M., & Che Ibrahim, C. K. (2016). The challenges of construction waste management in Kuala Lumpur. *Jurnal Teknologi*, *78*(5–3), 115–119. https://doi.org/10.11113/jt.v78.8522
31. *Offsite construction waste management: Lessons from Hong Kong*. (2020). 2020. https://doi.org/10.1016/j.resconrec.2012.09.007.Contact
32. OECD. (2005). *A framework for biotechnology statistics organisation for economic co-operation and development* (vol. 52).
33. *OECD Membership and the Values of the Organisation*. (n.d.).
34. Development Alternatives, & GIZ. (2015). ***Resource efficiency in the Indian construction** sector*. *1*, 94.
35. Yu, A. T. W., Wong, I., Wu, Z., & Poon, C. S. (2021). Strategies for effective waste reduction and management of building construction projects in highly urbanized cities—A case study of Hong Kong. *Buildings, 11*(5), 1–14. https://doi.org/10.3390/buildings11050214

A Short Review on Impact of Urbanization on Surface and Ground Water Quality in India

Surendar Varadharajan

1 Introduction

Urbanization impacts the earth's environment. Urbanization is connected with the expansion of air temperature, hot days and the reduction of relative dampness, wind speed along with vegetation. Developing populaces and relocation toward fabricated regions is driving area utilize change as urbanization over the planet and by two thousand and fifty about seventy percentage of the total populace do need to live in metropolitan regions. Urbanization carries with it a scope of ecological difficulties for both the neighborhood, territorial and more extensive condition as an immediate consequence attribute biochemical and substantial variations to geophysical frameworks [1]. Alteration of the area amid urbanizing construct variation in one and other, i.e. the sort together with greatness through overflow forms. Mentioned above progressions product from flora expanse, clay compression, dumping, depleting together with area coverage along impenetrable rooftops and streets. Urbanization additionally causes variation in area along with river bank. Loss of land along river bank living spaces diminishes the capacity through the water parting to channel supplement along with residue [2]. The misfortune in past area decreases invasion toward soil, while the presentation about counterfeit waste reinstate regular route. The present mix do for the most part considered to have significant impact on the hydrological reaction of a region to precipitation. Regardless of the present and potential development in urban zones couple of logical investigations have given careful consideration to the hydrological effects of across the board urbanization on already country zones. The four interconnected but discernible effect of land exploitation variation on the geophysics of a zone.

1. Variation in crest stream qualities.

S. Varadharajan (✉)
Sustainability Cluster, School of Engineering, University of Petroleum and Energy Studies, Dehradun, Uttarakhand, India
e-mail: vsurendar1990@gmail.com

N. A. Siddiqui et al. (eds.), *Advances in Construction Safety*,
https://doi.org/10.1007/978-981-19-4001-9_19

2. Variation in complete spillover.
3. Variation in characteristics of aqua.
4. Geophysical comfort change.

2 Urbanization Trends

Of all land exploitation variation influencing the geophysics of a zone, urbanization is through fluke the mightiest. Mural variation related along urbanization, especially sprawl, acted noteworthy amid the last 50 years and is required to proceed through the following decades. Changes in atmosphere can likewise majorly affect the geology of an aqua shed. Examinations have indicated increments in mean stream because of increments in precipitation and decreases in stream because of increments in temperature [4]. Surface aqua departure perhaps specifically influenced by aqua nature debasement, wells supply perhaps cut off for clean reasons and additionally for surveillance sense (surges dangers).The primary effect of area utilization on residue originates from introduction of the dirt to storm overflow. This happens principally when uncovered ground is uncovered amid development. It is notable that dregs creation is delicate to arrive slant. Silt yield from urban territories has a tendency to be bigger than in UN urbanized zones regardless of whether there are just little and generally distributed groups of exposed soil in metropolitan area [5]. Surface seas have consumed in excess of 550 billion tons of CO_2 from air, roughly 30% of the aggregate anthropogenic CO_2 outflows. Ingestion of CO_2 from the air has profited mankind altogether by decreasing the ozone harming substance levels in the air. In any case, when anthropogenic CO_2 is consumed via seawater, compound responses happen that diminish seawater pH, grouping of carbonate particle [6]. While particular sources are hard to distinguish, potential contaminations in urban conditions incorporate road litter, final results from petroleum derivative ignition, elastic and metal dissolved from vehicles, erosion of electrifies roofing materials, pet squanders and composts and pesticides connected to gardens [7]. Quick and unstructured or somewhat controlled metropolitan development of Bhagyanagar city has created colossal weight on the regular assets and condition. This examination will give the establishment to detail the strategies to counteract corruption of groundwater and will be useful in feasible advancement in India [8] (Fig. 1).

3 Urbanized Ecosystem

An environment can be characterized as "an arrangement of collaborating animal categories and their neighborhood, non-organic condition working together to support life". On account of the urban condition, it is both conceivable to characterize the city as one single environment or to see the city has made out of a few individual biological communities, e.g. parks and lakes. The environments are road trees, parks,

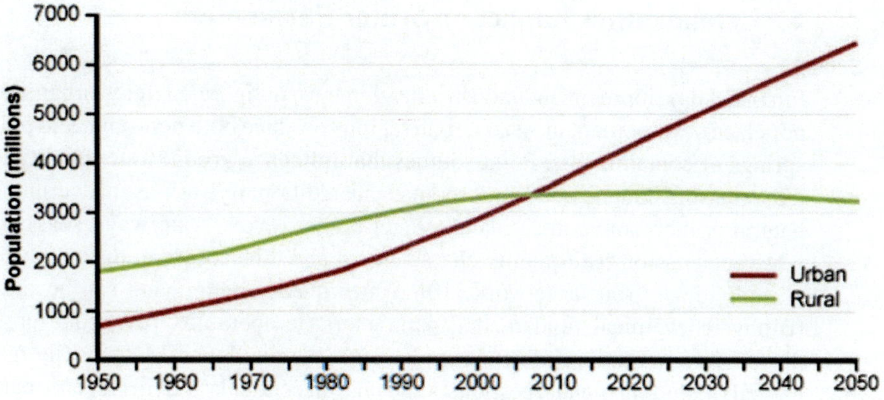

Fig. 1 Global urbanization trend [3]

urban backwoods, developed land, wetlands, lakes/ocean and streams. Among this, wetlands comprise different sorts of bogs and marshes. Lakes/ocean incorporates the vast water territories while streams alludes to streaming water [9]. Safe drinking water is fundamental human right and essential for solid life. In numerous parts of the world, clean water has turned out to be restricted and is in shortage. It is being anticipated that in the following century, it will turn out to be considerably all the more restricting because of expanded populace, urbanization and environmental change (Fig. 2).

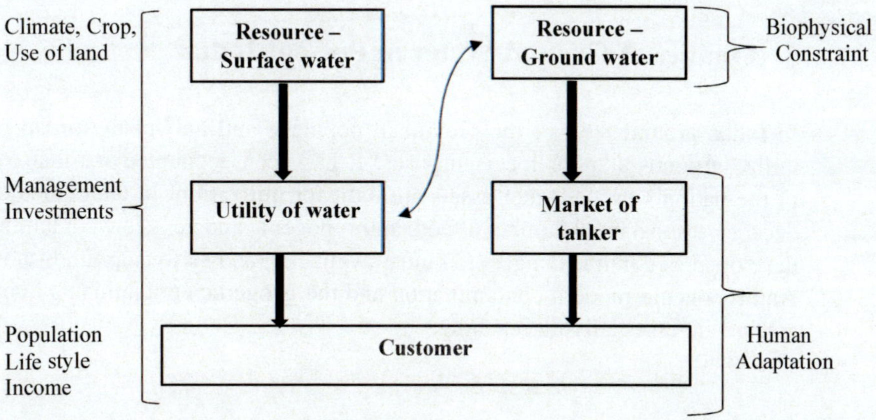

Fig. 2 Land water relation

4 Urbanization Impact on Water Parameters

Financial development and industrialization have invigorated quick urbanization and populace development in Asia. Urban regions produce both non-point and point well-springs of contaminants. Point sources that influence groundwater quality incorporate releasing underground storerooms, and additionally various inadvertent spills of natural or inorganic contaminants. Water bodies (lakes, waterways, seas) are assets where possession isn't plainly characterized and thus access to them is available to all individuals from a network [10]. Water quality parameters can be ordered by (i) physicochemical fundamental parameters (temperature, pH, broke up oxygen, disintegrated natural issues and supplements, (ii) smaller scale toxins (inorganic and natural) including metals, pesticides and pharmaceuticals and (iii) organic parameters with pathogens microorganisms and cyanobacteria [11]. The effects of urbanization on water quality have been very much recorded. Many contextual investigations show that urban streams display expanded levels of phosphorus, N_2, TSS, BOD, metals, oils and fecal coliform bacteria [7]. The quick development of urban territory has two fundamental consequences for ground water assets, for example, impacts on normal energize of aquifers because of fixing of ground with cement and contamination of ground water because of spillage from seepage and mechanical wastage and effluents. The expanding interest for water because of urban development powers, the specialists to import the huge volumes of water from outside the city. The extended urban communities later at that point additionally create expansive volumes of waste water.

5 Quality of Ground Water in Present India

In India, around 80% of the provincial populace and half of the urban populace utilize groundwater for local purposes. It has been accounted for that over 33% of the nation's groundwater assets are unfit for utilization. In the course of recent decades, a substantial number of both anthropogenic and geogenic contaminants has developed as genuine dangers to human wellbeing and serious groundwater utilize. Anthropogenic modern contamination and the geogenic contaminants fluoride and arsenic are especially far reaching.

5.1 Discharge by Industries

Groundwater contamination because of modern effluents is a typical issue in significant Indian urban communities, including Delhi, Mumbai, Kolkata, Ludhiana and Kanpur. In Kolkata, modern waste from the generation of the bug spray Paris-Green (copper acetoarsenite) by a nearby plant brought about the as tainting of hand tube wells utilized for drinking. In excess of 7000 individuals were influenced and a large number of these people built up the dermatological indications of perpetual as harming. In these and different instances of modern contamination, shallow tube wells are more defenseless to tainting than profound tube wells. There are around 2.6 million little scale enterprises (SSIs) in India, moved to a great extent in urban agglomerations. The majority of these SSIs don't treat their mechanical discharge.

The modern city of Ludhiana, Punjab depends totally on groundwater that is sullied in numerous regions with hexavalent chromium (Cr [VI]), a presumed cancer-causing agent and cyanide (three—Twenty seven thousand μg/L). Because of the absence of a surface water asset for squander transfer, a few processing plants in this city have been discovered straightforwardly releasing modern emanating into groundwater through splash pits. In an overview in Chennai, the capital of Tamil Nadu, very nearly 60% of the groundwater was observed to be unfit for utilization. The unregulated exercises of tanneries in the city have brought about genuine groundwater chromium tainting in four regions. Overwhelming metals, including Cr, nickel (Ni), cadmium (Cd) and mercury (Hg) have additionally been found in the groundwater of 43 regions of 14 states in India. The normal dumping of risky synthetic substances by a Union Carbide industrial facility in Bhopal, Madhya Pradesh polluted nearby groundwater [12].

Over the top misuse of groundwater assets because of urbanization in numerous spots all through the nation has prompted a relentlessly declining water level, which can adjust nearby water quality. It has been accounted for from Bangladesh that water system pumping assumes a job in the preparation of as to groundwater. With an end goal to empower expanded state command over groundwater withdrawal through state enactment, the Central Groundwater specialist flowed the Model Bill to direct and Control the Development of Groundwater, an exertion that was first endeavored in 1970.

6 Effect of Urbanization on Surface Water

The single central purpose behind water contamination in India is urbanization at an uncontrolled rate. The expanded interest for water from the developing populace can put included pressure officially extended assets. In and around urban areas, water is regularly hard to come by and subject to expanding rivalry by various clients. Individuals utilize the waterways to discard every one of their losses from homes, ventures and business organizations. Any progressions to the nature of surface water likewise influences groundwater in light of the fact that they are connected by the procedures of the water cycle so toxins from the surface will penetrate down and sully soil and groundwater also. It is evaluated that urban communities with populaces of in excess of one lakh individuals produce around 16,662 million liters of wastewater in multi day. For some odd reason, 70% of the general population in these urban communities approach sewerage offices. Urban communities and towns situated on the banks of Ganga produce around 33% of wastewater created in the nation.

An ongoing report on lakes titled 'Wetlands: Treasures of Bangalore' uncovered the accompanying outcomes because of quick urbanization.

- 90% of the lakes were sewage-nourished because of maintained stream of untreated sewage and mechanical effluents, dumping of strong squanders and building flotsam and jetsam.
- Water quality examination of 80 lakes found that half of the lakes were much dirtied.
- None of the lakes had water that was fit for drinking as indicated by guidelines set by CPCB.
- 79% lakes fell under Class E classification ordered by CPCB as reasonable for water system, mechanical cooling or controlled waste transfer.
- 29% lakes could be grouped under Class E and D as appropriate for angle culture and untamed life proliferation.
- Only one lake Myla Sandra 1 and 2 fell under Class A that was observed to be appropriate for drinking purposes as indicated by the CPCB characterization.

The Central Pollution Control Board (CPCB) has set up a network of monitoring sites on rivers across the country to keep an eye on pollution. People in 26 states and 5 territories across the country use the network. There are 870 stations spread across the country. Monitoring data of major rivers shows that organic pollution is the main source of pollution, and almost all surface water sources are contaminated to some extent by Coliform Group of Bacteria that make them unfit for human consumption unless they are cleaned. Sabarmati, Godavari, Satluj, Yamuna, Cauvery, Ganga, Krishna, Tapi, Mahanadi and Brahmani are the rivers with the most organic and bacterial pollution. Mahi, Narmada, Brahmaputra and Beas are the rivers with the least organic and bacterial pollution. The water quality standard is used to integrate the integration of a designated use for a certain water body with the water quality rules that protect it. Recreation, drinking water, shellfish propagation and harvesting are among some of the things that are permitted [13].

Other studies also proved that urban recharge component of groundwater have been more than ten times higher than the natural recharge in Hyderabad region.

A study on water quality of Tapi River in Surat at different locations (Chowk, Navadi and Chowpati) revealed the following results. BOD is 49 mg/l, 59 mg/l and 56 mg/l where it shall be 30 mg/l, COD is 101.3 mg/l, 119.6 mg/l and 113.6 mg/l. TDS is 3630, 4250 and 4810 mg/l where it should be 2100 mg/l [14]. Different investigations additionally demonstrated that urban energize segment of groundwater has been in excess of ten times higher than the normal revive in Hyderabad district.

7 Conclusion

Urbanization is a standout among the most great and obvious anthropogenic powers on Earth. The procedures of urbanization and improvement are connected to natural corruption, they are in no way, shape or form and homogenous or uniform as far as their express spatial examples or suggestions to biophysical working. In this examination, we propose that urbanization can be portrayed by the four distinct measurements of scene structure, scene setup, arrive utilize force and network. The quick urban sprawl has brought about a lot of developed terrains being supplanted with building lands. Be that as it may, from the audit, it is presumed that many urbanized water bodies are getting the flood of contaminations which are in charge of changing its water quality. The expanding human settlement and industrialization in closeness of numerous waterways is among the conspicuous reasons for changes in water quality. Groundwater quality administration must be put inside a coordinated structure that incorporates surface and groundwater, farming, residential supply, ecological conservation, mining and industry. Groundwater quality administration must be put inside an incorporated system that incorporates surface and groundwater, farming, household supply, natural protection, mining and industry. A considerable lot of the states in India are not ready to maintain a strategic distance from contamination of ground and surface water because of the accompanying reasons: absence of ordered aptitudes in many state contamination control sheets, lack of asset, labor and mindfulness. It is less educational and persuading to recognize the urbanization issues, observing consistence with administrative cutoff points, set release limits, set up needs for contamination control endeavors and execute suitable control measures.

References

1. Miller, J. D., et al. (2014). Assessing the impact of urbanization on storm runoff in a peri-urban catchment using historical change in impervious cover. *Journal of Hydrology, 515*, 59–70.
2. Alberti, M., et al. (2007). The impact of urban patterns on aquatic ecosystems: An empirical analysis in Puget lowland sub-basins. *Landscape and Urban Planning, 80*(4), 345–361.
3. United Nations. (2001). World urbanization prospect; The 1999 revision. United Nations, Washington, DC. Online at http://www.un.org/esa/population/pubsarchive/urbanization/urbanization.pdf. Accessed on October 20, 2014.
4. Claessens, L., et al. (2006). Effect of historical changes in land use and climate on the water budget of an urbanizing watershed. *Water Resources Research, 42*(3).
5. Leopold, L. B. (1968). Hydrology for urban land planning: A guidebook on the hydrologic effects of urban land use.
6. Feely, R. A., et al. (2010). The combined effects of ocean acidification, mixing, and respiration on pH and carbonate saturation in an urbanized estuary. *Estuarine, Coastal and Shelf Science, 88*(4), 442–449.
7. Carle, M. V., et al. (2005). Patterns of watershed urbanization and impacts on water quality 1. *JAWRA Journal of the American Water Resources Association, 41*(3), 693–708.
8. Wakode, H. B., et al. (2018). Impact of urbanization on groundwater recharge and urban water balance for the city of Hyderabad, India. *International Soil and Water Conservation Research, 6*(1), 51–62.
9. Bolund, P., & Hunhammar, S. (1999). Ecosystem services in urban areas. *Ecological Economics, 29*(2), 293–301.
10. Ramachandraiah, C., & Prasad, S. (2004). *Impact of urban growth on water bodies: The case of Hyderabad*. Centre for Economic and Social Studies Hyderabad.
11. Delpla, I., et al. (2009). Impacts of climate change on surface water quality in relation to drinking water production. *Environment International, 35*(8), 1225–1233.

12. Chakraborti, D., et al. (2010). *Examining India's groundwater quality management*. ACS Publications.
13. Trivedi, R., et al. (2007). Biological monitoring of water quality in India–needs and constraints. In *Proceedings of Taal2007: The 12th World Lake Conference*.
14. Kamal, P., et al. (2014). Impact of industrialization and urbanization on water quality of river Tapi Surat, Gujarat, India. *Journal of Environmental Research and Development, 9*(2), 306.

Specifying Safety Instrumented System Response Time: Case Studies from Oil and Gas Facilities

G. Unnikrishnan

1 Introduction

Controlling process hazards in facilities

Minimizing and eliminating hazards in process plants are a constant endeavour for Process Safety. The concept of functional safety and the added layer of protection called Safety Instrumented System (SIS) evolved from the above concept. Figure 1 shows the context of the SIS layer of protection that comes into action when other layers before it fails. The requirements of SIS and the related International Electrotechnical Commission's Standards IEC 61508 and IEC 61511 are well known to the industry [1].

Several issues need to be addressed while designing the SIS and specifying the safety instrumented function (SIF) loop, which consists of the sensor, the logic solver, the Final Control Element and its interconnections.

This paper's main objective is to discuss certain key issues that came up during the application of the SIS in an oil and gas facility and a long-distance pipeline. These cases illustrate that in real world applications, the situation can become complex and in spite of the well-developed theoretical foundations, and it becomes difficult to determine if adequate safety has been provided.

Further content is in four sections. Section 2 will summarize the basic concepts of SIS. Section 3 will focus particularly on the topic specifying the SIS response time. Case studies highlighting certain key factors while specifying SIS response time will be in Sect. 4. Section 5 will provide concluding remarks.

G. Unnikrishnan (✉)
Kuwait Oil Company, Ahmadi, Kuwait
e-mail: ukrishnan77@yahoo.com

N. A. Siddiqui et al. (eds.), *Advances in Construction Safety*,
https://doi.org/10.1007/978-981-19-4001-9_20

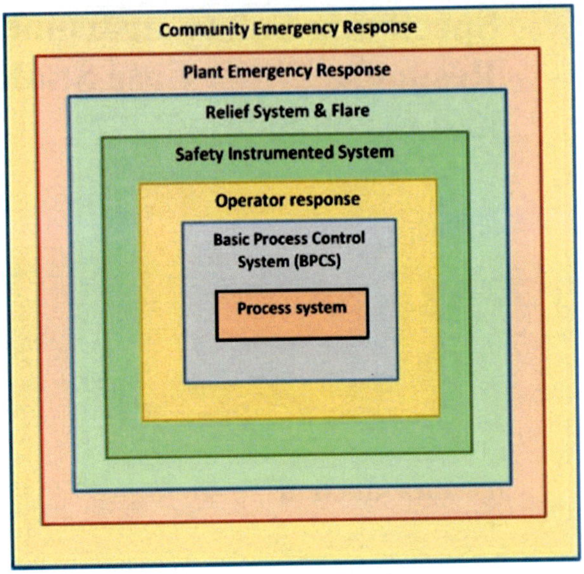

Fig. 1 Layers of protection concept

2 Basic Concepts of Safety Instrumented System (SIS)

The seventies and eighties saw major growth of industries in general and particularly process industries. The rapid industrialization caused series of accidents and fatalities too. Major accidents at Seveso and Bhopal challenged the engineers and scientists to come up with better safety systems to protect the employees and environment. Building on efforts of several individual organizations, International Electrotechnical Commission (IEC) in 1998 published the generic IEC Standard 61508. When electronic systems and Programmable Logic Controllers were developed, functional safety also incorporated the same. IEC 61511 published in 2003 is specific to the process industry sector and incorporates the concept of functional safety and layers of protection.

The SIS concept adds a completely independent layer of protection with high reliability in addition to the layers already available for the process system. The reliability of the SIS is expressed as Safety Integrity Level (SIL) and measured in Probability of Failure on Demand (PFD). The design specifications of the Safety Instrumented System (SIS) are collectively called Safety Requirements Specifications (SRS).

3 Safety Instrumented System (SIS) Response Time

Functional safety standards require that the overall SIS response time be determined and documented in the Safety Requirements Specifications (SRS) [1].

3.1 SIS Response Time

IEC 61508 mentions SIS response time as "the time within which it is necessary for the safety function to be completed." and IEC 61511 defines it as "the time required for the Safety Instrumented System to bring the process to a safe state."

From the above, it is possible to visualize the time required for the SIS action to be completed before the process hazard blows up into a disaster. The situation is given in Fig. 2.

Looking at Fig. 2, which shows the process variable values versus time available for action, the following is clear [2].

a. As soon as a process upset is initiated, the first layer of protection to react is the Basic Process Control System (BPCS). It tries to control the process variable (PV) back to the normal set point. When it fails, the process upset continues and reaches the alarm limit.
b. On hearing/seeing the alarm, operator action is expected. It could be that there is no action from the operator or that he/she could not control the upset situation. The process upset now reaches the trip point of the SIS.
c. The SIS has to act once the process variable (PV) reaches the Trip set point, which means that the following actions need to be completed:

   – Detection of the process PV that has reached abnormal situation (Trip set point)

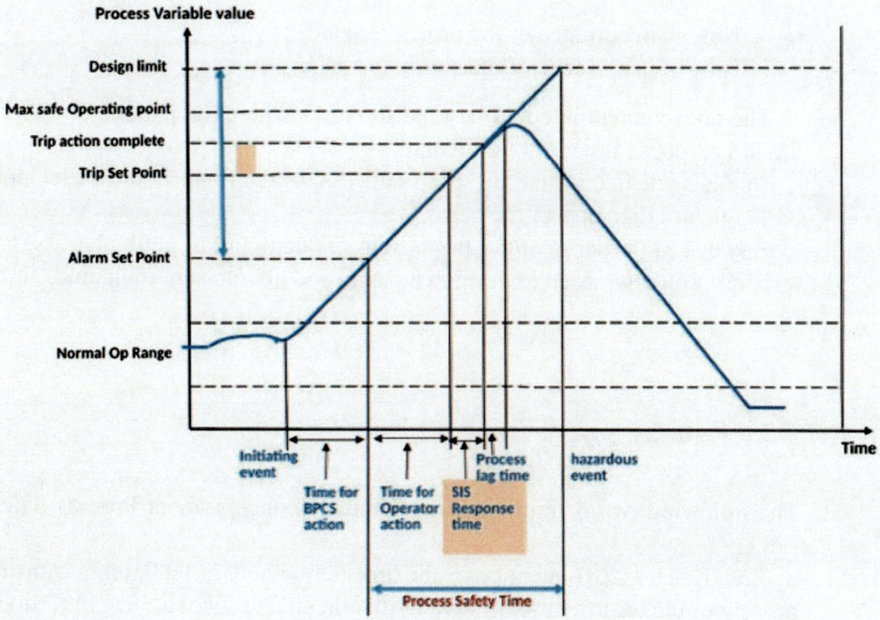

Fig. 2 Time available for action-Process Safety Time (PST)

- Transmission to the logic solver
- Computation of the corrective action by the logic solver
- Transmission to the Final Control Element
- Action of the Final Control Element to complete the SIS (Trip) action.

The important factor is that the SIS has to act before the process variable (PV) breaches and goes through the protection layer. Thus, what matters here is the speed of action of SIS or the response time.

On the other hand, the SIS response time cannot be too fast because it can cause a surge pressure upstream on fast closing of the valve, in the case of incompressible fluids. For compressible fluids, there could be other issues.

Now the response time of the SIS loop can be obtained from the manufacturers, which is possible only after the design specifications are complete. However, how do we know that it is fast enough to prevent the hazard event or too fast to cause surge pressure? In order to understand if the response is fast enough, we need to know the speed of propagation of the process upset. Here is where the Process Safety Time (PST) becomes important [3].

Before reviewing PST, let us see certain thumb rules in practice in the industry for an initial estimate of the SIS response time. That is the SIS response time should be less than half the PST.

$$\text{SIS response time} \leq \frac{1}{2}(t_{\text{haz}} - t_{\text{al}}) \tag{1}$$

where

t_{hz} Time of occurrence of the process hazard
t_{al} Time of occurrence of the alarm

The above equation could be used for cases where the parameters, specifically, the SIS response time is not critical.

Another practice is that the speed of action of the Final Control Element should be 1 s/inch of diameter of the valve.

However, in order to study the system in detail and to get better assessment of response times, we need to develop a model and do transient (dynamic) simulations [4].

3.2 *Process Safety Time*

The following two definitions gives definitive concept about Process Safety Time (PST):

IEC 61511-1:2016 mentions: “the time period between a failure occurring in the process or the basic control system (with potential to give rise to a hazardous event) and the occurrence of the hazardous event if the safety instrumented function is not performed” [1].

CCPS Guidelines for Safe and Reliable Instrumented Protective Systems: "the time between a failure occurring in the process or its control and the occurrence of the hazardous event" [5].

Clearly, PST is dependent of the system/equipment geometry, the dynamics of the process and external disturbances. The PST will be affected by different external disturbances as can be seen below from case studies. Therefore, to protect the equipment under consideration, the independent protection layer (IPL) should act before the hazardous event, that is, its action should be completed within the PST. In addition, there could be multiple independent protection layers for an equipment, in which case, assigning of priority of activation of the IPLs need to be analysed. Another factor is the Process Lag Time, which is basically the inertia of the system to change. Process Lag Time also has to be considered while determining the PST.

From Fig. 2, the following relationships can be derived:

$$\begin{aligned}\text{PST} &= \text{Time between initiation of the alarm for} \\ &\quad \text{process upset and occurrence of the process hazard} \\ &= \text{Time of occurrence of the process hazard}t_{\text{hz}} \\ &\quad - \text{Time of occurance of alarm for upset condition}t_{\text{al}}\end{aligned}$$

$$\text{PST} = t_{\text{haz}} - t_{\text{al}} \tag{2}$$

Taking the corresponding process variable values as PV_{al}, PV_{hz} and velocity of the PV as $\text{PV}/\text{d}t$

$$\text{PST} = (\text{PV}_{\text{hz}} - \text{PV}_{\text{al}})/(\text{PV}/\text{d}t) \tag{3}$$

The equation connects all the variables that determine PST.

3.3 Calculation of Process Safety Time (PST)

Since PST is strictly a property of the process system, it has to be calculated from the details of process parameters. Following examples illustrate the features of the issues.

(i) If it is large vessel, it is relatively easier. In oil and gas facilities, the crude storage tank with large cross-sectional area is a good example. Taking the example of a key PV such as liquid level for the tank, it can be seen that this PV raises slowly and so there is ample time for the SIS to act. (Notwithstanding the fact that several accidents involving tank overflows have happened in spite of the high PST available for controlling the same). In such cases, manual calculation or judgement of experts could suffice. Please see Fig. 3.

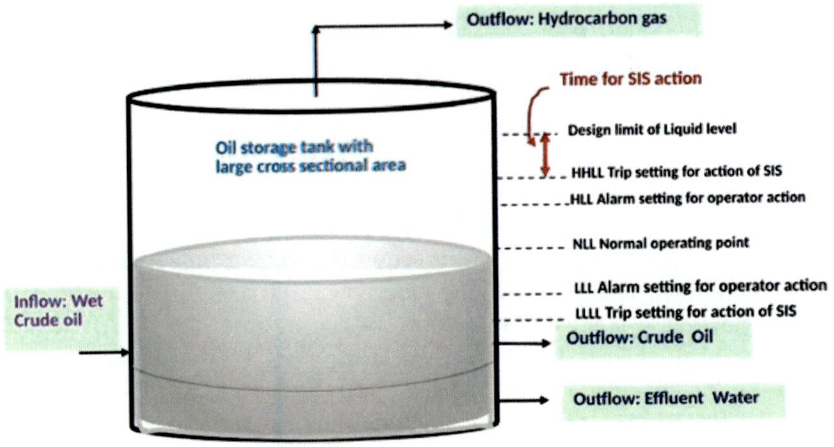

Fig. 3 Oil storage tank having large cross-sectional area

(ii) Another example is the case of an inlet two phase crude oil/water separator. Please see Fig. 4.

The following is the more common scenarios considered as hazards in oil and gas separator. Abnormal values of the parameters often termed as High-High or Low-Low will cause shutdown of the Emergency Shut down Valves (ESDVs) around the equipment, thereby preventing further progress of the abnormal situation.

– High-High pressure in the separator

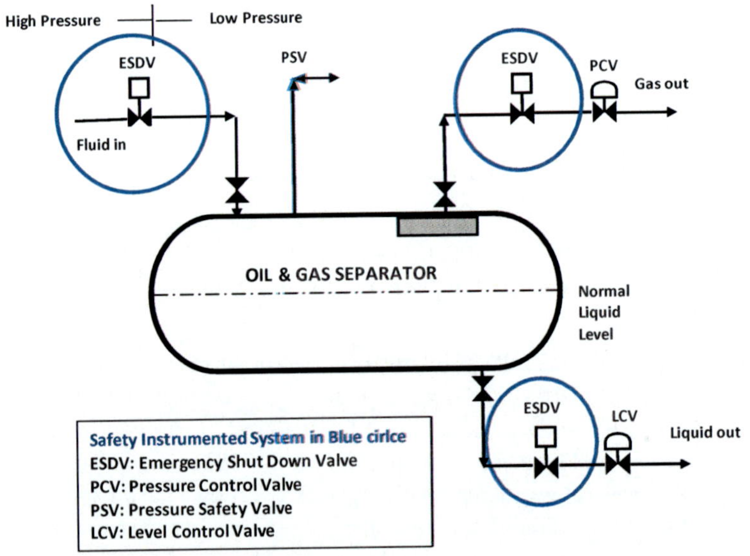

Fig. 4 Oil and gas separator with protection layers

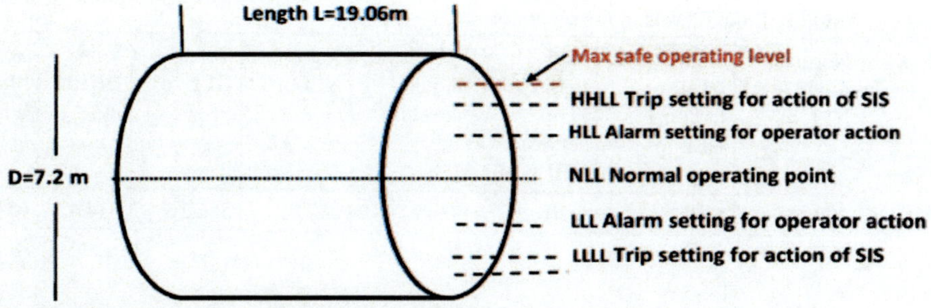

Fig.5 High-High liquid level hazard and design limit in oil and gas separator

- High-High liquid level in the separator
- Low-Low liquid level in the separator

There could be other parameters also such as Low-Low oil–water interphase level, but they are not considered here. The protections—Independent Layers of Protection (IPL)—listed above, are shown in Fig. 5.

The hazardous situations in oil and gas separator are generally not that fast, since the separator is a large vessel with high cross-sectional flow area. It takes time, often in minutes for the PV to reach abnormal situations. A dynamic model of the process is required to simulate the movement of PV from normal to abnormal as well as the action of the SIS.

Further, in some cases, we have to consider arrival of slugs into the vessel as a result of disturbances from upstream. The slugging phenomena in multiphase flow has a complex behaviour and several factors such as the well fluids composition, multiphase flow regime, flowline and header sizes, profile of the piping, mode of operation, etc. The scenario requires complex models as can be seen from the case study given below.

4 Case Studies

4.1 Case Study 1

SIS response time for the oil and gas separator for the hazard condition of High-High level in the separator. Please see Fig. 5.

a. Hazard condition: The Flow Control Valve at the liquid outlet fails closed fully due to malfunction. We consider the partial volume of the liquid in the separator for this upset condition when the liquid level rises.

Related separator details, process information and calculated values are given in Table 1:

Diameter of separator: 7.2 m and length of separator: 19.06

Table 1 Liquid levels in the separator

Normal liquid level NLL m	High liquid level. (Alarm) HLL m	High-High liquid level (Trip action) HHLL m	Max safe liquid level m	Vol at NLL m3	Vol at HLL m3	Vol at HHLL m3	Vol at Max safe liquid level m3
3.60	5.10	6.20	6.45	387.82	556.16	634.43	655.20

All liquid levels are from the bottom of the vessel.

Normal inlet volumetric flow rate = 0.20 m^3/s

From the above, the time taken for each action can be calculated. Same is given in the Table 2.

As can be seen, there is 6.52 mins for operator to take action from High liquid level (Alarm point) to the High-High liquid level. But once that opportunity is missed, then there is only 103.88 s (1.7 mins) available for the Trip or SIS action to close the inlet and outlet Emergency Shut Down Valves (Final Control Element), before the upset liquid level goes through the safe limit. The above 1.7 mins is easily achievable by SIS loop, particularly by the Final Control Element, the Emergency Shut Down Valve.

Based on equation, the Process Safety Time can be obtained by adding the time taken by level to rise from alarm (HLL to HHLL) and the time taken for level to rise to maximum safe limit. That is 391.33 + 103.88 = 495.21 s

With the calculated value of PST, we can see that the SIS response time is well within half of the PST (247.60 s).

If the level rises above the maximum safe limit, the liquid will flood the mist eliminator and will go through the gas outlet line, with further serious consequences.

All looks safe as of now. In fact, this was the original design of the facility.

b. After the facility has been operating for many years, it was noticed that there was increasing high-level upsets and shutdowns due to the same. However, the shutdowns initiated by the SIS were late, happening after reaching the hazard conditions. By that time, the liquid level had reached the mist eliminator and flooded it. **All the same, the control system and level control valve at the liquid outlet was working fine.**

The reason for frequent **high-level upsets was the onset of slug flow in the inlet** piping. Although the design of the separator had incorporated certain volume to accommodate slug flow, **that was based on assumptions during the design done 12**

Table 2 Time taken for corrective actions

Time taken for level to rise from NLL to HLL	841.71 s (14.03 min)
Time taken for level to rise from HLL to HHLL	391.33 s (6.52 min)
Time available for completing Trip (from HHLL to max safe liquid level = SIS response time)	103. 88 s (1.73 min)

Table 3 Slug arrival & Revised time taken for corrective actions

Time taken for level to rise from NLL to HLL	112.23 s (1.87 min)
Time taken for level to rise from HLL to HHLL	52.18 s (0.87 min)
Time available for completing Trip (SIS) action)	13.85 s

years back, since at that time, the reservoir was producing oil without any flow issues. Slug flow phenomena developed over the period of time as the reservoir matured, and the volume given in the original design for the separator became insufficient to handle the slug volume.

A flow assurance study was taken up to determine if there is slug flow in the current conditions and if it is there, what can be done about it with minimum disturbance to operations.

The flow assurance study established presence of slug flow and an increasing trend for more frequent and increased volumes of slug flow.

When we consider a case of slug arrival at the rate of 1 m^3/s, then the time available for corrective actions changes dramatically. Please see the Table 3.

The above table gives that the SIS has to act much faster, within 13.85s to protect the process system.

After analysis, it was decided to change the existing SIS's Final Control Element unit to a valve with faster response characteristics. Higher slug arrival rates will eventually require a slug catcher before the separator.

4.2 *Case Study 2*

This case study describes a High Integrity Pressure Protection System (HIPPS) positioned at the inlet of a long-distance pipeline of 48-inch diameter transporting crude oil.

HIPPS is a SIS designed and built as per IEC 61508 and 61511 with the specific application of preventing overpressure in production flowlines, piping systems within a process facility and pipelines. HIPPS has the same components as a SIS but will often contain more redundancies. Figure 6 shows schematic of a HIPPS with electronic logic solver, typically used in oil transportation pipelines, from Health and Safety Executive, UK [6].

The HIPPS loop detects high pressure in the pipeline and closes the Final Control Elements (ESD Valves) to protect the downstream pipelines, which are designed for a lower pressure rating. HIPPS is completely independent standalone system. HIPPS has three sensors and a voting logic of 2 out of 3 for the valves to close. Each sensors/transmitters and its loop is independent even within HIPPS. There are two Final Control Elements also. All of the above reduces the failure rate of the system.

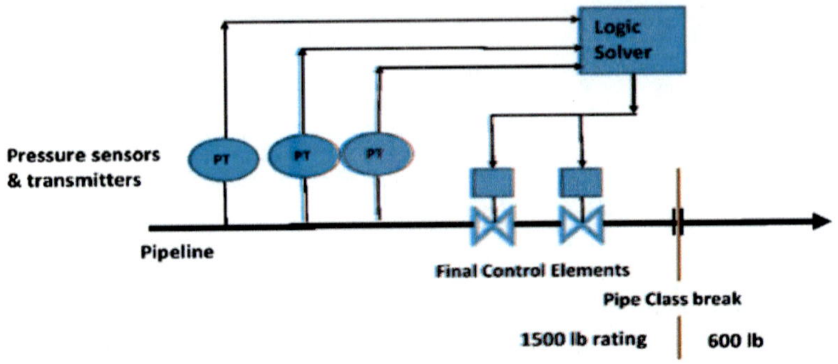

Fig.6 Typical HIPPS loop

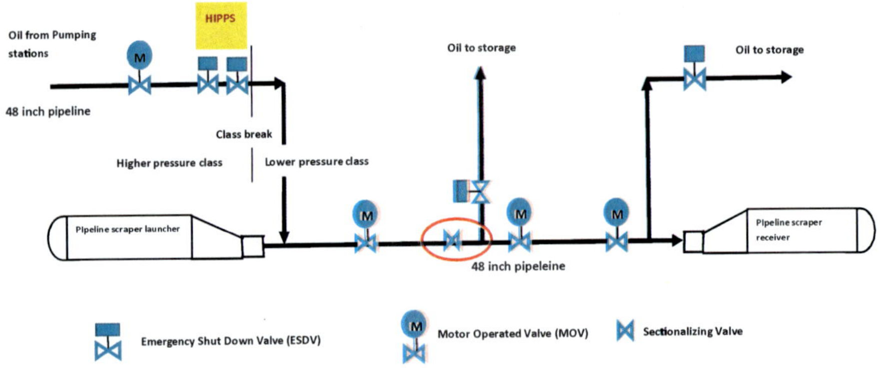

Fig.7 Simplified schematic of pipeline system

Let us now review a typical long-distance pipeline. Figure 7 is a simplified example of the pipeline system. Details of pipeline are given in Table 4.

The pipeline has another branch pipeline going out to another intermediate storage. There are sectionalizing valve stations at every 15–20 km kilometres as per code requirement. These are gas-operated valves (GOVs) since power is not available on the route of the pipeline.

The GOVs close automatically on detection of any leak upstream or downstream of the location of the GOV. GOVs do not act that fast, for a 48-inch line, it is typically in the range of 120–180 s.

There are several cases of the pipeline operations that can result in surge pressure in the line. Scenarios that have impact on the HIPPS valve are noted below: (This is not an exhaustive list. Interested readers may refer to several books available on the subject [7]).

- Motor-operated valve (MOV) closure at the inlet and outlet of the pipeline.
- Closure of the sectionalizing valve at the midpoint of the pipeline.
- Pumps trip

Table 4 Basic details of the pipeline system

Diameter	48 inches
Distance	150 km
Fluid	Stabilized crude oil
Normal pumping pressure at inlet	18.6 barg/(270psig)
Receiving pressure at end of pipeline	4.2 barg/(60 psig)
Pipeline design pressure	58.6 barg/(850 psig)
Provision of HIPPs	Yes
Pipeline code	ASME 31.4 pipeline transportation for liquids and slurries
HIPPS	Provided at inlet and end of the pipeline. Set point originally set at 40 barg, reduced to 38 barg HIPPS response time: originally set at 10 s, later reduced to 8 s

– Pumps start up and flowrate ramp up.
– Pipeline operation at part capacity

For the purposes of the paper, we will focus on the MOV and sectionalizing valve closure.

On modelling and doing a transient simulation of the system, it was found that, the closure of MOVs within a time span of 100 s gives rise to a surge pressure within the design limits of the pipeline. Nonetheless, the HIPPS valve acts at 40 barg (original HIPPS set point) to close the valve.

However, the closure of the sectionalizing valve at the midpoint in the line, in a time span of 160 s gives rise to a surge pressure 70 bar at the inlet of the pipeline after 25 s. This is happening despite of the fact that the HIPPS valves closing off the inlet flow or source of pressure. Please see Fig. 8.

After several iterations of the transient flow simulation model, it was found that reducing the set point of HIPPS to 38 barg and reducing the response time of HIPPS from 8 to 6, kept the rise in surge pressure to within the design pressure of the pipeline. The above was considered as the optimum solution for the problem.

The above case studies highlight the need to study the response time of SIS in certain scenarios.

5 Concluding Remarks

SIS are provided in process systems/units to protect the same from reaching hazardous conditions. The SIS provides a highly reliable and independent system that will automatically act when the BPCS operator actions fail.

However, while specifying the parameters of the SIS, the Process Safety Time and SIS response time need to be analysed to see if the response time is sufficient when

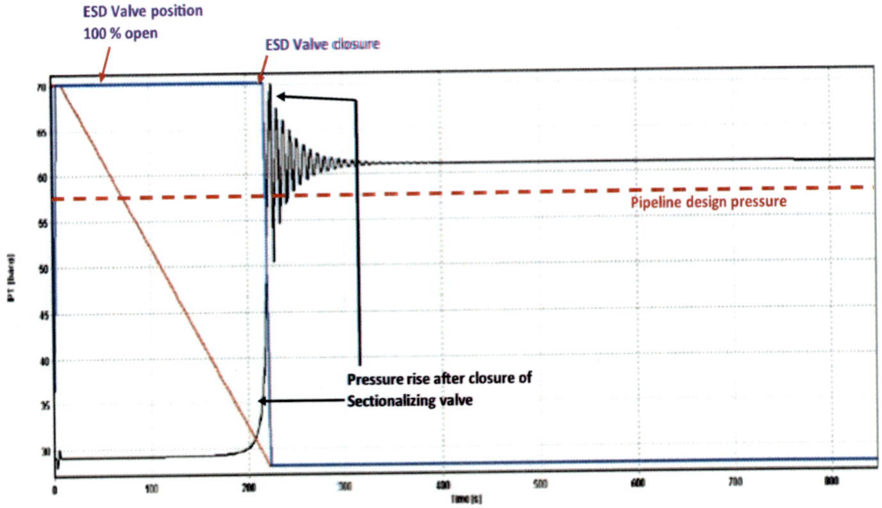

Fig.8 Surge pressure rise on closure of sectionalizing valve

compared with PST. The above case studies illustrate that in oil and gas facilities where multiphase fluids are handled, slug flows are to be studied for its impact on SIS response time. Sometimes the action of SIS ESD valve can create surge pressures in the piping/pipelines. The above situation calls for transient analysis of the fluid system and several iterations will be required to fine-tune the SIS response time and/or the process system/unit's design parameters to ensure safe operation of the SIS.

References

1. O'Brien, C., & Bredemeyer, L. D. (2009). *Final elements & the IEC 61508 and IEC 61511 functional safety standards*. Exida.com LLC.
2. Barnard, G., & Creel, W. (2015). Impacts of process safety time on layer of protection analysis. *Process Safety Progress, 34*(4), 383–388. https://doi.org/10.1002/prs.11759
3. Chastain-Knight, D. (2019). Confirming the safety instrumented system layer of protection. *Process Safety Progress, 39*(1). https://doi.org/10.1002/prs.12079
4. Akintoye, A., & Harrow, S. (2016). Determination of Alarm Safety Response Time. *Hazards 26, Symposium Series No 161*. IChemE. U.K https://www.icheme.org/media/11753/hazards-26-paper-16-determination-of-alarm-safety-response-time.pdf
5. CCPS. (2011). *Layer of protection analysis: Simplified process risk assessment*. Wiley
6. Health and Safety Executive. (2011). Safety instrumented systems for the overpressure protection of pipeline risers. HSE.GOV.UK. https://www.hse.gov.uk/foi/internalops/hid_circs/technical_osd/spc_tech_osd_31.htm
7. Pickford, J. (1969). *Analysis of Surge*. Macmillan & Co.

Understanding the Critical Conditions During a Backdraft—A Review

Sakshi Dubey and Akshi Kunwar Singh

1 Introduction

The occurrence of the fire is the first step and the primary source of occurrence of the backdraft. Every situation where fire occurs does not lead to backdraft, but there is a chance that even in such a situation where backdraft cannot be expected, backdraft can happen. The possibility of fire occurs when the three main elements of the fire triangle come together at one place, which is "Ignition Source, Fuel, and Oxidant". The oxygen plays a vital role in the events resulting in the backdraft. When the fire develops from flammable liquids or even ignitable solids with more oxygen demand than the amount of oxygen present in the area, the fire will starve. During this stage, fire starves due to the lack of oxygen. When a small opening is created in the space for the oxygen to flow through could lead to the sudden expansion of the fire, resulting in a fire ball engulfing the whole space. Deflagration may also be the final result of such a phenomenon [6].

The primary condition responsible for the occurrence of the backdraft is the presence of starving fire in an enclosed place, assuming that some flammable material is present in a closed space, such as a room or a less ventilated space, which came in contact with a heat source or an ignition source. This contact of fuel with the ignition source in the presence of oxygen or air causes fire. As the space is less ventilated, the oxygen demands of the fire won't be met. It does not extinguish the fire, but the ventilation controls it, and the behaviour of the fire is highly dependent on the shape and size of the opening created [10]. Opening a door or breaking a window results in the inflow of the air currents, also known as the gravity currents. These gravity currents provide the extra oxygen required to the fire, resulting in a rapid increase in the temperature of the space or room, pushing out the Fuel–air mixture through the opened door or window into the open space [7]. The fuel–air mixture will

S. Dubey · A. K. Singh (✉)
School of Engineering, UPES, Dehradun, India
e-mail: singh.akshi85@gmail.com

N. A. Siddiqui et al. (eds.), *Advances in Construction Safety*,
https://doi.org/10.1007/978-981-19-4001-9_21

combine with the gravity currents at the interface and ignite, increasing the temperature rapidly and causing a fireball or even deflagration. The increase in temperature and the pressure due to the rapid ignition causes the sudden expansion.

Many experiments have been conducted to identify the critical conditions of the backdraft and the effective preventive measures since it is the most dangerous phenomenon and poses a severe threat to the firemen in rescue operations. After conduction, many investigations, analysis, and experiments, the main reason for the occurrence of the backdraft is the fraction of volume of the ignitable gases present in the air in the enclosure, which varies for different combustible gases or materials or fuels [8]. Further studies have found that the backdraft occurrence also depends on the lower explosion limits of the combustible fuels. When the fraction of volume of the combustible gases favours the backdraft occurrence, then the backdraft occurs. So, the preventive measure taken can now be concentrated on increasing the volume fraction or decreasing it such that it's not in the range which favours the backdraft.

Few studies were conducted on finding out the critical condition of the backdraft, which includes experimental studies, theoretical studies and also simulation conducted by [3, 5, 11]. Each study has a different approach in finding the critical conditions, but all studies conducted state that the fraction of volume of combustible gases is one of the critical conditions leading to the occurrence of the backdraft. Each method taken to conduct the study and the results found is compared and discussed in the paper.

2 Review

2.1 Experimental Study

In the experimental study conducted by [11] to identify the critical conditions of the backdraft, they considered a small compartment as their experimental setup with specific dimensions having two small openings, designed in such a way that it'll withstand the high pressures and temperatures of the backdraft and also used methane as their fuel source. Temperature and pressure readings were taken using thermocouples and electronic pressure transducer, respectively, in the compartment. Water mist was used as a protection measure, and also concentration measures of O_2, CO, and CO_2 were taken. The flow rate of the fuel entering into the compartment and also the air currents are recorded. The flame was ignited in the compartment, and after some predetermined time, the fuel inflow was seized to make the fire extinct, and then the experiments were conducted to record the results.

After performing the calculation, it has been found out that the mass fraction of the unburned methane used as the fuel is almost equal to the critical mass fraction required for the backdraft to occur. It has been found that the total mass flow of gases entering the compartment is more significant when compared to the mass flow of gases exiting the compartment after opening the hole in the compartment before igniting the fuel. The experiment was repeated many times with different mass fractions of oxygen, carbon dioxide and carbon monoxide, where only a few times witnessed the backdraft. The main factor is the fraction of mass of the methane, which is the fuel, which is not burnt, and the critical value of the fraction of mass for the backdraft to occur is determined to be 9.8. If the fraction of mass of the unburnt fuel is more than 9.8, then the backdraft would occur, and if the fraction of mass of unburnt fuel is less than 9.8, then the backdraft would not take place.

The result is illustrated using a flammability diagram of the methane, nitrogen, and oxygen mixture [2]. A flammability diagram is a triangle with a concentration of methane, oxygen, and nitrogen, which are considered in this case, on each side marked from 0% mass fraction to 100% mass fraction—plotting the lower flammability limits of the mixture on the diagram and showing all the possible mixture combinations that can be formed when the opening is created in the compartment. An envelope can be observed on the flammability diagram, and all the mixture concentrations inside the envelope form flammable mixtures. The ends of the envelope formed in the flammability diagram give the minimum required mass fraction for the occurrence of the backdraft. All the mixtures beyond this limit would lead to a backdraft. It has been stated that if the flammability diagram is accurate, we can obtain a critical fraction of the mass of the unburnt methane or the fuel for the occurrence of the backdraft.

2.2 Simulation

The simulation study is done by [5], and done on *"Fire Dynamics Simulator"* by using the code which is developed by *"National Institute of Standards and Technology"*, following the flame index, which is a scalar product of oxidant gradient and fuel, and LELFM (*"Large eddy laminar flow model"*), which is used mainly to solve the Navier–Stokes equation [4, 9]. Considering two scenarios where the gases are mixed beforehand and the other, the gases are not mixed.

After being analysed, the simulation results led to the division of the backdraft phenomenon into five stages.

2.3 Ignition Stage: The gravity currents from the opening created in the compartment reach the source of ignition and ignite. The results have shown that the premixed gases create a flame front that travels downwards to the surface.

2.4 Free Spherical Propagation: The flame front propagation towards the created opening was identified in this stage. It is found that the upstream creates a mixture where the flame propagates, creating a thermal expansion, increasing the pressure, driving the mixture of oxygen, and other combustibles towards the opening in the compartment.

2.5 Plane Front Propagation: In this stage, the concentration of oxygen near the surface was observed to be three times more when compared to the concentration near the upper surface, and the fuel concentration which was pushed downwards from the top surface was seven times more than the fuel concentration from the lower surface. The mixture near the surface was practically observed to be unburnt when compared to the other oxygen starving fuel because the mixture formed was not a flammable mixture. During this process, it is observed that the turbulence in the lower part, near the exit, is more, and it spreads more rapidly than the mixture near the top surface.

2.6 Flame Front Stretching: Due to the turbulence created, speed up to 13 m/sec were reached by the flame front, which accelerates towards the exit and 40 m/sec speed were reached during exit.

2.7 Fireball is the final stage where the accelerated flame front deforms after losing momentum and forms a ball structure rising upwards. This flame front, when mixed with fresh air outside, ignites viciously and creates a backdraft fireball.

3 Theoretical Analysis

Chen et al. [3] conducted the theoretical analysis along with an experimental study to identify the critical conditions of the backdraft. This study states that backdraft is possible only if the volume fraction is more than the necessary amount, and this critical amount varies for different materials [1]. The study is carried out by assuming certain assumptions, which are (1) ideal gas law is applicable, (2) chemical reaction is the only heat source, (3) there is no difference in densities, so there would be no buoyancy and (4) solid phase and gaseous phases are homogenous.

Experiments have been conducted using wood and heptane as fuel sources, and it was observed that the fire in the enclosure reaches the smouldering stage after the opening, which was created in the box, is closed. The temperatures in the box spike as the backdraft occurs. The experiments state that, while free burning, the volume fraction of oxygen, which was used continuously is reduced abruptly, and the CO_2 volume fraction increases. It was observed that when the opening was made, volume fractions of oxygen and carbon dioxide decreased rapidly and increased before the backdraft occurred and rapidly increased and then decreased before the backdraft.

The results state that, for wood, the critical value for volume fraction is 8.7% and 2.5% for heptane. Since, the backdraft also depends on the lower explosion limits of the combustible materials, the ratio of volume fraction to the lower explosion limit is taken into consideration. Results show that if the combustible material is wood, then the ratio of the fraction of volume to lower explosion limit is 1.43 and 1.37 for heptane, at the time when the opening is created in the compartment, and 7.69% is observed to be as the critical mass fraction for the occurrence of a backdraft. In a theoretical study, 9.3% identified as a critical volume fraction for the possibility of a backdraft which is similar to 10. An average value of the volume fraction ratio to the lower explosion limit is 1.4, which is observed to be similar to the ratio of every fuel. When the ratio reaches that critical value, which is 1.4, the backdraft will occur; it is stated that if the ratio of the fraction of volume to the lower explosion limit is more than 1.84, it results in a fireball. The size of the fireball formed is proportional to the ratio of the fraction of volume to the lower explosion limit.

4 Conclusion

We have witnessed many different methods for finding out the critical conditions for the occurrence of backdraft even though all the results point towards a single direction.

(a) An experimental study done by Weng and Fan [11] provides that the condition required for the backdraft to occur depends on the fraction of mass of the unburnt fuel. In their experiment, the fuel was methane, so it was observed that the critical fraction of mass of unburnt methane is 9.8. If conditions in the compartment prevail, so that the fraction of mass of fuel, which is methane, burning is more than 9.8, then backdraft will occur, or else the backdraft does not occur.

(b) The simulation study done by Ferraris shows the detailed phases and changes that are undergoing in the compartment during the occurrence of the backdraft. The backdraft process was divided into five stages: ignition stage, free spherical propagation, plane front propagation, flame front stretching and finally, the fireball.

(c) Theoretical analysis done by Chen states that the condition for the backdraft to occur depends not only on the critical mass fraction but also on the lower explosion limit of the fuel that is undergoing combustion. The ratio of the critical fraction of volume to the lower explosion limit is also considered stating that, if the ratio of critical volume fraction to the lower explosion limit is more than 1.43 for wood and 1.37 for heptane, then the backdraft would occur. It has been observed that all types of fuels, which when combusting in an enclosure, pertaining a condition such that the value of the ratio of critical volume fraction to the lower explosion limit equal to or more than 1.4, then the backdraft will occur and a fireball backdraft phenomenon will occur if the value is more than 1.84.

References

1. Anderson, M. K., et al. (2000). Ignition signatures of a downward smolder reaction. *Experimental Thermal and Fluid Science, 21*(1–3), 33–40.
2. Beyler, C. (2016). ***Flammability limits of premixed and diffusion flames*** (pp. 529–553). Springer.
3. Chen, A., et al. (2011). Theoretical analysis and experimental study on critical conditions of backdraft. *Journal of Loss Prevention in the Process Industries, 24*(5), 632–637.
4. Cook, A. W., et al. (1997). A laminar flamelet approach to subgrid-scale chemistry in turbulent flows. *Combustion and Flame, 109*(3), 332–341.

5. Ferraris, S., et al. (2008). Large eddy simulation of the backdraft phenomenon. *Fire Safety Journal, 43*(3), 205–225.
6. Fleischmann, C. (1994). Backdraft phenomenon. *National Institute of Standards and Technology*. NIST-GCR-94–646.
7. Fleischmann, C. M., & McGrattan, K. (1999). Numerical and experimental gravity currents related to backdrafts. *Fire Safety Journal, 33*(1), 21–34.
8. Gottuk, D. T., et al. (1999). The development and mitigation of backdraft: A real-scale shipboard study. *Fire Safety Journal, 33*(4), 261–282.
9. McGrattan, K., et al. (2004). Fire dynamics simulator (version 5), technical reference guide. *NIST special publication, 1018*(5).
10. Utiskul, Y., & Quintiere J. (2003). Wall-vent compartment fire behavior under limited ventilation. *Chemical and Physical Processes in Combustion* 37–40.
11. Weng, W., & Fan, W. (2003). Critical condition of backdraft in compartment fires: A reduced-scale experimental study. *Journal of Loss Prevention in the Process Industries, 16*(1), 19–26.

Advanced Technologies in Health Safety and Environment in Construction Industry

Abhishek Nandan, P. Mondal, Bikarama Prasad Yadav, and K. Sai Bharadwaj

1 Introduction

The construction company is responsible for increasing the economy of the countries across the globe, contributing around 29 $10 trillion to the worldwide Gross Domestic Product (GDP) (McKinsey Global Institute 2017). In 2015 and 2016, the event business contributed around 6.2% of the US GDP with quite $650 billion, while this commitment keeps on developing [5].

With the changes in the economic growth of the world, there are a lot of changes in the development processes of the companies. One such and large change is the implementation of digital technologies all across the globe. Along with other IT companies, construction companies have also integrated digital platforms in their systems. These digital techniques are implemented in almost all the sectors of the construction field like building models, communication purposes, managing the projects construction and maintenance processes, engineering infrastructures, and above all is to produce a united digital platform for the construction industry. The digital economy has been increased across the entire globe with an increase in labor productivity and opens many new dimensions in this field. The construction industry has been considered as the most complex having a large number of technological processes for fulfilling the needs of humans. This research is outlining various areas of digitization for the construction industry.

Implementation of electronic format for planning urban procedures, utilizing the electronic database for managing and storing the urban data for various activities, collection of digital data for processing and publishing the data extraction easily from different systems, and developing various reference platforms. Figure is showing the various dimensions of digitalization in the construction industry (Fig. 1).

A. Nandan (✉) · P. Mondal · B. P. Yadav · K. Sai Bharadwaj
Department of Health Safety Environment and Civil Engineering, UPES, Premnagar, Dehradun 248007, India
e-mail: abhisheknandan24@gmail.com

N. A. Siddiqui et al. (eds.), *Advances in Construction Safety*,
https://doi.org/10.1007/978-981-19-4001-9_22

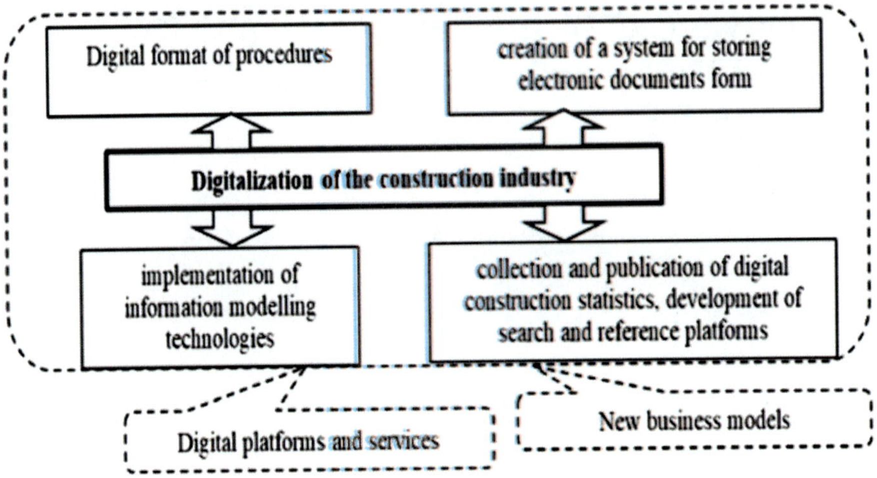

Fig. 1 Dimensions of digitalization in construction industry [40]

The main task is to improve the construction company with the help of modern and technical implementations for enhancing the quality of the construction industries. This research is suggesting the BIM technologies as the key features for quality improvement in the industrial field (Fig. 2).

Along these lines, the event labor pool is profoundly significant worldwide and to the US economy. The improved efficiency and quality of the work can be prompted with the help of the government assistance of labor pool for potentially genial work. Creating superior structures also as common works augment the experience and all things are supporting the economy of the country. Paradoxically, any negative test confronted by the event labor pool influences their efficiency, the character of their

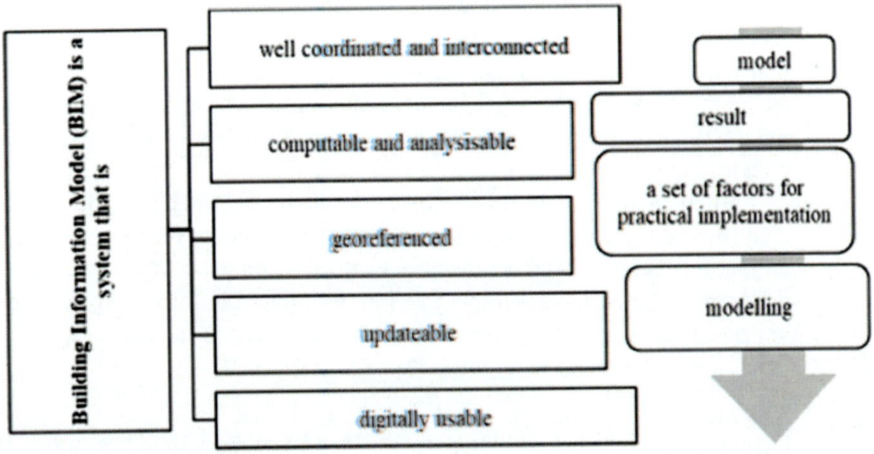

Fig. 2 BIM technologies in construction area [40]

work, and well within the end of the day causes undesired results on the economy and therefore the experience of the overall population. Likewise, the government assistance of the event labor pool needs to be amplified. Nonetheless, such a labor pool faces special difficulties that contrarily impact their government assistance, especially their actual security. The health and safety of the laborers are very important at the workplace and due to environmental fatalities, where most of the laborers lost their lives. Within the US development industry, 970 cases of the fatalities were reported in 2017 yielding a casualty pace of roughly 10 laborers killed annually for every 100,000 full-time workers in development. As indicated by the International Labor Organization (ILO), development laborers in created nations across the world are 3 to multiple times sure to have a deadly mishap at work than different ventures. These number increases multiple times in non-industrial nations. Limiting the high number of labor environment including the wounds and fatalities, various practices approaches are executed within the development industry that territory from social to designing wellbeing.

Social methodologies [7, 21, 42] underscore improving laborer mindfulness regarding the perils and therefore the utilization of the best number of wellbeing insurances within the work environment. Designing controls incorporate receiving wellbeing best practices (e.g., watchman and security rail frameworks) to three forestall laborer openness to risks. Notwithstanding security execution for the businesses development remained very low. Researchers ceaselessly search for elective techniques and practices that would altogether improve wellbeing execution within the development business. An assessment of the foremost late distributions on development security uncovers a faraway from on the usage of innovation for wellbeing administration. The innovation took place mainly during the twentieth century and still continues, and this innovation has been roughly multiple times from the mid-2000s to the mid-2010s. This rate has been supported through the foremost recent decade (2010–2019). The construction company has to face many challenges due to advanced technology and emerging changes in the competitive market. Therefore, the risk identification and hazard assessment are very important. Figure 3 below is showing some schemes for hazards identification (Fig. 4).

For Risk assessment, following three steps should be followed.

Zhang et al. [62] stated the sensor-based mechanization is available for improving safety protocols and standards concerning the construction industry. This technology has become so advanced which improved data collection, transformation, and processing, and thus, it helps in the modernization of safety management in the construction industry. Sensor-based technologies have made the work easier and facilitated the construction sector. Regarding this, sensor-based technology consists of vision-based, location-based, and wireless sensor networks. In locating sensor-based technology, various sensors are used which easily detect location include the radio frequency identification (RFID), global positioning system (GPS), Zigbee, ultra-wideband (UWB), wireless local area network (WLAN), and ultrasound. There are different sensors used in this technology such as light sensors, temperature sensors, pressure sensors, and optical fiber sensors. The most important part of this new technology is wireless network is the wireless sensor network (WSN), and it is

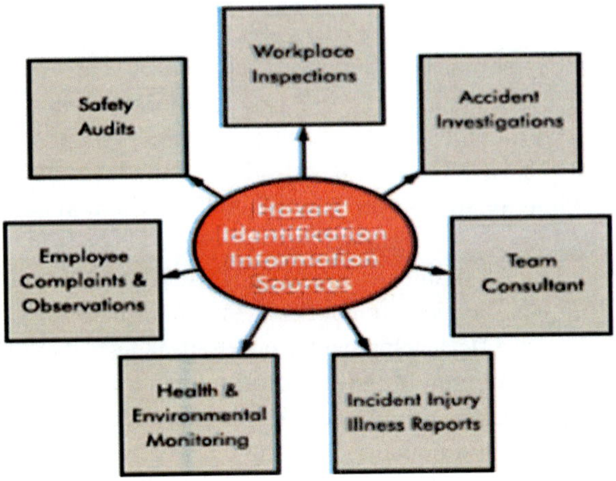

Fig. 3 Hazard identification

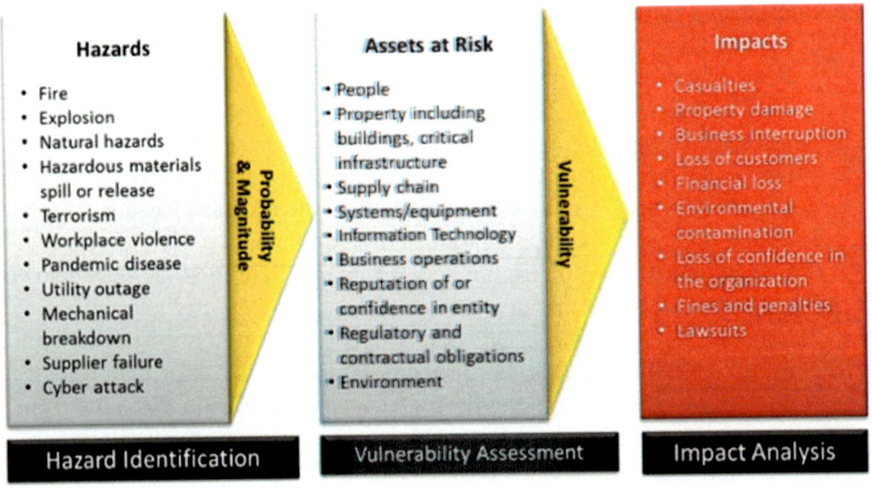

Fig. 4 Risk assessment

mainly composed of a communication module, central processor, and sensor nodes. These sensors monitor environmental or physical conditions and transfer the data to the main unit. Various algorithms are used for this purpose and it is according to the requirements of the safety management construction industry including locating and tracking the object, noise removal or error removal, and classification. There are many applications of this technology in construction safety management such as safety route planning and prediction, integrated safety management, and accident informing system [62].

Tay et al. [61] depicted the 3D (three-dimensional) painting for the construction industry, which is also known as additive manufacturing. It is a type of manufacturing process which produces automatically complex geometric shapes without the help of any dies, tooling, and fixture. In industry of construction, many new techniques have been applied for facilitating the workers.

On one hand, it has made the working easier and on the other hand, it replaces the old methods and also decreases the need for human resources. Thus, this innovation in the construction industry has played a very important role by minimizing manual work and hence, reducing the high capital investment. Therefore, many construction industries are now starting to use this technology, and it has given a very productive effect on the environment. By implementing 3D printing, in this, industry can also decrease the additional form of work and can also improve the printing quality of the present system by limiting the harms produced by manual working [61].

At the site of workers, the safety of employees is considered a top priority by most construction industries. The workers have to be exposed to many hazards at a location which results in serious injuries or fatalities. For this purpose, various safety measures and approaches have been adopted by the construction industry such as traditional and old strategies and methods that have been replaced by new computer-aided technologies. These new methods include virtual reality, mixed reality, computer-generated simulations, augmented reality, and serious games. The findings of this study were found to be very effective and greatly favor the implementation of the latest technology. This study has measured the effectiveness in terms of injury rate reduction and knowledge acquisition. The outcomes of this study have shown that the use and applicability of computer-aided technology have confirmed to support its efficacy, and the statistical evidence also favored this new technology. The study has compared both old and new methods and shows that both techniques are best, but in the world of today, the engagement of new models and designs is necessary for coping with present challenges of time. However, some challenges and issues have to face by this computer-based technology. But this new method is found to be superior in many technical features as compared to old tools [10].

Exploration on using innovation for wellbeing and wellbeing also (Nnaji the executives) has expanded in light of advances can make different security benefits by perceiving working environment risks that aren't ordinarily achievable for laborers and removing work environment perils right off the bat within the venture lifecycle. Generally, this investigation means to draw upon flow research endeavors on this subject and amplify the uses of innovation for word-related wellbeing and wellbeing (OSH) the executives within the development industry. These innovations are very helpful for protection of the laborers. The protection measures like wearable detecting gadgets and exoskeletons are very important for upgrading the systems and easiness of the workers. Likewise, the essential capacity of Building Information Modeling (BIM) is to enhance with the help of the useful data and different other modern techniques. Artificial intelligence is a very new concept being used for tackling many issues for environmental protection. AI is utilizing many simple and easy-to-handle strategies that can be handled by the health, safety, and environmental professionals for their daily life routine. All the new research and models are based on AI techniques, a tool named AI assistant that helps the HSE professionals and experts in tackling different protection measures. This tool is a very innovative technology and is the best alternative to the existing search engines. AI can answer human nature and questions in response to the data and then respond accordingly. There are many other applications of the HSE AI like controlling the stock in warehouses and managing the hazardous waste at different disposal sites [30].

2 Background

For the accomplishment of any development project, overseeing OHS has been considered essential parameters. Security and wellbeing episodes adversely sway timetable, quality, and cost of the task just as worker confidence, organization notoriety, protection expenses, etc. Such negative results couldn't just affect development laborers, and their association, however, could likewise influence the encompassing local area or whole society in a ruinous way. The chain of command of controls is regularly utilized in development for OSH the board. The order of controls is a framework involving various levels of control to relieve working environment risks and oversee OSH. To be explicit, the chain of command of controls contains five levels [elimination, replacement, designing, organization, and individual defensive hardware (PPE)] going from the most compelling to the most unsuccessful as far as alleviating work environment risks. Disposing of and subbing the dangers, e.g., subbing unsafe emanating materials with zero producing materials are viewed as best since they truly eliminate all or part of the risks). The designing and control tools like machine guarding, guardrails, blockades, and various frameworks are considered as the second-best as indicated for the control chain. They separate specialists from the actual risks yet don't allocate with or lessen the actual security. At last, regulatory controls (e.g., security signage and preparing) and PPE (wellbeing footwear and eyewear) are viewed as least successful regarding moderating working environment dangers. Authoritative controls improve specialist attention to the dangers yet do not lessen the actual perils, and PPE is utilized to limit the effect of the risks if there should be an occurrence of an episode without relieving the actual perils itself. The following segment will examine how unique development advances could be utilized for wellbeing the board in development. These innovations can give various kinds of risk control, going from end to authoritative, contingent upon the type, and usefulness of the innovation utilized. Different applications of the latest technologies for the construction companies are very important for the development of businesses and have got a lot of success in the past few years. The greater part of the advancements was at first received and used to either improve the nature of the result or proficiency of the development cycle, the two of which at last lead to decreased expenses and improved benefits. Nowadays, it is clear that increasingly more development innovations are as of now being utilized for the wellbeing and wellbeing of the executives. These advancements can be utilized in various development applications to moderate working environment dangers. As referenced, a few innovations could be utilized to help train laborers to perceive working environment dangers by giving cases like genuine situations. For preparing the development laborers and gear administrators, blended reality recreation has been utilized for on recognizing and alleviating work environment risks of development undertakings and machine activity. Li et al. created a multiuser-accommodating virtual climate preparing apparatus that development businesses could use to prepare their laborers for a safe methodology for tower crane erection and destruction. The created apparatus give a bit by bit methodology to perform erection and destroying of a pinnacle crane safely. The entirety of the preparation is empowered through a computer-generated experience climate which opens to insignificant to zero-hazard when finding out about erection and destroying the system of pinnacle cranes. A particular device is a successful authoritative control to help development laborers comprehend dangers hindered in pinnacle crane erection and destroying measures. Importantly, using innovation for wellbeing, the executives could give more viable controls (e.g., designing controls) than basically preparing laborers on the most proficient method to recognize risks (i.e., managerial controls), and created wellbeing instruments to distinguish and location potential development dangers from the get-go in the undertaking lifecycle, which is applicable during the planning stage. Similarly, built up a security decide checking stage that analyzes the structure frameworks and consequently recognizes any potential work environment dangers. When the risk is recognized and sorted, predictor measures are created by the stage to dispense with the dangers from the plan or alleviate the risks during development tasks. The instrument and stage depicted overuse BIM to plan for development specialist wellbeing before starting development. Planning for development laborer security guarantees that a critical bit of the actual risks are taken out from the development cycle which is the best strategy for peril alleviation as per the chain of command of controls. Another innovation arising in the development business is wearable wellbeing gadgets (WSDs); WSDs are accepting significant consideration inside both industry and the scholarly world. WSDs are little wearables that specialists can append to their body, outfits, or assistants to screen their wellbeing, and additionally. These gadgets are exceptionally viable, simple to utilize, and cheap wellbeing apparatuses [6, 55]. To be explicit, WSDs are utilized in development to forestall musculoskeletal issues of field staff, forestall work falls, survey actual remaining burden and exhaustion degree of field laborers, assess risk acknowledgment capacities of the two specialists and supervisors, and screen laborer mental status [2, 28]. Different other innovation applications for security the executives exist yet, for curtness, they are not portrayed in the current investigation. Readers intrigued by more data on advances utilized by and by for wellbeing and wellbeing administration are encouraged to audit applicable articles in the reference list [2, 14, 25, 28, 33, 45].

This can help to improve safety in construction in many ways. Since the past few years, it has been seen that much attention has been paid to the safety measures at the construction sites. It should be the topmost priority of the companies to provide full safety to their employees. The five main technological areas that help to improve safety at the workplaces are the digitization of safety processes and integration with mobile apps, sharing of the content with the help of apps and digital platforms, monitoring of the dangerous activities, and management of crises with the help of technology. The analysis and identification of the different aspects concerning safety are very important on a monthly or weekly basis [10].

Nowadays, artificial intelligence (AI) and machine learning (ML) have now been implemented in almost all sectors including safety, risk prediction, and risk mitigation. The main AI risk dimensions are strategy, trust, ethics, compliance, finance, technology, and people.

Strategy
Strategic planning is very essential for the success of the project. The main risk being associated with this includes the wrong selection of the project, lack of digital platforms, and mismanagement from the executive team.

Trust
The poor user experience and lack of trust in employees are the main causes for risks.

Ethics
Ethical considerations are a very essential part of a successful business, and the main AI risk dimensions for ethics are biased in data, bias in algorithms, and human unintentional behaviors.

Compliance
The risk dimensions include violation of privacy, lack of monitoring and controlling mechanism, and proper analysis of the risks.

Finance
The proper management of finance is very important for the development of the country. The main risk included in this scenario is an imbalance between the resources and their implementation, over budgeting, and the higher cost of the infrastructure. Therefore, proper planning and prediction analysis are very essential before the project execution.

Technology
The main risks associated with the technology are lack of skills among employees, data issues, less integration of the ML and AL latest models, and reliance on vendors [8].

2.1 The Objective of the Study

As stated before, this examination intends to expand on the ebb and flow research endeavors concerning the crossing point between innovations utilized in the development of the executives and wellbeing, and augment the utilization of innovation for OSH the board. Such augmentation is preposterous without understanding the advantages and restrictions just as the boundaries of receiving innovation for

OSH executives. Although a few investigations exchange limits and boundaries to address factors that discourage the utilization of innovation in the development business. These terms are not continually interchangeable. Boundaries are elements that keep associations or people from embracing innovation. Barriers are regularly pre-appropriation factors. Then again, limits are factors that restrict the all-encompassing execution of an innovation—ordinarily saw at the post-selection period of innovation combination. To guarantee the qualification between hindrances and limits, this examination characterizes "boundaries" as the elements that forestall the reception and utilization of a wellbeing and wellbeing innovation, and "restrictions" are factors that limit the utilization of innovation effectively received. The elements that keep individuals from embracing advancements utilized for security and wellbeing.

The purpose of this paper is to identify risk mitigation using AI. ML and AI are very latest techniques, and advanced analysis is very helpful for event interpretation, events automation, and various actions for decision-making plans. This will discuss the different safety and other risk mitigation for the construction companies. AI, compliance management, safety management, and incident management are the focus of this research.

Its administrations and consistent use and use post-reception ought to be recognized for the fruitful execution of advances on a development project for security and wellbeing administration. In order accomplish the examination objective introduced over, the three essential goals are set as follows: (1) recognize accessible innovations for OSH the board, (2) distinguish and rank the advantages and restrictions of advances for OSH the executives, and (3) distinguish and measure hindrances to receiving advances for OSH the board and propose answers for defeat such obstructions.

3 Research Methodology

For addressing the three goals of this investigation, a multi-strategy has been proposed. The methodology comprises an organized audit and substance examination of the accessible writing on the theme and a review of industry experts. The two methodologies utilized are depicted in detail beneath.

Initially, an integrative audit of existing writing was directed to recognize the innovations as of now utilized in the business to improve laborer wellbeing and wellbeing, their applications, advantages, and constraints. An integrative audit of writing is an extensive methodological methodology of surveys that joins information from observational and hypothetical writing to build up a theoretical model, survey proof-based discoveries, and investigate concerns related to a specific subject. For recognizing the audit cycle that has been received by past development-related examinations, significant elements were hired to put together concerning wellbeing execution [36]. Besides, the survey cycle has been executed broadly as a device for distinguishing boundaries and advantages related to utilizing security and wellbeing advancements in wellbeing-related fields. The survey measure in this investigation was embraced from and includes six stages, in particular, (1) setting up the managing questions, (2) testing the writing, (3) gathering information, (4) examining included investigations, (5) talking about the outcome, and (6) introducing integrative survey. The depiction of each stage is prohibited in this composition yet could be found in Souza et al. (2010) for quickness.

The four essential inquiries guided the audit interaction, specifically, "What advances are utilized for OSH the board in the development business?" "What are the advantages of utilizing these advances in OSH the executives?" "What are the impediments related to utilizing these advances in overseeing specialist security and wellbeing on a development project?" and "What are the obstructions that keep organizations from embracing these advances for OSH the board?" Various datasets taken from different Web sites and distributers like Google Scholar, Scopus, American Society of Civil Engineers (ASCE), Taylor and Francis, Emerald, and Elsevier were utilized for datasets due to their for recognizing valuable distributions by utilizing watchwords related with the directing inquiries (e.g., "development administration", "development wellbeing gadgets", "wellbeing advancements", and "development advances", and "laborer wellbeing and wellbeing"). For example, while looking through Scopus, the accompanying inquiry code was used: (TITLE-ABS-KEY ("Construction Management" OR "Development" OR "Structural Engineering" OR "Fabricated Environment") AND TITLE-ABS-KEY ("Worker Safety" OR " Worker Health" OR "OSH" OR "Word related Safety and Health" OR "OSH") AND TITLE-ABS-KEY ("Technology" OR "Advancement" OR "Gadget")) PUBYEAR > 2000 AND PUBYEAR < 2019. Afterward, the recognized distributions were screened by zeroing in generally on the title, theoretical, and ends, just as the figures and tables. If a distribution was considered important (that is, it contained a conversation on the use of wellbeing/wellbeing advances on a development project), a further nitty–gritty assessment of the substance was performed to distinguish possible advantages, impediments, and obstructions of embracing and utilizing security and wellbeing advances. Distributions with restricted data on the utilization of wellbeing and wellbeing innovation or use of the innovation for an alternate reason other than OSH, the executives were ignored.

4 Results and Discussions

Figure 5 indicates that even more than 50 per majority of people were pleased with the general standards for safety at the workplace, and 22 per majority of people were neither pleased nor unhappy with existing conditions. A significant 13% of respondents claimed that the standards on the worksite were unsatisfying. The economic prospects of this area have changed with various countries (clients and authorities), making protective steps necessary for regional subcontractors and suppliers. These findings are consistent with the regional study conducted for Saudi Arabia, which emphasizes that while changes are made to safe working practices, there is space for progress and sufficient pro-activeness.

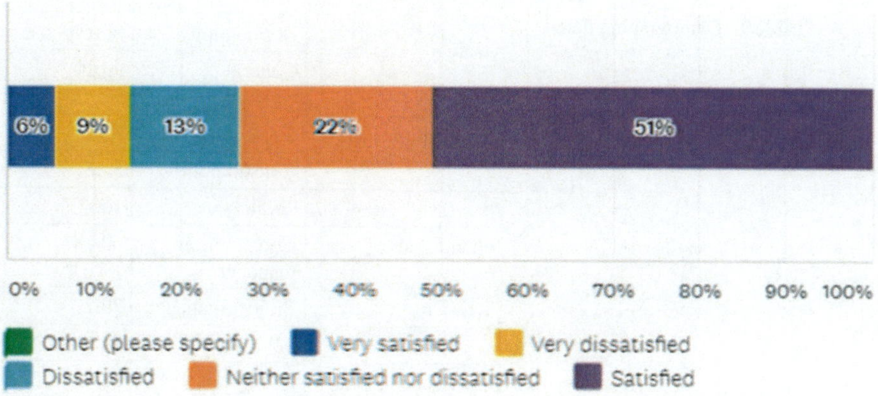

Fig. 5 Satisfaction level of safety measures in workplace

4.1 Survey Participants and Demographics

About 157 participants from US has been taken for this study, and the review has been noticed that helped the researchers for quality data and to deduce the exact results for the research study. Next, stepping away from all replies from individuals with fewer than five years of instruction program. Next, the studies have emphasized that only those employees who are competent as their 273 organizations have adopted OSH management technology and those with job titles that infer. Eleven participation in governance was included. Finally, researchers removed replies that displayed signs of straight-line and delivered answers well below the normal response time.

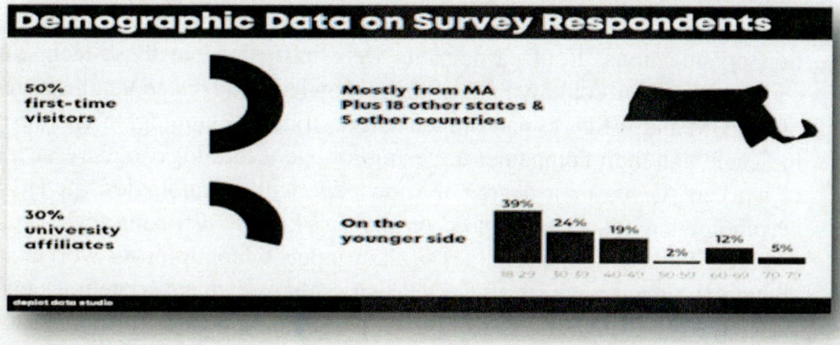

After the consistency reviews, 102 answers were found appropriate. These controls increased the efficiency and feasibility of the analysis by excluding unnecessary responses or comments from individuals with no expertise on the subject. Respondents came primarily from the general project team (90.20%). For the business revenue, 33% of the members come from small enterprises, 59% have been

Table 1 Demographic data

	Demography	% response	*D*
	General contractor	90.20	92
Organization Type	Sub-contractor	3.92	4
	Consultants	5.88	6
	Total	100	102
Job type	Construction manager	60.78	62
	Project manager	39.22	40
	Total	100	102
Experience in year	5–10 years	58.82	60
	10–20 years	31.37	32
	More than 20 years	9.81	10

from full-back businesses (USD 11 million to USD 1 billion), and 8% are from major corporations. California, Florida, and New York provided 63% of the replies provided. Table 1 summarizes the demographic data of participants. Both respondents were connected to these technologies at some ability, but roughly 86% of respondents suggested that they are currently using BIM and WSDs as part of their security strategic planning. Fifty-nine people mentioned that their companies are using robotic technology to increase the security of workers. Despite their degree of expertise with devices, the reply of the researchers tends to be focused on first-hand knowledge using these innovations, thus improving the accuracy of their contributions.

Simply by asking stakeholders regarding their views on the advantages and disadvantages of applying the defined innovations, interviewees were told to show whether was acquainted with the techniques used as part of their protection and health production organizations. Both participants were introduced to these technologies to a certain extent, but roughly 86% of the respondents suggested that they are currently using BIM and WSDs as part of their safety strategic planning. Fifty-nine interviews indicated that their companies are using robotic technology to increase the security of workers. Given their degree of knowledge with technologies, the answer of the stakeholders tends to be focused on the first knowledge using these technologies, therefore improving the accuracy of their inputs which improve worker safety. The above mentioned analysis suggested that employees have sometimes endured and suffer greatly from various workplace accidents, industrial disorders, psychiatric, and environmental issues. This call for action derives from the highest importance typically provided to terms and conditions of employment by Ghanaian businesses. Studies by small and medium-sized businesses in Ghana show that financing, sustainability considerations, and skills growth are their targets, and workplace health and safety training was nowhere near their primary concern; again. This lack of safety consideration is contradictory to study results that now the management contribution to safety is among the most significant quantitative measurements of the environment and/or community of protection in an enterprise. Except for the suggested portfolio management steps, it needs attitudinal reform between government leaders reflecting Ghana's involvement in the oil and gas industry, corporates representatives of the oil corporations itself, and workers employed on the oil rigs. Regarding the possibility that applied therapists should be interested in HSE management, I would like to ask the following question: who else is best qualified to cope with risk attitudes, safety environment, safety performance, safety purpose, behavior-based protection weather safety incentive, safety behavior, behavior-based safety measures, and accident-prone personal traits that decide the degree of conformity with safety regulations. This study offers valuable knowledge that could action plan organizations that are considering using these innovations to handle OSH. Third, this research did not define the causal association among advantages, obstacles, hurdles, and the entity or organizational tendency to implement the software. An evidential questionnaire should be performed to obtain additional observations in this respect. Using Structural Equation Modeling (SEM) and other similar methods, efficiency could be evaluated. Similarly a process (SEM) has been utilized for identifying the relationship. Table 1 gives information about organization type, experience, and job title.

Figure 6 illustrates relation between age and job title. A very relation may be tested using modeling approach modeling (SEM) or related approaches. Third, a related mechanism (SEM) may be followed to define the correlation here between the usage of these innovations and the effect on the trailing and stagnating metrics. Eventually, this study centered specifically on the viewpoint of the management team, considering the essential role they play in the decision to buy and introduce

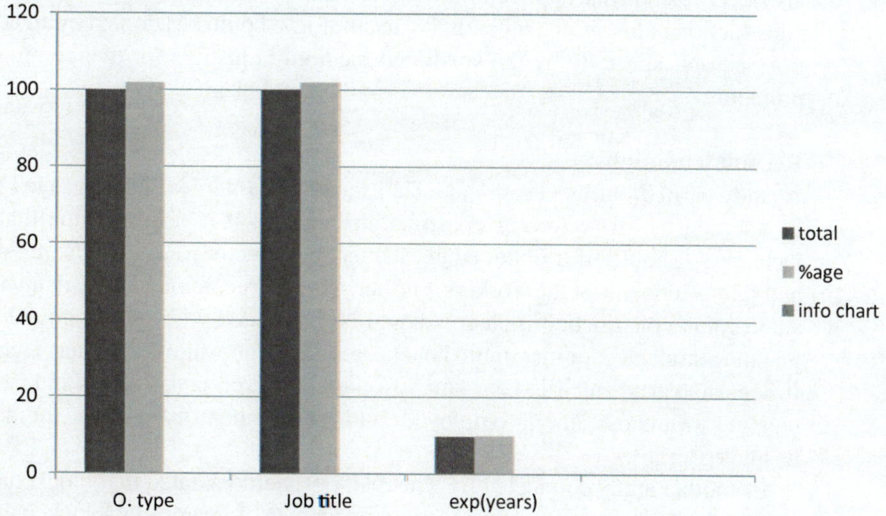

Fig. 6 Relation between age and job title

technologies in light of protection. Provided that prior research indicates that effective technology implementation is a combination of a highest and bottom-up method [56], future research should focus on the perspective of agricultural workers to create useful data to enhance the expertise given in this article.

5 Conclusion

It can be deduced that new and innovational technology has played an important role regarding the safety of employees in the construction industry. The article has clearly developed an understanding about the new technology and the advancements concerning three factors health, safety, and environment (HSE) in construction sector. Machine learning and artificial intelligence are the two latest algorithms which are now being used for safety and health concerns in the industrial industry. In past years, technology is not much advanced and the aim of present study is to fill this gap by introducing new technologies and algorithms in construction sector. Many studies have been carried out for knowing the importance of advanced technology used in construction industry concerning the HSE factors. The building practitioners and technology experts were evaluated for knowing the methods of implementing new technology in construction sector. Furthermore, the findings of this study depict that by executing various new techniques in industrial sector helped the academic researchers and industrialists to solve the problems and also support them in coping with current issues and problems. However, various barriers have to be faced by introducing these new methods, and it is also expected that it will also help in overcoming all the problems caused due to adaptation of new technology. Furthermore, the study also recommends that applications of blockchain technology should be introduced in construction industry, and other digital technology should be adopted by all construction sectors especially by the small and medium industries for overcoming all the problems.

Recommendations:
A study carried out by Ghosh et al. [26] stated that Internet of Things (IoT) should be adopted by all sectors of construction industry. It is need of time that digital technology should be introduced in establishment sector for making sure the health and safety concerns of the workers. Further, it is also recommended by many research studies that construction industry should be facilitated with proper guidelines and planning strategies for maximum benefits based on innovation and latest technology. It is particularly important for small to medium-sized construction industries, for creating awareness among employees and management team about the advanced technology [26].

In another study conducted by Kiu et al. [39] demonstrated that blockchain technology should be introduced in construction industry. Recently, the blockchain digital technology lacks proper designs and concrete framework for carrying out proper functions in industrial sector. This study proposed that real life block-chain technologies and applications should be adopted by industries. Therefore, it is also the need of time to introduce such digital technology in construction industry for reducing risk at work site and thus ensuring the safety of workers [39].

References

1. Acar, E., Kocak, I., Sey, Y., & Arditi, D. (2005). Use of information and communication technologies by small and medium-sized enterprises (SMEs) in building construction. *Construction Management and Economics, 23*(7), 713–722.
2. Ahn, C. R., Lee, S., Sun, C., Jebelli, H., Yang, K., & Choi, B. (2019). Wearable sensing technology applications in construction safety and health. *Journal of Construction Engineering and Management, 145*(11), 03119007.
3. Al-Gahtani, S. S., & King, M. (1999). Attitudes, satisfaction, and usage: Factors contributing to each in the acceptance of information technology. *Behaviour & Information Technology, 18*(4), 277–297.
4. Asanka, W. A., & Ranasinghe, M. (2015). Study on the impact of accidents on construction projects. In *6th International Conference on Structural Engineering and Construction Management*, 58–67.
5. Associated Builders and Contractors ABC. (2019). ABC News Release. https://www.abc.org/News-Media/News-Releases/entryid/9801/constructions-contribution-to-u-s-economy-highest-in-seven-years. Accessed 27 November 2019.
6. Awolusi, I., Mark, E., & Hallowell, M. (2018). Wearable technology for personalized construction safety monitoring and trending: review of applicable devices. *Automation in Construction, 85*, 96–106.
7. Azeez, M., Gambatese, J., & Hernandez, S. (2019). What do construction workers want? A study about representation, importance, and perception of us construction occupational rewards. *Journal of Construction Engineering and Management, 145*(7), 04019040.
8. Bhatti, B. (2020). 7 Types of Ai Risk and How to Mitigate their Impact. https://towardsdatascience.com/7-types-of-ai-risk-and-how-to-mitigate-their-impact-36c086bfd732
9. Bioservo. (2019). Return on Investment—ROI. https://www.bioservo.com/professional/return-on-investment-roi. Accessed 27 November 2019.
10. Branthonne, D. (2017). 5 Ways Technology Can Improve Safety in Construction. https://www.novade.net/5-ways-tech-can-improve-safety-construction/
11. Center for Disease Control and Prevention CDC. (2020). Hierarchy of Controls NIOSH. https://www.cdc.gov/niosh/topics/hierarchy/default.html Accessed 9 January 2020.
12. Chan, A. P. C., Darko, A., Olanipekun, A. O., & Ameyaw, E. E. (2018). Critical barriers to green building technologies adoption in developing countries: The case of Ghana. *Journal of Cleaner Production, 172*, 1067-1079.
13. Chan, P., & Leicht, R. (2014). The role of integrated project delivery elements in adoption of integral innovations. In *EPOC 2014 Conference*
14. Cheng, J., Chen, K., & Chen, W. (2019). State-of-the-art review on mixed reality applications in the AECO industry. *Journal Construction Engineering and Management, 146*(2), 03119009. 744
15. DataInsider. (2018). The FFIEC cybersecurity assessment tool: A framework for measuring cybersecurity risk and preparedness in the financial industry. 746
16. Delgado, J. M. D., Oyedele, L., Ajayi, A., Akanbi, L., Akinade, O., Bilal, M., & Owolabi, H. 747 (2019). Robotics and automated systems in construction: Understanding industry-specific 748 challenges for adoption. *Journal of Building Engineering, 26*, 100868
17. ENR. (2019). A Match Made in Cyber Hell. https://www.enr.com/articles/46832-750construction-cybercrime-is-on-the-rise. Accessed 2 Nov 2019. 751.

18. Esola, L. (2018). Construction market flooded with devices seeking to cut injuries in hazardous 752 sector. https://www.chubb.com/microsites/chubb-construction/wearable-753technology-in599construction.aspx. Accessed 07 September 2019. 754.
19. Foster, L. L. (2008). Legal issues and risks associated with building information modeling 755 technology. Doctoral dissertation, University of Kansas
20. Gambatese, J. A., Michael Toole, T., & Abowitz, D. A. (2017). Owner perceptions of barriers to prevention through design diffusion. *Journal of Construction Engineering and Management, 143*(7), 04017016.
21. Gambatese, J. A., Pestana, C., & Lee, H. W. (2016). Alignment between lean principles and practices and worker safety behavior. *Journal of Construction Engineering Management*. https://doi.org/10.1061/(ASCE)CO.1943-7862.0001209,04016083
22. Gambatese, J. A., Lee, H. W., & Nnaji, C. A. (2017b). Work zone intrusion alert technologies: Assessment and practical guidance (No. FHWA-OR-RD-17-14). Dept. of Transportation. Research Section.
23. Ganah, A., & John, G. A. (2015). Integrating building information modeling and health and safety for onsite construction. *Safety and Health at Work, 6*(1), 39–45.
24. Gao, Y., Gonzalez, V. A., & WingYiu, T. (2019). The effectiveness of traditional tools and computer-aided technologies for health and safety training in the construction sector: A systematic review. *Computers & Education.*
25. Gheisari, M., & Esmaeili, B. (2019). Applications and requirements of unmanned aerial systems (UASs) for construction safety. *Safety Science, 118*, 230–240.
26. Ghosh, A., Edwards, D. J., & Hosseini, M. R. (2020). Patterns and trends in Internet of Things (IoT) research: future applications in the construction industry. *Engineering, Construction and Architectural Management*.
27. Hallowell, M. R., Hardison, D., & Desvignes, M. (2016). Information technology and safety: Integrating empirical safety risk data with building information modeling, sensing, and visualization technologies. *Construction Innovation, 16*(3), 323–347.
28. Hasanzadeh, S., Esmaeili, B., & Dodd, M. D. (2018). Examining the relationship between construction workers' visual attention and situation awareness under fall and tripping 774 hazard conditions: Using mobile eye tracking. *Journal of Construction Engineering and Management, 144*(7), 04018060.
29. Hayne, G., Kumar, B., & Hare, B. (2014). The development of a framework for a design for safety BIM tool. *Computing in Civil and Building Engineering*, 49–56.
30. Hojageldiyev, D. (2018). Artificial Intelligence in HSE.
31. Howard, J., Murashov, V., & Branche, C. M. (2018). Unmanned aerial vehicles in construction and worker safety. *American Journal of Industrial Medicine, 61*(1), 3–10.
32. Inan, F. A., & Lowther, D. L. (2010). Factors affecting technology integration in K-12 classrooms: A path model. *Educational Technology Research and Development, 58*(2), 137–154.
33. Jebelli, H., Choi, B., & Lee, S. (2019). Application of wearable biosensors to construction sites. I: Assessing workers' stress. *Journal of Construction Engineering and Management, 145*(12), 04019079.
34. Karakhan, A. A., Rajendran, S., Gambatese, J., & Nnaji, C. (2018). Measuring and evaluating safety maturity of construction contractors: Multicriteria decision-making approach. *Journal of Construction Engineering and Management, 144*(7), 04018054.
35. Karakhan, A., & Alsaffar, O. (2019). Technology's role in safety management. *Professional Safety Journal of the American Society of Safety Professionals*, 43–45.
36. Karakhan A., Xu, Y., Nnaji C., & Alsaffar O. (2018) Technology alternatives for workplace safety risk mitigation in construction: exploratory study. In I. Mutis, T. Hartmann (Eds.) *Advances in Informatics and Computing in Civil and Construction Engineering*. Springer
37. Kim, K., Kim, H., & Kim, H. (2017). Image-based construction hazard avoidance system using augmented reality in wearable device. *Automation in Construction, 83*, 390–403.
38. Kim, S., Moore, A., Srinivasan, D., Akanmu, A., Barr, A., Harris-Adamson, C., Rempel, D.M. & Nussbaum, M.A., (2019). Potential of exoskeleton technologies to enhance safety,

health, and performance in construction: industry perspectives and future research directions. *IISE Transactions on Occupational Ergonomics and Human Factors*, 1–7.
39. Kiu, M. S., Chia, F. C., & Wong, P. F. (2020). Exploring the potentials of blockchain application in construction industry: a systematic review. *International Journal of Construction Management*.
40. Kudryavtseva, V. A., & Vasileva, N. V. (2020). On the development of a united digital platform in the construction. *IOP Conference Series: Materials Science and Engineering*.
41. Lai, P. C. (2017). The literature review of technology adoption models and theories for the novelty technology. *JISTEM-Journal of Information Systems and Technology Management, 14*(1), 21–38.
42. Langford, D., Rowlinson, S., & Sawacha, E. (2000). Safety behaviour and safety management: Its influence on the attitudes of workers in the UK construction industry. *Engineering, Construction and Architectural Management, 7*(2), 133–140.
43. Lee, S., Yu, J., & Jeong, D. (2013). BIM acceptance model in construction organizations. *Journal of Management in Engineering, 31*(3), 04014048.
44. Leung, M. Y., Liang, Q., & Olomolaiye, P. (2015). Impact of job stressors and stress on the safety behavior and accidents of construction workers. *Journal of Management in Engineering, 32*(1), 04015019.
45. Li, X., Yi, W., Chi, H. L., Wang, X., & Chan, A. P. (2018). A critical 814 review of virtual and augmented reality (VR/AR) applications in construction safety. *Automation in Construction, 86*, 150–162.
46. Lu, W., Huang, G. Q., & Li, H. (2011). Scenarios for applying RFID technology in construction project management. *Automation in construction, 20*(2), 101–106.
47. Manuele, F. A. (2005). Risk assessment and hierarchies of control. *Professional Safety, 50*(5), 33–39.
48. Manuele, F. A. (2008). Prevention through design addressing occupational risks in the design and redesign processes. *Professional Safety, 53*(10), 28–40.
49. Meyer, J. (2011). Workforce age and technology adoption in small and medium-sized service firms. *Small Business Economics, 37*(3), 305–324.
50. Mitropoulos, P., & Tatum, C. B. (1999). Technology adoption decisions in construction organizations. *Journal of Construction Engineering and Management, 125*(5), 330–338.
51. Mitropoulos, P., & Tatum, C. (2000). Forces driving adoption of new information technologies. *Journal of Construction Engineering and Management, 126*(5), 340–348.
52. Nnaji, C., Lee, H. W., Karakhan, A., & Gambatese, J. (2018). Developing a decision-making framework to select safety technologies for highway construction. *Journal of Construction Engineering and Management, 144*(4).
53. Nnaji, C., Gambatese, J., & Lee, H. W. (2018). Work zone intrusion: Technology to reduce injuries and fatalities. *Professional Safety, 63*(04), 36–41.
54. Nnaji, C., Gambatese, J., Lee, H. W., & Zhang, F. (2019). Improving construction work zone safety using technology: A systematic review of applicable technologies. *Journal of Traffic and Transportation Engineering* (English edn.).
55. Nnaji, C., Gambatese, J., Karakhan, A., & Eseonu, C. (2019). Influential safety technology adoption predictors in construction. *Engineering, Construction and Architectural Management, 26*(11), 2655–2681. https://doi.org/10.1108/ECAM-09-2018-0381
56. Nnaji, C., Okpala, I., & Kim, S. (2019). A simulation framework for technology adoption decision making in construction management: a composite model. In *Computing in civil engineering 2019: visualization, information modeling, and simulation* (pp. 499-506). American Society of Civil Engineers.
57. Oesterreich, T. D., & Teuteberg, F. (2016). Understanding the implications of digitisation and automation in the context of Industry 4.0: A triangulation approach and elements of a research agenda for the construction industry. *Computers in Industry, 83*, 121–139.
58. Okpala, I., Nnaji, C., & Karakhan, A. (2020). Utilizing emerging technologies for construction safety risk mitigation. *ASCE Practice Periodical on Structural Design and Construction Journal*. https://doi.org/10.1061/(ASCE)SC.1943-5576.0000468

59. Park, C. S., & Kim, H. J. (2013). A framework for construction safety management and visualization system. *Automation in Construction, 33*, 95–103.
60. Rogers, M., & Lopez, E. (2002). Identifying critical cross-cultural school psychology competencies. *Journal of School Psychology, 40*(2), 115–141.
61. Tay, Y. W., Panda, B., Paul, S. C., Mohamed, N. A., Tan, M. J., & Leong, K. F. (2017). 3D printing trends in building and construction industry: a review. *Virtual and Physical Prototyping, 12*(3).
62. Zhang, M., Cao, T., & Zhao, X. (2017). Applying Sensor-Based Technology to Improve Construction Safety Management. *Sensors, 17*(8).

Review of Earthquake Resilience and Safety in Building Construction

Geetanjli Rani, P. A. Arun, Umar Muktar, Nitty Ann Abraham, and Shariq Ansari

1 Introduction

The expanding pace of advancement and urban growth in developing countries makes them more vulnerable to natural disasters. There is an increasing level of risk due to factors such as urban sprawl, building density. Earthquakes are inescapable yet the outcomes can be in part controlled utilizing a compelling risk reduction framework, since risk is dependent on various elements, which if, managed properly will help effectively reduce the risk. The impacts of tremors on a poorly arranged community have demonstrated to be destroying with long-term effects. Seismic risk reduction (SRR) looks into the measures that can be taken to forestall or decrease impact of an earthquake, essentially, before one occurs. Preparedness is a significant segment of risk reduction, involving taking measures beforehand to caution and educate communities on ways to be prepared in the event of an earthquake, and sensitizing people [1]. The city of Dehradun lies in ZONE 4, which is second most active region in terms of earthquake. Figure 1 shows the MCT and MBT passing through Dehradun District, which makes the study area most dangerous and prone to the earthquakes [2]. This paper contains the structural review and recommendation towards safety in building construction of Dehradun city (Fig. 2).

G. Rani (✉) · P. A. Arun · U. Muktar · N. A. Abraham · S. Ansari
University of Petroleum and Energy Studies, Dehradun, India
e-mail: grani@ddn.upes.ac.in

N. A. Siddiqui et al. (eds.), *Advances in Construction Safety*,
https://doi.org/10.1007/978-981-19-4001-9_23

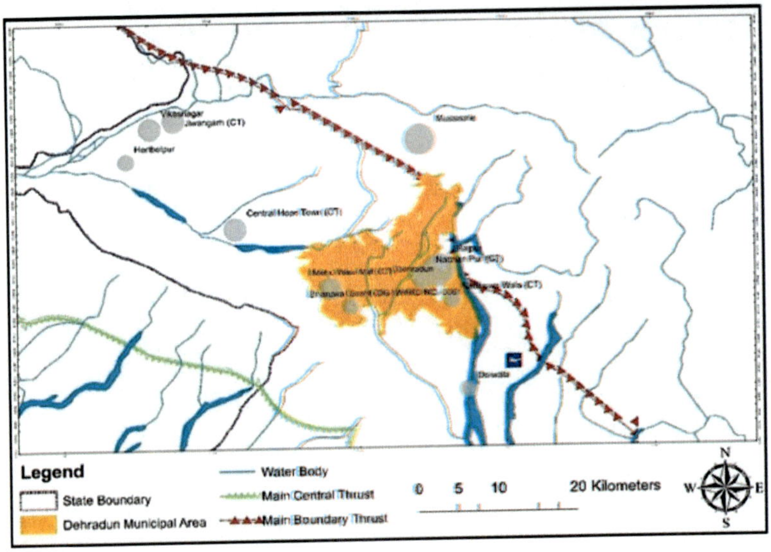

Fig. 1 MCT and MBT crosses along Dehradun [3]

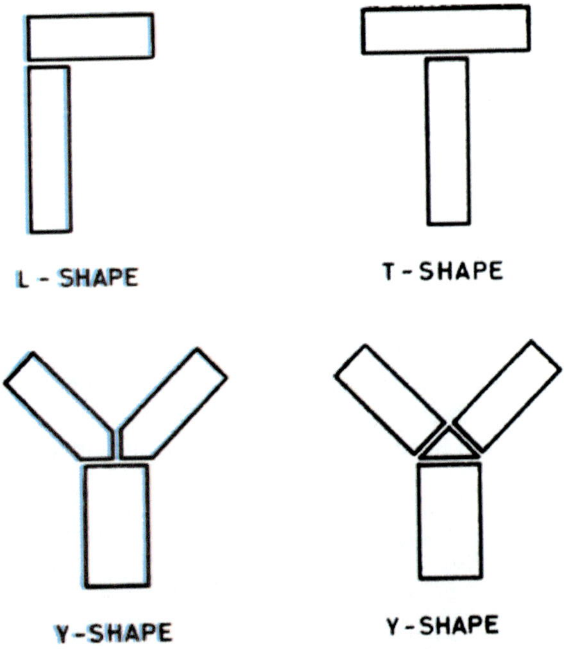

Fig. 2 Shapes of building formations [4]

2 Analysis

2.1 Acts

The disaster managements in building are followed with proper rules and regulation. The disaster management act, 2005 this is to manage the disaster and for matter connected with it or identical accidents that are happened previously. This act came into president on 23rd December, 2005. This act has total 11 chapters that cover all disaster management. Chapter covers national, state and district authority, etc. The state disaster management plan discusses about the hazards by nature and their safety measures.

2.2 Norms and Standards

The building should be constructed according to the Indian Standard IS 4326: 1993. This deals with the earthquake resistant design and building construction including masonry units, Timber construction and prefabricated structures. The seismic tremor power is a component of mass, structure should be light weighted that can easily handle the earthquakes especially roof and upper storey of the building.

General principles of the buildings are followed as below:

1. Most of the buildings catch fire after the earthquake, fire safety requirements should be done according to the Indian Standards, IS 1641:1988, IS 1642:1989, IS 1643:1988, IS 1644:1988 and IS 1646:1986 [5].
2. All structural testing reports like soil instability/earthquake zone classification should be submitted to Public Works Department (PWD). This implies for new building/ extension works, etc.
3. All masonry buildings are to be built on earthquake proof line, i.e. with single unit foundation and stringer courses at each floor level and eave level except the ground floor.
4. Structural Engineer shall prepare structural drawings as per the norms and shall furnish the certification on the structural plans.
5. It id preferred to have simple rectangular design and be symmetrical with respect to both weight and rigidity so that building coincides with each other in which case no separation sections other than expansion joints are necessary.
6. If the symmetry structure is not possible height or mass, the arrangement will be made for tensional and different impacts because of quake powers in the auxiliary structure or the pieces of various rigidities might be isolated through fold segments. The length of such structure between separation segments will not ideally surpass multiple times the width.

Small length of buildings forming *L*, *T*, *E* or *Y* shapes are not to be provided with separate section.

Table 1 Building categories for earthquake resistance [7]

S. No.	Building categories	Range of α_h
1	*A*	Less than 0.05
2	*B*	0.05 to 0.06 (both inclusive)
3	*C*	More than 0.06 and less than 0.08
4	*D*	0.08 to less than 0.12
5	*E*	Equal to or more than 0.12

1. In special cases, structure should not exceed 15–20% of total dimension of the building [6].
2. Minimal damage to the frame structure can lead to the deformation of non-structural elements [6].

Seismic Strengthening- Prior to seismic strengthening/retrofitting of any existing structure, Empanelled Structural Engineer shall carry out evaluation of the existing structure as regards structural vulnerability in the specified wind/seismic hazard zone. If as per the evaluation the seismic resistance is assessed to be less than the specified minimum seismic resistance as given in the note below, action will be initiated to carry out the upgrading of the seismic resistance of the building as per applicable standard guidelines (Table 1).

- For masonry buildings reference shall be made to IS 4326 and IS 13935.
- For concrete buildings and structures, reference shall be made to IS 15988: 2013 Seismic evaluation and strengthening of existing RCC buildings [8].

3 Resilience Structure Technology

Many kind of technology is implemented in earthquake resistant in building that prevents from catastrophic accidents. The building room height should not exceed 11 feet 4 inches in special cases according to section 3 of IRC it can go up to 12 feet. It reduces the motion and controls inertial force acting horizontal wave flow in building. Technologies are developing every day in order to perform a safe environment. Pradhan Mantri Awas Yojana (PMAY) has constructed 119 constructed and 4876 under construction; Ravij Awas Yojana (RAY) has constructed 1406 and 1188 start's construction soon. These buildings are constructed according to masonry unit (PCC, MS, RCC) [9] (Fig. 3).

3.1 Shear Wall

Shear wall is structural material used to resist lateral force, it works in form of cantilever principle. These wall are constructed in vertical direction that resist the

Fig. 3 Resilient structure technologies and programmes, [10]

horizontal force. Walls, floors and rooftops to the ground towards a path corresponding to their planes. Models are the strengthened solid divider. Parallel powers brought about by wind, tremor and lopsided settlement loads, notwithstanding the heaviness of structure and tenants, make power twisting tension. Shear walls are significant in elevated structures subject to horizontal breeze and seismic powers. They are plane/flanged in section, it has more strength in closed wall compared to intersection and openings [11] (Fig. 4).

3.2 *Bracing Wall*

Brace Wall boards as comprehended by designers and code authorities are simply territories of encircled divider containing no entryway or window openings and which have allowed in supporting, askew board sheathing, or a code-endorsed sheet material to solidify the structure against racking. Much of the time, the main contrast between a shear divider and propped divider is the consideration of hold-down sections on the shear wall (Fig. 5).

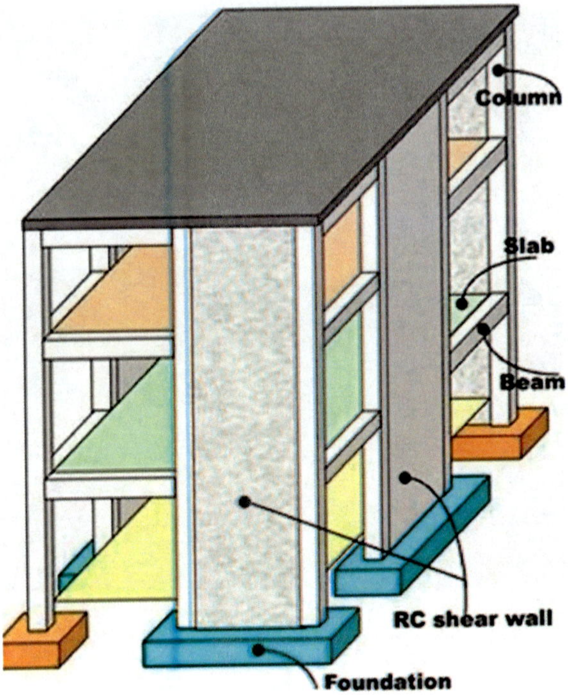

Fig. 4 Shear wall [12]

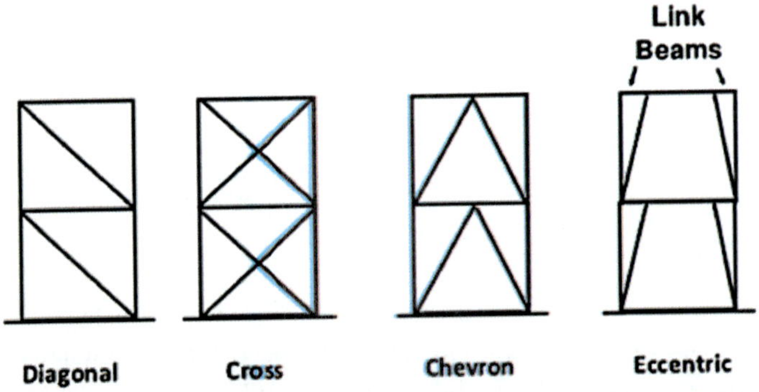

Fig. 5 Bracing wall [13, 14]

3.3 Dampers

The frameworks are planned and fabricated to ensure basic upright natures, control basic harms, and to forestall wounds to the occupants by engrossing seismic vitality and lessening misshapenness in the structure. It has a Shock Absorber for the buildings that reduce the tension, few methods of damping are viscous, friction, yielding,

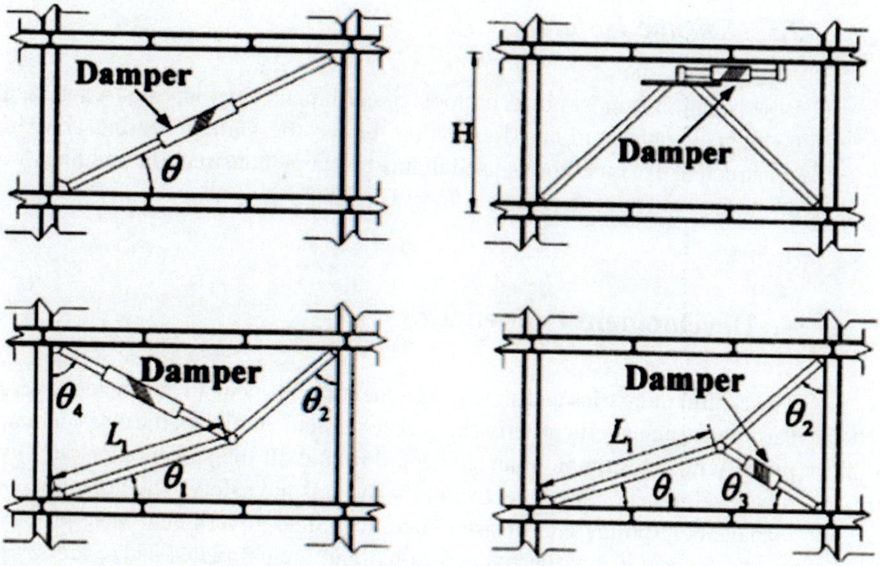

Fig. 6 Dampers [15]

magnetic and tuned mass dampers. It is the common method used in constructing building earthquake zone [15] (Fig. 6).

3.4 Rollers

The gap in between a building's vertical wall in the foundation pit. A small gap is filled with roller at time of earthquake the vertical wall of pit may hit the building when the ground moves. Disasters cannot be stopped but can control and minimize the damage (Fig. 7).

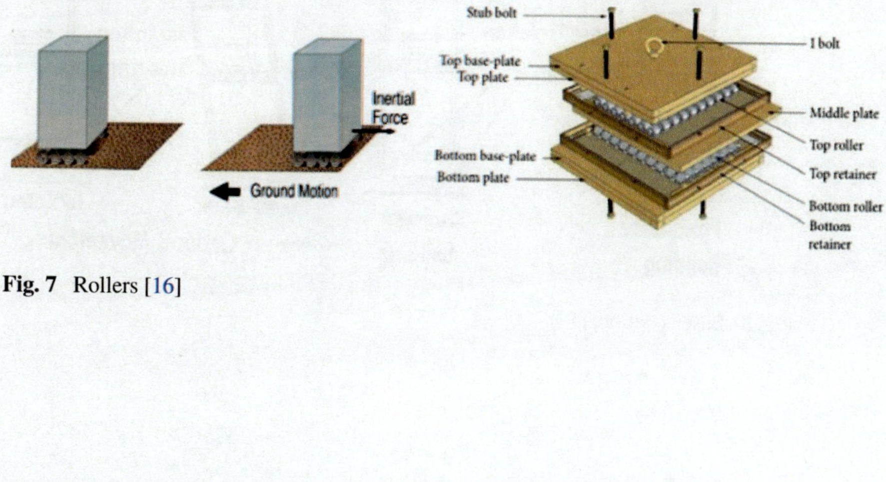

Fig. 7 Rollers [16]

3.5 *Seismic Isolation*

Base isolation is most powerful tools of earthquake engineering. A spherical sliding bearing over the building foundation, above the sliding bearing column base is constructed. This method is used in most of large construction buildings with small displacement between adjacent floors (Fig. 8).

4 Development Guidelines

The central ministries and state government department and agencies have appointed earthquake management activities for the implementation of the disaster management plans, with the suitable management skill and all the stakeholders are involved in this development. The objective is for the seismic safety. The guideline is done by management policy and disaster specifics. State government have provided based on the criteria it may be five-year plan and annual plans. There are six pillars of earthquake management in effective seismic safety, the six pillars are:

i. Earthquake Resistant Construction of New Structures.
ii. Selective seismic strengthening and retrofitting of existing priority structures and lifeline structure.
iii. Regulation and Enforcement.
iv. Awareness and Preparedness.
v. Capacity Development (Education, Training, R&D, Building Capacity and Documentation).
vi. Emergency Response.

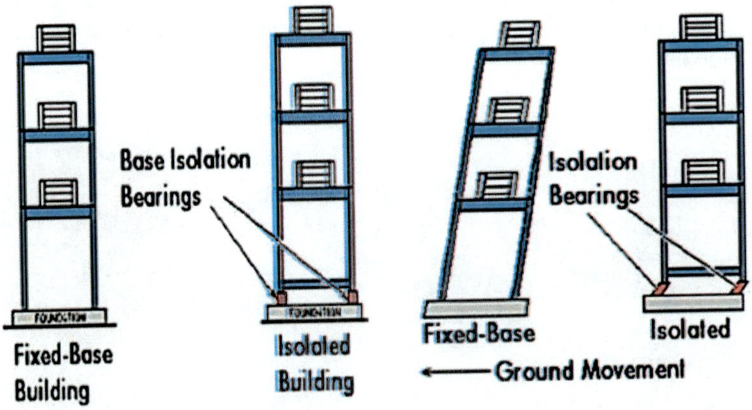

Fig. 8 Base isolation, [17]

Table 2 Previous milestone for the implementation of the guideline [18]

S. No.	Item	Commencement	Action and date of completion
Phase-I implementation of the guidelines			
1	Development of detailed action plans for each phase-I activity	With immediate effect	Complete by 30 June 2007
2	All activities of phase-I	With immediate effect	Underway by 1 July 2007
3	Mid-term monitoring and correction of implementation plans of all phase-I activities	With immediate effect	Complete by 31 December 2007
4	Completion of phase-I activities	With immediate effect	Complete by 31 December 2008
5	Major review of all action plans of all activities of phase-I	With immediate effect 1 January 2009	Complete by 31 December 2009
Phase-II implementation of the guidelines			
6	Identification of activities to be undertaken in phase-II and development of detailed action plans for the same	Initiated by 1 July 2009	Complete by 31 December 2009
7	Implementation of all phase-II activities		Underway by 1 July 2010

The previous guideline phase-I was done by 2009 and phase-II was started by 1 January 2010. Seismic risk reduction is done by mobilising effective stakeholder participation [18]. (Table 2).

4.1 Building Byelaws

The guidelines broadly provide the quality and design specifications of houses as well as materials used for housing, particularly for roofing and walling, have a bearing on the vulnerability of houses to earthquakes, landslides, floods and fires.

To ensure disaster resilient built form in order to mitigate the structural losses due to vulnerability of the urban form, the building byelaws has incorporated 'risk-based classification' of buildings with parameters for building classification on the basis of risk level, the building heights, floor area or F.A.R., the slope angle for construction activities and also based on the experience of the design and building team itself (Table 3).

There are situations when these buildings byelaws are not followed by some construction agencies and then these buildings become vulnerable in the times of a disaster. Thus, it is required to monitor each type of construction activity at its

Table 3 Classification details of disaster resilient built [8]

Parameters		Risk level		
		Low risk	Medium risk	High risk
Building classification		Low hazard occupancies as defined in NBC-2005 (non-assembly buildings)	Moderate hazard Occupancies as defined in NBC-2005	High Hazard occupancies in NBC-2005 (closed assembly buildings)
Height	Plains	Up to 9.0 m	>9.0 m up to 21.0 m	>21.0 m
	Hills	Up to 7.5 m	>7.5 m up to 9.0 m	>9.0 m
Floor area		Covered area on each floor $\leq$ 350 m^2	Covered area on each floor > 350 m^2 and $\leq$ 500 m^2	Covered area on each floor > 500 m^2
Slope		$\leq 10°$	$>10°$ and $\leq 26.5°$	$>26.5°$
Experience of the design and building team		More than 3 (three) buildings	Fewer than 3 (three)	No previous experience

several stages of construction for which the Building byelaws of Uttarakhand have provisions for inspection of building construction to ensure its compliance to the norms. The table below shows various types of inspections that are followed during an inspection survey and the time at which it should happen according to building completion stage:

4.2 *Training of Masons*

The training of labours are properly done on site and off site, in offsite they are explained by showing the working model or 3-D visual of the structure. They are explained with the usage of earthquake materials and how to handle them in construction site. Most of earthquake constructions are implemented in small structures of simple masonry unit (RCC, PCC, MS structures). Construction is done by experienced structural engineer guidance. Workers are guided and trained in every stage of construction, concrete work are done as per the Indian Standards. These structures are constructed for small residential buildings and government office by trained workers (Table 4).

5 Conclusions

The Dehradun city is into three categories risk areas green, yellow and red. These areas are plotted with proper rules and regulations, according to construction byelaw. The government of Uttarakhand is no violation in buildings that are constructed in

Table 4 Inspections w.r.t. building completion, [8]

Name of inspection	Time of inspection	Risk category of building		
		Low	Medium	High
Preliminary inspection (as per sanctioned drawing)	At completion of plinth level	Self-inspection/certification by supervision engineer or architect	Self-inspection/certification by supervision engineer or architect	By sanctioning authority
Intermediate inspection	At completion of $\leq$ 15 m height	Not required	Third party inspection by structural engineer	Third party inspection by structural engineer
Final inspection	At building completion	By sanctioning authority	By sanctioning authority + Fire Deptt. + other Line deptts	Joint inspection- (third party inspection by structural engineer Fire Deptt. and other Line Deptts.)
Surprise inspection	At any time	Not required	Minimum one inspection by sanctioning authority	Minimum two inspections by sanctioning authority
Complaint-based inspection	At any time	On receipt of complaint		
Periodic occupancy renewal certificate	Mock drill inspection	Not required	Once every 5 years	Once every 3 years

high (red) and medium (yellow) risk zones that does not mean to safe for a disaster like an earthquake; these building can withstand disaster with some minor injuries rather than total collapse, which may lead to loss of life.

Every hill region construction site should undergo slope stability analysis, location of the plot should not come under the frequent landslide zone at time earthquake or any natural disaster and it is recommended to build a retaining wall in slope region which helps in strengthening of surface and reduce landslide this can be done for existing as well as new structures. All buildings should be constructed with safe distance 15–20% of the total structure.

If the region comes under the high (red) risk zones buildings can be constructed using resilience technology like dampers, bearing wall (timber structure). The building should not exceed more than 2 floors or less than 9 m (hill region)/21 m (plain) with floor area more than 500 m^2. For medium (yellow) risk zones technologies like dampers, shear wall, base isolated system can be implemented in the construction of 2 or 3 floors with a floor area of 350–500 m^2. The low (green) risk zone that as the slope area less than 10 m can be implemented with economic and resourceful technologies like base isolation system, shear wall, dampers, etc. It can be implied on both hill and plain region.

Existing building structures can be renovated with shear wall technology that also depends on the age and type of construction. Various seismic strengthening and structural resilience technologies are described in the study, which can be, incorporated appropriately in construction and retrofit works. This study also highlights that there must be a system to regularly review the consistency and viability of the codes, rules and regulations. There also ought to be an organisation to ensure the compliance of construction works to building byelaws. Engineers and contactors must be licensed to do works in these areas, which are under seismic risk.

Organizations must be created liable for guaranteeing compliance of byelaws. Proper Land use regulations must be reviewed and enforced. Deployed of licensed engineers and contractors for earthquake resilient structures in seismic hazard prone areas. Inclusion of Disaster Risk Reduction expert in project approvals. Implement an Engineers Act in order to make them progressively liable for the works they do. Adopt ward level earthquake risk reduction plan and awareness measures some of which are recommended in community strengthening section of this report.

References

1. National Disaster Management Authority (NDMA). (2019). *Design for national seismic risk management programme, Inception Report.* NDMA, Government of India.
2. Sharma, M. L., & Lindholm, C. (2012). Earthquake hazard assessment for Dehradun, Uttarakhand, India, including a characteristic earthquake recurrence model for the Himalaya frontal fault (HFF). *Pure and Applied Geophysics, 169*, 1601–1617.
3. Geetanjli, R., Arun. P. (2020). Sustainable City Planning Strategy Review for Next Level Technology, Earthquake-Resistant Buildings for Dehradun City. *National Conference on Sustainable Management of Environment & Natural Resource through Innovation in Science and*

Technology SMTST, 1–9. Dehradun: Springer.
4. IS.4326. (1993). Earthquake Resistant Design and Construction of Building-Bye Laws of Practices.
5. Bureau of Indian Standards (BIS). (2005). *National building code of India.*
6. IS. 4326. (1993). *Earthquake resistant design and construction of building-bye laws of practices.*
7. IS 13828: 1993. (1993). Improving Earthquake Resistance of Low Strength Masonry Buildings–Guidelines.
8. TCPD, Uttarakhand Housing and Development Authority. Building Bye laws. Town and Country Planning Department, TCPD.
9. GOI, Urban Development Department, Government of Uttarakhand. Uttarakhand—Disaster resistant technology. http://mohua.gov.in/upload/uploadfiles/files/3-%20Uttarakhand.pdf
10. Ravij Awas Yojana (RAY), G.O.I. (2018).
11. Anshumn, S., Dipendu, B., & Bhavin, R. (2011). Solution of shear wall location in multi-storey building. *International Journal of Civil Engineering, 9*(2), 493–506.
12. IIT-BMTPC, Earthquake Tips 23, Learning Earthquake Design and Construction.
13. Roslida, A. S., Muhammad, F. K., Muhammad, H. D., Gamil, G. A., Abu, B. F. (2019), Comparative Structural Performance of Diagrid and Bracing System in Mitigation of Lateral Displacement, https://doi.org/10.1088/1755-1315/220/1/012025
14. Angel, A., George, C. C., Boris, B. (2017). Experimental study on eccentrically braced frames with a new type of bolted replaceable active link.
15. Duggal, S. K. (2013). *Earthquake—Resistant design of structures.* Oxford University Press.
16. Yeats, R. S. (1967). ***Living With Earthquakes in the Pacific Northwest.***
17. Talikoti, D. R., & Thorat, M. V. (2014). Base Isolation In Seismic Structural Design. *International Journal of Engineering Research & Technology (IJERT).*
18. National Disaster Management Authority (NDMA). (2019). *Earthquake Disaster Risk Index.* Ministry of Home Affairs, Government of India.

Land Pooling To Be a Tool for Land Development and Management

Vibhor Goel, Rahul Silori, and Harsh Jha

1 Introduction

Urban land pooling is a strategy for coordinating the servicing and sub-division of independent landholdings in urban periphery areas to facilitate planned urban growth. Because it incorporates these procedures, it is also called "urban land consolidation," "land readjustment," "land re-plotting," and "land redistribution" in specific nations. In Japan, South Korea, and Taiwan, as well as several places in Australia and Canada, it is extensively used. In several Indian cities, a similar method called plot reconstitution is applied under land pooling schemes [1]. Land as a resource is a vital factor for the social and economic development of any country. Whether it is about managing biodiversity to ensure good quality of life or just to boost the happiness index of people, we need to have a sound management plan for land resources development to achieve all these objectives with an overall sustainable perspective [2]. This socio-economic development of a country is directly related to the level of urban and rural infrastructure available within the nation [3]. Over 400 million people still do not have access to electricity, 300 million do not have access to safe drinking water, and 1.5 billion do not have access to basic sanitation [3]. Infrastructure development is a requirement for long-term economic growth. If the area wants to continue its growth momentum, reduce poverty, and respond to climate change, developing countries would need to invest \$26 trillion in infrastructure from 2016 to 2030, or \$1.7 trillion per year. Especially in Southeast Asia, approximately 5.7% of the GDP of a nation

V. Goel (✉)
Fichtner Consulting Engineers Private Limited, Noida, India
e-mail: ar.vibhor@outlook.com

R. Silori · H. Jha
School of Engineering, University of Petroleum and Energy Studies, Dehradun, India
e-mail: rsilori@ddn.upes.ac.in

H. Jha
e-mail: 500063075@stu.upes.ac.in

N. A. Siddiqui et al. (eds.), *Advances in Construction Safety*,
https://doi.org/10.1007/978-981-19-4001-9_24

is required to be invested in infrastructure development. India spends around 4% of GDP on infrastructure projects and it needs to spend at least 1.5% more annually to meet UN sustainable development goals [4].

This paper highlights how exactly this financial pressure could be abated and how positive public financing could lead to the development of infrastructure projects in small but growing cities-towns where rural-to-urban transition could provide an opportunity for a well-planned greenfield development as well as helps to manage the peripheral urban growth in big cities.

The best land development model to fund almost all sorts of infrastructure and development projects is land pooling, specifically for greenfield development [5]. Few states in India use land pooling as a primary land development model while planning for greenfield areas, Gujrat and Maharashtra being one of them. Also, these states have the maximum successful greenfield land pooling projects [6, 7]. Today the traditional way of land development, the "land acquisition model" has become a bigger problem. Whenever relocation and rehabilitation are carried out, most of the landowners are not happy with the decision of the administration on relocation sites and the amount of compensation paid to them [8]. Whereas inland pooling, landowners provide a portion of their land willingly to bring development projects in the area. In return, their land value increases, and compensation is also paid for the same [9]. This land parcel is used to provide infrastructure facilities in the area, and the rest is sold for capital gains to invest in the same project.

2 Literature

2.1 *Urbanization Trend and Need for Land Management*

In India, as per the 2011 census, one out of three people are living in urban areas [10]. This number is estimated to grow from 300 million in 2011 to 814 million by 2050, which means one out of every two people in India will be living in urban spaces. Also by 2030, the number of cities with populations of more than 1 million will grow from 42 to 68 [11]. Four of India's cities, namely Ahmedabad, Bangalore, Chennai, and Hyderabad with currently 5–10 million inhabitants are projected to become megacities in the coming years, for a total of seven megacities projected in the country by 2030 [12]. All this urban population, especially in small but growing cities where the rural-to-urban transition, will demand an immense amount of urban infrastructure ranging from small parks for recreational purposes to growth centers like central business districts (CBDs) for economic growth. However, convincing the landowners to give up their land for development has always been a tedious task do to. Most of the large-scale projects like housing projects, commercial projects, railway lines, metro projects, flyovers, bridges, hydro, and solar power plants require huge parcels of land. The two important land development models that can be used for developing these large-scale infrastructure projects are land acquisition and land pooling models.

Capital for investment and land for these projects are the two most essential resources required to provide these infrastructures. Both can be arranged well by the most undervalued land development model, Land Pooling. It is an old saying in India that what comes from land goes back to it and land is indeed the only resource that we all need. Sadly, it is not available for everyone. Therefore, the planning should be in such a way that everyone has access to it until it lasts. Land pooling can be used to grow this very concept.

2.2 Land Procurement Becoming an Obstacle for Infrastructure Development in India.

Since Independence, the Indian administration and government are aware of the fact that a country like India with such a huge population needs infrastructure projects to take care of their needs. For this, the government has continuously increased the investment of GDP in infrastructure projects. Apart from huge capital investments, the second most common obstacle faced by the decision-makers in the land transfer from local owners to the development authorities resulting in years-long delayed projects. According to a report of 2018 submitted in the parliament, more than 435 major infrastructure projects got delayed because the primary step of land acquisition could not be completed on time. The major cause of this delay was the unwillingness of the landowners to transfer land rights to the government because the terms of land acquisitions did not meet the requirements of the owners. However, these numbers were quite high before 2013 when the process of land acquisition was based on a 120-year-old statute, the "Land Acquisition Act 1894." Later, this act was replaced by the "Right to Fair Compensation and Transparency in Land Acquisition, Rehabilitation, and Resettlement Act 2013" [13]. Yet even today, many metro projects, rail projects, and flyovers are stuck and waiting for the approval of landowners to give up the land and accept the compensation given to them.

To learn more about the limitation of land acquisition, it is important to understand this concept and why this is used majorly as a land development tool in India. There are numerous definitions of land acquisition and the simplest enough to understand the concept is-it can be defined as the measures and procedures taken by the government to acquire the required land from its owner for a project planned to cater to the public need. Sometimes, these parts can be a small percentage of the land but in the majority of the cases, and these acquisitions are conducted on a huge scale resulting in taking away the entire villages, neighborhoods, tribes, and communities. Just because it is simple and easy to implement, this model has been widely used in India for decades before independence. The two major identifiable limitations from this Land Acquisition Act 1894 which was leading to delays in land procurement were:

- Rehabilitation and resettlement of the public and community displaced by the acquisition program were not considered.
- The compensations made to the landowners were based on market value, and future land value increments were not considered while compensation.

These were addressed by the LARR act 2013, however, the higher land compensation, consent of affected families, and rehabilitation and resettlement cost by this act had exponentially increased the project cost and time taken to finally start the implementation, making it difficult for the decision-makers to implement such projects in high numbers. Also, the compensation and resettlement done are not up to the social cost of the landowners. That is why this model of land development is still not beneficial for the stakeholders of the projects. To tackle these issues, decision-makers are bringing up land pooling policies to boost faster development. Land pooling policy, approved by these authorities, removes the bottleneck of land acquisition and provides an advantage to landowners of getting the developed land in return.

2.3 *Understanding Land Pooling*

Land pooling or land readjustment is a land development model which provides the required infrastructure and promotes efficient, sustainable, and equitable land development in an area by public financing of the project [14]. The process is carried out by preparing town planning schemes for the area and then promoting them among the landowners so that they hand over their land parcels to the government collectively for infrastructure development. These land parcels are legally consolidated and transferred to the agency responsible for carrying out the land pooling. Once the development of the project is completed, the land is handed over to the original owners, after deducting some portion as the cost for the same [15]. This increases the land value by at least 3–4 folds [16]. In addition, land pooling entitles the original titleholder to the ownership, and no rehabilitation and resettlement are carried out.

E.g., a Land pooling mechanism was selected by the Delhi Development Authority for the construction of housing societies and other commercial complexes in Delhi. In case the landowner has more than 20 Ha of the land, the pooling percentage will be 40%, which means 60% of the land will be returned to the owner after development, and the land price increment with this infrastructure will be at least two three times the current land rates. Else if the landowners have 02–20 Ha of land, then the land returned to the owner will be 48% of the total share [17].

3 Case Study of Land Pooling in the Area of Navi Mumbai

In the case of Town Planning Scheme 8 in Navi Mumbai under the NAINA Scheme [18], the plan was prepared for 11 town planning schemes in the interim development plan proposed by CIDCO special planning authority. TPS 8 was entirely rural and had a total population of 3000 with no or little development in the region. The objective was to develop this area with appropriate infrastructure to cater to the needs of at least 30,000 people. As per the policy, 60% of the land was taken from the owners to provide infrastructures like roads, schools, health services, public amenities, open spaces, and growth centers. The rest 40% was returned to the original landowners with increase land values.

Figure 1 shows the khasra map of the land parcels where the plots are unorganized with no or limited infrastructure services. Figure 2 shows the final land-use proposal on the area by applying TPS with final plots allotted to all the landowners. This is how land pooling can be implemented through a town planning scheme. The cost of the project was incurred by selling the commercial spaces and semi-public areas

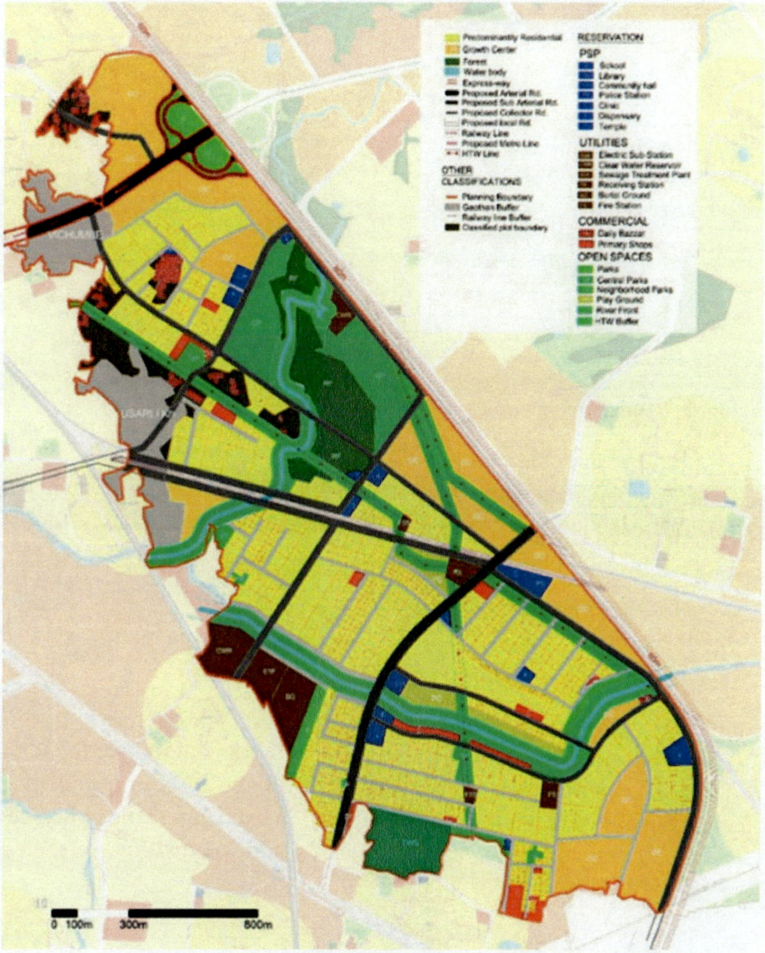

Fig. 1 Plan after implementation of land pooling policy

Fig. 2 Unorganized plots before land pooling

to private players at increased land prices, which was approximately thrice the land price before planning the project. The sold land percentage is almost equal to 40% of the total land pooled.

3.1 Merits and Demerits of Land Pooling

Land pooling schemes, which were created as an alternative to the time-consuming Land Acquisition Act, are proving to be a huge success in Indian states. With the clear benefit of receiving developed land in exchange, landowners are more willing to choose this strategy over the outdated Land Acquisition Acts.

Advantages

Most of the lands remain positioned with the original land. Only a small part of the land is used for the planned urban growth by the development agency. This also ensures the minimization of the negative impacts of urbanization on farmers, i.e., the original owners. In case the land is sold or developed, the increment from the development accrues to the original owner. It promotes equitable sharing of costs and benefits.

The strategy is frequently funded and implemented by landowners, who stand to earn handsomely from the initiative. It also helps to avoid the costly and unusual bureaucratic process of obtaining the land.

The readjustment of land helps the land owners to get back the large portion of the land. The requirement for working in conjunction between landowners and the state administration necessitates strong collaboration during the project's implementation. It provides infrastructure and service providers with a good opportunity to recoup their ongoing expenditures and gain access to land. Further, systematic administration could lead to increased land distribution equality.

Usually in developing countries, plots in urban areas are smaller, irregular, and not-so-accessible. Land pooling provides an excellent opportunity for potential development and infrastructure. It also obliterates the difficulties of land and population density.

Disadvantages

The procedure prescribed for the preparation and implementation of land pooling might be complicated and cumbersome. It could lead to the entire thing being a little time-consuming.

The betterment charges levied on the finalization of the scheme may not meet the cost of the infrastructure assessed at the beginning due to a not-so-quick process.

The current system does not force-land development, leading to the development being slow. The capital value appreciation might take time as well.

3.2 Problems with the Current Practices Acts and Policies Affecting Land Pooling

As an urban planner, I find it essential for the actual public financing of the projects to be done only when both the decision-makers and the public work together to plan and implement such projects. Certainly, what we see today is that public participation is done only to follow the norms and minimum standards for project implementation. In most cases, it is carried out in closed room stakeholder meetings. In Arnstein's ladder of public participation, we as a country are still at the degree of tokenism, where informing and basic consultations are the only steps for participating. The affected landowners never even get a chance to lay down their opinion on these projects. Sadly, most of them are not even aware of a project being planned on their land. Only a public notification is published in the official gazette. We are still stuck at the 4th step of the ladder of participation. To achieve direct citizen control in decision-making, we need to climb 4 more steps.

The second crucial requirement is to promote land development models like land pooling in development acts of the respective states. As land is a state subject, the state administration has a right to govern any developmental project on, over, and under the land in their jurisdiction. Many states in India do not even have a land pooling act to implement these policies. The decision-makers need to amend the town and country planning acts of their respective states.

Also, provisions to develop an area by creating a town planning scheme using land pooling as a tool for development must be added to the respective acts. The definition of "development" should promote sub-division of land, so that town planning schemes can be carried out efficiently as per these acts. References can be taken from Gujarat and Maharashtra Town planning Acts. Today, Maharashtra, Gujarat, Delhi, Andhra Pradesh, Tamil Nadu, and the western State of Rajasthan have popular land pooling schemes. Recently, Bihar has also stepped into the game and has proposed TPS schemes for development in Patna.

4 Conclusion

Today, every state has its own development rules, and all of them have a different perspective of land development. Some prefer land acquisition while some conduct land pooling as Town Planning Schemes. The rest choose the suitable model as per the requirements of the respective projects. It is equally necessary to have educated and trained personnel to carry out these projects. Currently, only a few institutions hire actual planners to plan the town planning schemes.

Therefore, involving the public in decision-making, realizing required amendments to town and country planning acts, and hiring planning experts could act as a leading step toward the implementation of effective public financed projects.

References

1. Archer, R. W. (1986). Readjustment technique to improve land development. *10*(4), 3975186. 0197–3975186
2. Beltr, G. (2013). Asia development bank's report India : Promoting inclusive urban development in Indian cities urban planning and land management for promoting inclusive cities, pp 2–30.
3. Yoshino, N., & Paul, S. (2019). *Land Acquisition in Asia: Towards a Sustainable Policy Framework*. Asian Development Bank Institute.
4. Majumdar, R. (2021). Budget 2021: Infra push can boost economy, improve competitiveness, *Business Today*.
5. Josh, R. (2004). Land reservations for the urban poor: The case of town planning schemes in Ahmedabad.
6. Archer, R. W. (1992). Introducing the urban land pooling/readjustment technique into Thailand to improve urban development and land supply. *Public Administration and Development, 12*(2), 155–174. https://doi.org/10.1002/pad.4230120204
7. Archer, R. W. (1988). Land pooling for resubdivision and new subdivision in western Australia. *American Journal of Economics and Sociology, 47*(2), 207–221. https://doi.org/10.1111/j.1536-7150.1988.tb02029.x
8. **Mathur, S. (2013). Self-financing urbanization: Insights from the use of town planning schemes** in Ahmadabad, India. *Cities, 31*, 308–316. https://doi.org/10.1016/j.cities.2012.09.004
9. Mahadevia, D., Pai, M., & Mahendra, A. (2018). Ahmedabad: Town planning schemes for equitable development-glass half full or half empty (p. 24). https://doi.org/10.13140/RG.2.2.12683.75040
10. Census, "Census," (2011). www.censusindia.gov.in
11. Sankhe, S. et al. (2010). *India' s urban awakening : Building inclusive cities , sustaining economic growth*. McKinsey Q. (pp. 1–33).
12. United Nations, "World Urbanization Prospects," 2014. ST/ESA/SER.A/366
13. GOI. (2013). The right to fair compensation and transparency in land acquisition, Rehabilitation and resettlement act (pp. 1–46). https://legislative.gov.in/sites/default/files/A2013-30.pdf
14. Sanyal, B., Deuskar, C. (2012). A Better Way to Grow?: Town planning schemes as a hybrid land readjustment process in Ahmedabad, India. *Value Capture and Land Policies Proceedings og 2011 Land Policy Conference*, 149–181. https://www.lincolninst.edu/sites/default/files/pubfiles/2180_1505_LP2011_ch07_Town_Planning_Schemes_as_Hybrid_Land_Readjustment_Process_In_India_0.pdf.
15. Mahadevia, D., Mahendra, A. (2018). Ahmedabad: Town planning schemes for equitable development—glass half full or half empty? https://doi.org/10.13140/RG.2.2.12683.75040
16. Ingram, G. K., & Hong, Y. H. (2012). *Land Value Capture and Land Policies*.
17. Joardar, S. D. (2006). Development mechanism in spatial integration of cities. In *ISOCARP Conference* (pp. 1–15).
18. **CIDCO SPA NAINA. (2013). Draft development plan for Navi Mumbai airport influence notified area (NAINA)** *40*, 627. https://cidco.maharashtra.gov.in/pdf/other/Combined_Report.pdf

Implementation of Behaviour-Based Safety Management in Achieving Inclined Driver Safe Behaviour

P. R. Sushmitha, P. A. Arun, and Madhuben Sharma

1 Introduction

All accidents are avoidable, if we have a comprehension of how they happen. It is also seen with inclined emphasis on safety has rendered great reduction in accident rate. The ideology on causes of the accidents has evolved over time. At earlier times, there was a misconception about the accidents that they are preordained to be or unavoidable [1]. All accidents are routed to two categories of causes: 1. Unsafe acts and 2. Unsafe conditions. The unsafe acts are committed by humans. Even unsafe conditions are indirectly rendered by negligent human behaviours. A researcher, Heinrich, sampled 75,000 accidents to bisect the causes of the accidents. Evidently he revealed that 88% of the accidents have occurred by unsafe human behaviour, and 10% of the accident has occurred due to unsafe conditions prevailing. Rest 2% of the accidents he reviewed had other unstoppable reasons [2]. Followed by Heinrich, Bird and George came up with the modern accident causal chain theory which also conveyed that the unsafe acts are the substantial ground for accidents [3]. The BBS concept which is centred on behavioural science was coined by Earnest and Palmer [4].

Profitably, applying a technique which will target correcting the unsafe behaviours will help circumvent most accidents irrespective of the field. In that way, behaviour-Based Safety is the limelight technique to deal with safety conscience. All humans cannot be expected to hold inherent safe attitude by themselves. Safe attitude is believed to be groomed and moulded through appropriate trainings [5]. The core methodology of BBS will be to observe/recognize the behavioural nature of the humans. These observations should be classified as unsafe and safe behaviours. Potential unsafe behaviours which be ranked. The unsafe behaviours with respect to

P. R. Sushmitha · P. A. Arun (✉) · M. Sharma
Sustainability Cluster, University of Petroleum and Energy Studies, Dehradun, Uttarakhand, India
e-mail: arun.pa@ddn.upes.ac.in

N. A. Siddiqui et al. (eds.), *Advances in Construction Safety*,
https://doi.org/10.1007/978-981-19-4001-9_25

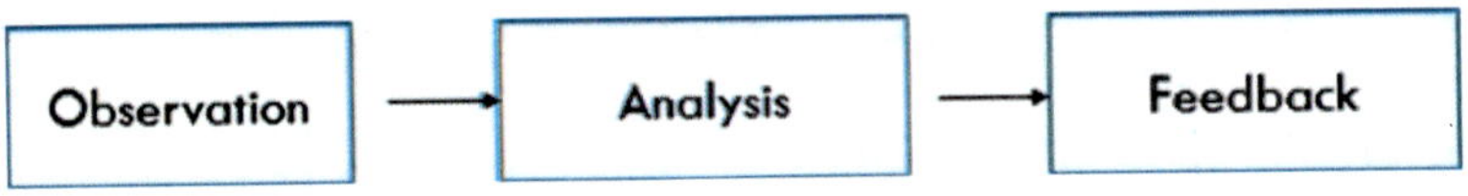

Fig. 1 BBS stages

each activity should be replaced with safe behaviour through positive and negative reinforcements [6]. The stages of BBS implementation follow the below fashion.

The success of BBS will be reflected when there is an attitude change. Attitude cannot be measures or observed but behaviours can be. When there is a repetition in behaviour correction, attitude will be changed. The BBS can be implied by fervent observation of behaviours. Dealing with these observations is the cardinal part of analysing them. Amendments to these identifies unsafe behaviour should be established through psychological reinforcements [7, 8].

BBS is implied with the ABC model. The ABC model comprises antecedents, behaviour and consequence. Antecedents and consequence are influencing factors of behaviours in way, where the antecedents act as the triggering points of a behaviour, meanwhile the consequences present us the likelihood of the behaviour recurring again [1]. The positive reinforcements can be parties, rewards and recognitions. The negative reinforcements can be fines, advisory letter and suspensions. Even though some safety professionals believe that BBS possesses some drawbacks like victim blaming and minimized focus point on immediate causes [10]. The leads of BBS overshadow the downsides [11] (Fig. 1).

BBS greatly aligns to the safety management systems. In order to minimize the likelihood and severity of the damages due to an accident, safety management systems should be in place which deals with the safety policies and how people are organized safely [12]. One of the biggest scorer of accidents is road injuries. It is estimated that almost 50 million are injured annually on road. Especially, India which is a low income country, there are economic and societal loss due to road accidents. From Word Health Organization-2018 reports, it is revealed that India holds road accident injuries triple the time high income countries does. Upon research, we know that negligence to follow road/vehicle safety precautions, unsafe hazardous physical conditions is the prime factors that leads to road accidents. It is evident that India had the major part to make road injuries as the second highest contributor of deaths in World Health Organization in 2002, world report on road traffic injury prevention and Geneva' World Health Organization. Among the contributors to road accidents, speeding needs more focus. From the statistics speeding multiplies, the probability of an accident which can even be fatal [14]. Speeding issues contribute almost 1/3rd of the fatal road injuries [15]. Speeding is an unsafe behaviour where the driver has to be observed, analysed and rectified with appropriate measures (like fines) which will reduce the chances of the drivers to commit speeding again. This is how the speeding behaviour of a driver can be changed over time through BBS. Researches explain that inadvertent driver behaviour stands behind 74% of the fatal road accidents. So, significance lies in grasping the driver behaviours [16]. In a sample of 15,000 licence holders, 2% of the sample was found to carry out irresponsible attitude and they were

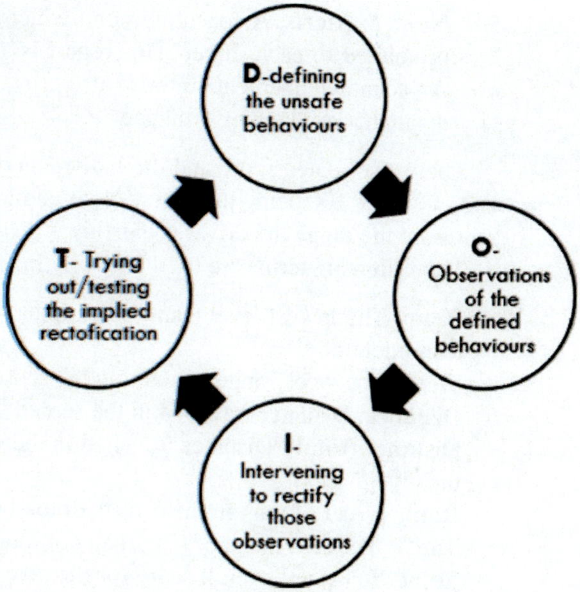

Fig. 2 DO IT process of BBS [9]

more prone to accidents. It is easy to seclude this population and change their attitude through behavioural change [17]. In this paper, we will discuss the outcomes of a project, where a smartphone application was designed and developed to monitor and gauge drivers' performance. About 300 drivers from a pharmaceutical company were enrolled in this project (Fig. 2).

About the App

This application also provides coaching insights on weekly basis. These insights help the drivers rectify and recognize the unsafe driving behaviour at individual level. This app is capable of distinguishing vehicle classes, where it measures the data compatible to the 2 wheeler or 4 wheeler vehicles. A "Vehicle Class" option is designed which enables the user to adapt to vehicle used currently. The mobile application calibrates drivers' behaviours like phone usage, jerking, swerving and cornering.

The app generates the following scores:

1. **Current Driving Score**: This score has a scale of 1–100 which will be updated upon completion of every ride after 10 min.
2. **Handling and Focus Scores**: This score reveals how smoothly the driver handles the vehicle. This score will be influenced by jerking and swerving.
3. **Team Ranking**: The employees (drivers enrolled in the app) from different units were grouped as teams. An average score of each team was shown in the app against the drivers' individual score.
4. **Coach**: This section facilitates the drivers to view their past weeks' score and analyse the score of each criterion measured.

5. **News & Alerts**: A coaching report which details the drivers' performance will be mailed to each driver. This report is also made available in the app itself. An acknowledgement button is also provided in the app to know if the driver is aware his driving performance.

The consolidated score and the individual driving behaviour metered between 1 and 100, with 100 being the best. Colour coding is also done on the score to visually represent the range the driver is existing.

The following terms are used basically in the application:

Team: The lowest level team (sometimes referred to as store or division) in the team hierarchy.
Week: The week range containing the scores.
Distance: Distance travelled in the scored week.
Distance (total): Distance travelled in the whole time range up to and including the given week.
Rank: Order of rank in the drivers of total score for the given scored week.
Top%: Per cent range of the drivers of total score for the given scored week.
Total: Your total overall score for the given week as a rollup of the lower driver behaviour scores of accel, brake, corner, jerk, swerve, phone and speed.
Handling: A rollup score consisting of braking, swerving and cornering.
Focus: A rollup score consisting of phone, jerking and swerving.
Eco: A rollup score consisting of braking, accelerating and speeding.

Behaviour Scores
Accel: Measures your tendency to accelerate hard relative to other drivers.
Brake: Measures your tendency to brake hard relative to other drivers.
Cornering: Measures your tendency to turn rapidly around curves or turns.
Jerk: Measures harsh or frequent speed changes.
Swerve: Measures harsh or frequent direction or lane changes.
Phone: Measures how much you are moving your phone around while driving, including picking it up or fiddling around with it.
Speed: Measures speeding infractions.

Key indicators in this dashboard and their colour coding are:
Scores >80 are GREAT
Scores 70–80 are GOOD
Scores 60–70 are AT RISK
Scores <60 are ACTION REQUIRED

Weekly Driver Scores
This score is generated to examine the driver behaviour on each type of driving traits listed above, and ranks are provided in the application within the teams. This score

is calculated on weekly basis. A date range option is also provided to certain view scores of a particular week.

Roll up Weekly Scores
This score is generated for regions. The users are ranked on the basis of geological regions. So from this section of the application, we can infer regional behaviour of the drivers. The time ranger of score generation will be weekly.

Driver Scorecard
In this section, we categorize the drivers of all regions and teams based on their scores from highest to lowest. The colour coding is applied on each driver, so we can navigate to the section of drivers who needs immediate attention. This score can be viewed weekly, monthly, quarterly and yearly basis.

Benchmark Scores
This score of driver is generated as a comparison to the similar scores with similar driving conditions of other universal drivers enrolled in the app from other companies.

Store Scorecard
This score is generated to compare consolidated score of a team.

For an instance, there are 5 units and 2 teams in each unit. A consolidated driver score of each team will be provided. The teams from all units will be ranked from highest to lowest fashion.

Weekly Score Trend
This section illustrates the aggregate weekly score of each team will be listed. From this section, we can know the increasing and decreasing trend of the score weekly.

2 Methodology

1. Driver nominations were received from various units of a pharmaceutical company.
2. Username and password were configured for each driver in the app.
3. The drivers were made to download the mobile application.
4. Their scores got generated after every ride against different driving criteria discussed above.
5. Behaviour-based safety management was implemented through consistent Microsoft Teams meeting.

In the meetings, drivers' performance was discussed. Drivers who have achieved "Great" categories were given appreciations and were asked to share how they earned good driving behaviour. Positive reinforcements were made through rewards like gift vouchers which boosts them to maintain consistent and great performance. Also, the inspiring talk from excelling talks kindled the interest of low performing drivers to accomplish better driving behaviour.

While the drivers who were in the "At Risk" category were shown and explained the consequences of unsafe driving traits. The drivers who were employed in the marketing of pharmaceutical goods used drive all day. Those drivers' performance was focused more. As part of negative reinforcement, fine system was put in practise whenever a driver's scores lower than 50.

The drivers who were in the "Good" category were given recommendation on which driving factors they have to focus on. Their driving reports were discussed to have a better comprehension of the feedback and to check the degree of awareness of the drivers in implementing the recommendations.

3 Results and Discussion

By executing BBS on driving behaviour, we can infer that positive and negative reinforcements can reflect improved safe driving behaviour in drivers.

Below are the illustration to the above statement, where the score have been recorded for a month with 4 weekly a scores (Fig. 3).

The above graphs are the total representation of the driving performance of all the drivers in the company.

Week 1: The scores were moderate when a general briefing was done that the mobile application is launched to monitor the driver performance. The drivers were conscious of their driving pattern. We aimed to achieve an attitudinal change even absence of a monitor. In the upcoming weeks, we screened their behaviour and driving traits. We analysed the scores of the drivers against each criterion.

Week 2: In the second week, meetings were conducted in clusters with the teams approximately 15–20 drivers at a time. In the first half of the meeting, we recognize the top ranker of the last week and encourage the safe performance with rewards. The best performance of each team was also asked to express his experience as a token of inspiration to the others drivers. The "At Risk and Action Required" drivers were focused more. Firstly, we appreciated these drivers in driving factor they scored well. Secondly, these drivers' reports were discussed in person and detailed counselling

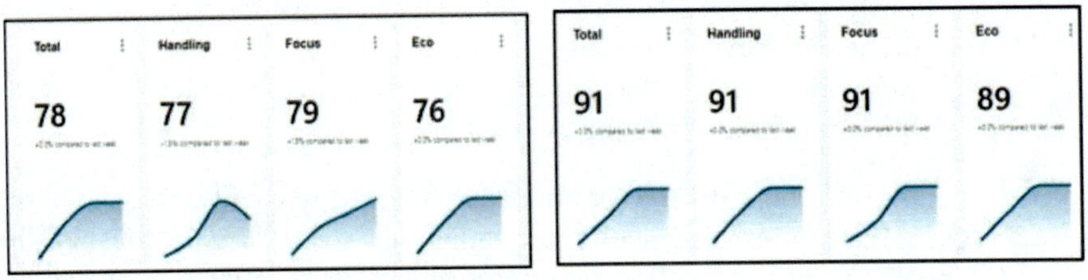

Fig. 3 Sample scorecard

were provided in a more individually customized way. For drivers whose score are lower than 50, fines were deducted and red cards were issued.

Week 3: The drivers' score either maintained or inclined. There was a gradual raise in the trend of the score. Especially, the "Action Required" drivers exhibited an improvement in the scores through negative reinforcements. However, few drivers' scores still needed attention.

Week 4: Positive and negative reinforcements were continuously put in practise. Inclined trend in the scores was evidently seen. To realize if there are actual attitude change, we notified them that the application will no longer monitor the driving performance and reviewed the driver behaviour to check if their behaviour is influenced by the surveillance only or there is an actual attitude change. The results in the portal of the application revealed that the drivers have adapted and basked in safe driving traits.

4 Conclusion

Behaviour-based Safety is a practically effective tool, where we can attain long term attitudinal change. Attitude cannot be measured however behaviour can be observed and measured. In the same way, behaviours of drivers were monitored through mobile application. Crucial driving factors like phone usage, sudden acceleration, swerving, cornering and jerking were calibrated. This project is an integrated idea of road safety and BBS. Real results were achieved after administering behaviour-based safety through personal counselling, rewards, recognition and fines. It is pivotal to effectuate long term attitude change than short term ones.

References

1. McAfee, R. B., & Winn, A. R. (1989). The use of incentives/feedback to enhance work place safety: A critique of the literature. *Journal of Safety Research, 20*, 7–19.
2. Heinrich, H. (1941). *Industrial accident prevention* (2nd ed.). McGraw-Hill.
3. Bird, F. E., & George, L. R. (1990). Practical loss control leadership. In Det Norske Veritas Inc.
4. John, A. (2006). An introduction to behavior-based safety. *Stone Sand and Gravel Review, 6*(2), 38–39.
5. Cooper, M. D. (1993). Goal-setting for safety. *The Safety & Health Practitioner, 11*, 32–37.
6. Scott Geller, E. (2001). *The psychology of safety handbook* (2nd ed.). CRC Press LLC.
7. Willam, R., & Holliday, C. S. P. (1999). Behavior based safety-safety professional's view, presented to: ASSE Georgia Chapter.
8. DSR. (2003). STOP for employees introduction. Texas: Du Pont (pp. 2–10).
9. Chen, D., & Tian, H. (2012). Behavior based safety for accidents prevention and positive study in china construction project. *Procedia Engineering, 43*, 528–534. https://doi.org/10.1016/j.proeng.2012.08.092

10. Pęciłło, M. (2010). Effectiveness of unsafe behavior modification programs—experiences of foreign enterprises, Bezpieczeństwo Pracy: nauka i praktyka, 11, 16–19 (in: Polish).
11. Skowron-Grabowska, B., & Sobociński, M. (2018). Behaviour based safety (BBS)—Advantages and Criticism. *Production Engineering Archives, 20,* 12–15. https://doi.org/10.30657/pea.2018.20.03
12. Fernandez-Muniz, B., Montes-Peon, J. M., Vazquez-Ordas, C.J. (2007). Safety management system: development and validation of multidimensional scale. *Journal of Loss Prevention, 20,* 52–68.
13. Peden, M., Scurfield, R., Sleet, D., Mohan, D., Hyder, A. A., & Jarawan, E. (2004).
14. Kloeden, C. N., McLean, A.J., Moore, V. M., Ponte, G. (1997). Travelling speed and the risk of crash involvement. Volume 1: findings. Report CR 172. Federal Office of Road Safety FORS, Canberra.
15. Wegmand, F., Aarts, L. (2006). Advancing sustainable safety: National road safety outlook for 2005–2020. SWOV Institute for Road Safety Research, The Netherlands
16. Singh, S. (2018, March). Critical reasons for crashes investigated in the National Motor vehicle crash causation Survey. *National. Highway Traffic Safety Administration*
17. Assum, T. (1997). Attitudes and road accident risk. *Accident Analysis and Prevention, 29*(2), 153–159. https://doi.org/10.1016/s0001-4575(96)00071-1

Human Attitude Improvement at Work Place in Oil and Gas Industry Through Training a Proposed Model for Accident Reduction

Mohammed Ismail Iqbal, Ibrahim Alrajawy, and Osama Issac

1 Introduction

From the yearly reports of seat stamped organizations from year 2016–2018, likewise from the contextual investigations accessible surmised 100 fiasco cases were examined and are realized that the conceivable danger during penetrating is bigger in number. The result demonstrates the adjustment of control estimates that is proposed to have protected and sound workplace at work place. A faculty working in the boring apparatus must be extremely focused and alerts during their functioning hours. These deadly occurrences feature the significance of the connection between the executives, frameworks, and human factors. Wellbeing the board frameworks all by themselves are not the appropriate response, but rather the discipline of human factors assists with uncovering holes in administration frameworks, assisting with guaranteeing their adequacy. In the oil and gas industry, contrasts exist in project workers' working methodology that occasionally display close to nothing consistency. Such non-standardized work measures lead to deviations by people and teams and trouble in preparing individuals as they move starting with one area or organization then onto the next. These deviations proceed to occur in the oil and gas industry and are normally veiled on a project worker's disaster report under the name "not following methods," which doesn't refer to legitimate restorative activities other than "follow the systems" to forestall comparable episodes.

According to the expert decision, accident don't just happen largely but they are caused due to unsafe acts and conditions. The innate intricacy of the cycles and the unpredictable idea of petrol items urge the petrol business to constantly look for and foster instruments and procedures to recognize, assess, and relieve potential

M. I. Iqbal (✉)
Research Scholar, Lincoln University College, Kota Bharu, Malaysia
e-mail: ismailiqbal@lincoln.edu.my; mdiqbal1988@googlemail.com

I. Alrajawy · O. Issac
Faculty of Business and Accounting, Lincoln University College, Kota Bharu, Malaysia

N. A. Siddiqui et al. (eds.), *Advances in Construction Safety*,
https://doi.org/10.1007/978-981-19-4001-9_26

dangers that can contrarily affect their interaction activities and more risk is related to drilling operation [1]. Mergs [2] above 80% work environment mishaps are connected to hazardous conduct, and every risk is associated with different criteria based on environment, based on nature of rock, [3] and each hazard is related with various rules dependent on climate, in light of nature of rock, and organizational factors [4]. There is an increasing demand and interest in relating the methodologies of risk assessment in operational phase [5] relating the methodologies of risk assessment in operational phase [5]. Operational risk assessment, real-time risk assessment, and dynamic risk assessment plays very important role in hydrocarbon industry. There are many provisions such as risk assessment which is used in industry [6].

Consistently many penetrating teams face testing and dangerous circumstance because of wellbeing concerns and wellbeing at locales (boring). Hence, penetrating tasks are viewed as three-fold perilous than [6]. The future examinations suggested that it should consider exploring information techniques like conduct perception. The other interceding components like enthusiastic fatigue, abilities, inspiration, and work pressure, could be viewed as which reinforce the results [7].

Woolfson [8] states in measure ventures domino hypothesis depicts mishap grouping as chain of five components or elements (dangerous demonstrations or conditions, issue of individual, mishap or injury, social, and natural) on the off chance that one factor falls, the other four elements would doubtlessly fall. Distinguishing freedoms to further develop hardware configuration includes deciding causal elements that can be followed back to problematic gear plan or client/gear interfaces. Introductory accentuation is put on exhaustive audits and investigations of exercises gained from heritage archetype or similar gear to help recognize and dispose of attributes in the new hardware. It is likewise implied dangers of major mishaps which have been exhibited by calamity like the fire and blast on the UK creation stage Piper alpha. Significant mishaps address a definitive, most terrible manner by which a seaward designing task can turn out badly. Mishaps cause passing, enduring, natural contamination, and interruption of business.

The oil and gas industry is comprised of various working organizations, administration organizations, sellers, workers for hire, and subcontractors—all interrelated what not expecting to interface flawlessly to work securely. With a labor force directed by a bunch of security management frameworks and cycles and systems remarkable to every association, working in arrangement—and working securely—is a test. We need to further develop how our individual organizations team up with one another. The Macondo very much was exploratory in nature, which means critical vulnerability stayed about the topography, the oil assets, and the development attributes that make the well simple or hard to penetrate. Overseeing such vulnerabilities eventually decides an organization's prosperity. The outcomes shown that oil laborers' that there is decline in dangerous conduct with commitment of security consistence and interest. In any case, mental condition was affirmed as mediator which helps in moderation. Thus, proposed that mental states of representatives and related variables should be considered to lessen perilous conduct [9]. Note [10] expressed helpless wellbeing society at work place is one of the main considerations that affects worker's security conduct at work place, which may straightforwardly bring about mishaps.

2 Problem Statement and Research Motivation

According to (HEMS) Health, Safety, Environmental administration frameworks in order to avoid the accident knowing the danger prior to making arrangements, knowing the danger area during drilling, human variables at work place are administering are to be taken into consideration. Nonetheless, it isn't generally comparative in a large portion of the cases. As indicated by WOAD information base in most recent couple of many years, around 6000 mishaps have occurred distinctly in seaward. For this study, a recording of mishaps occurred between year 2016–2018 is taken. At the point when you utilize the rehashed calamities occurring, it offers a hint of caution of disappointment in wellbeing the executives to guarantee legitimate boundary set up is the normal main driver. Insufficient preparing, absence of management and capability are the reasons are inability to discovering the dangers and hazard appraisal. The new patterns show that the vast majority of the mishaps happen in oil industry are generally in boring and creation space.

The accidents are largely in drilling from the Figs. 1, 2, and 3 and to minimize the accidents, training is considered to be as the tool which can help the employee understand the need of organization and the needs of work place which acts as a barrier. Further with various finding, a proposed barrier model is suggested.

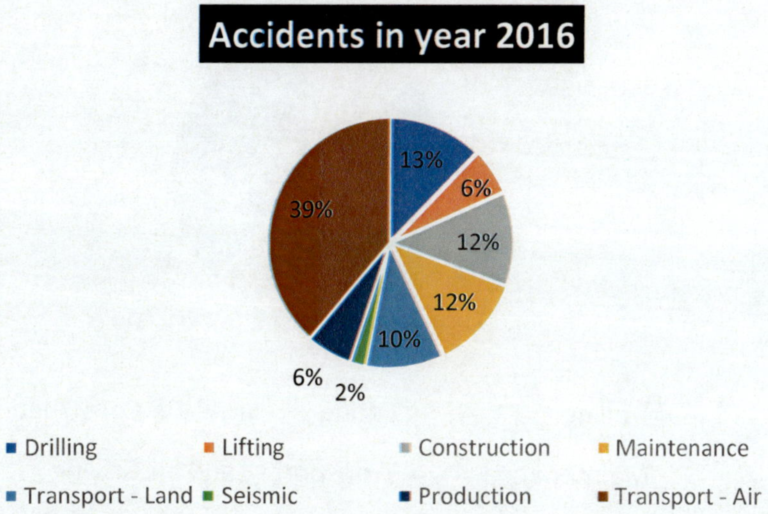

Fig. 1 Accidents in domain area in year 2017 [11]

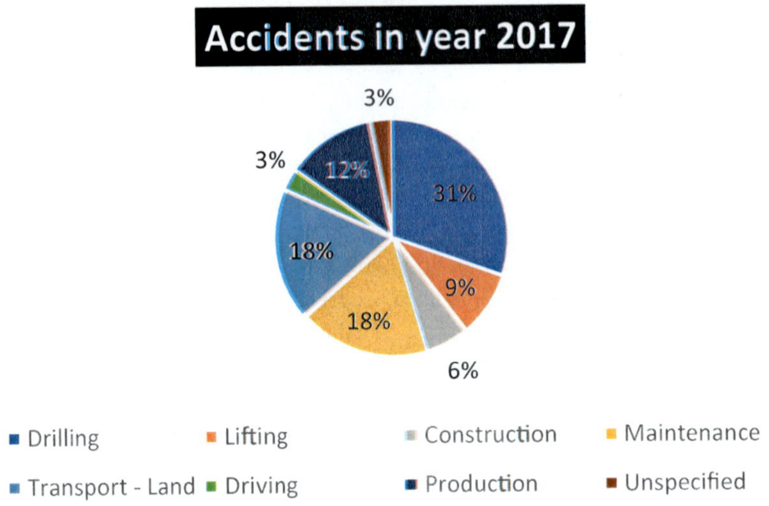

Fig. 2 Accidents in domain area in year 2017 [12]

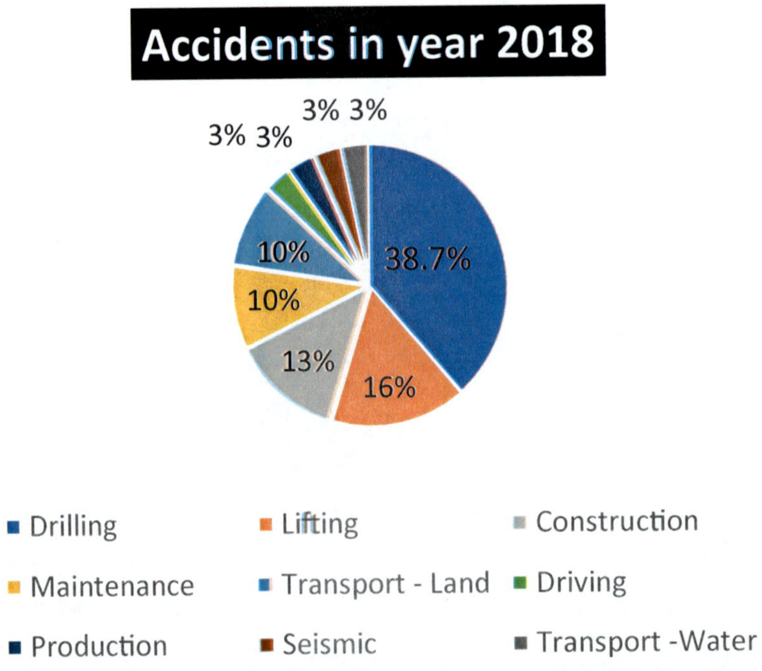

Fig. 3 Accidents in domain area in year 2018 [13]

3 Concept of Training

Al-Mughairi [14] states that preparation coordinates projects and plans which gives a worker another diverse expertise, which gives an improvement in proficient advancement openings and noticed execution at association. Likewise, when organization attempts to put resources into innovation which requests to give representatives preparing who are existing Van Wart [15]. The lone method of upgrading abilities, portrayal, and capacities of an individual is through preparing as it were. Hence, [16] determines that there is a significant increase from how to know in the individual staff when proper training is given in proper procedure. Having said that there must also upgradation in capacities of employee board which must inspire them from well arrange to well mannered. Laing [17] describes training as a step to enhance skills, learning, and understanding different viewpoints of an individual which also expands association with various groups. Training is an operative procedure in empowering utilization of individual potential to the maximum. Mzimela et al. [18] Poor and conflicting human execution keeps on being an issue that impacts the oil and gas industry's wellbeing execution. Comparative episodes happen across the business with natural main drivers named as human mistake. This "cause" lines up with the basic reason/impact relationship where the business has truly depended upon, yet which has not accomplished the ideal result of absence of repeat [19]. Subsequently, the competitive atmosphere in the market is obtained only via proper training and development [20]. Another wellbeing the executives practice to foster a well-working security culture is successful security preparing. Practically, all oil organizations in China are legally necessary to lead wellbeing preparing. Required information and capability associated with security for laborers are the critical substance of security preparing [21].

4 Major Accident Theories

Mishap causation model is orderly strategy for deciding the reasons for mishap. The incident of different exercises which is perplexing in a solitary existence is called as mishap. Assurance of makes driving it is troublesome since such countless factors are engaged with it.

According to Ismail et al. [25], working on various models for accident reduction is being considered important and much research work in done in developing new frameworks. There are various theories of accident causation which are listed are Heinrich, human factor, accident, epidemiological, systems, and behavior. In order to have good analysis of accident behavior, theory is considered.

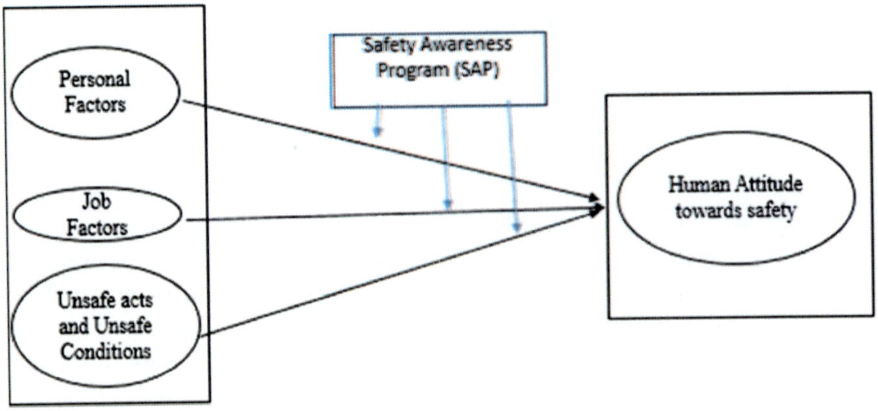

Fig. 4 Conceptual framework

5 Conceptual Framework

The effect of it is finished by directing variable wellbeing preparing. Coming about because of such target identified with BBS, the accompanying exploration theory are created and are proposed to be tried (Fig. 4).

The various assumptions that are made to for hypothesis of this research are below:

Hypothesis: *H(1): The relationship between personal factors and attitude of human towards wellbeing or safety is significant.*

H(2): The relationship between job factors and attitude of human towards wellbeing or safety is significant.

H(3): The relationship between unsafe act & conditions and attitude of human towards wellbeing or safety is significant.

H(4): The is moderating effect of training on human factors (i.e.: personal factors, job factors, unsafe acts and condition) influence human attitude of an individual.

6 Research Methods

6.1 Sample and Data Collection

As the data is collected from participants, the data screening has to be done for multivariate analysis which is crucial. It is required to get a meaningful outcome from the analysis as the corruptive data which is either insufficient or incomplete can be removed. One of reason not to get good outcome, it is because of the data quality. After collecting the data from the respondents, the researcher had to formulate the

Table 1 Reliability test of Cronbach alpha for each variable's question

Variables	Number of questions asked for each variable	Cronbach alpha
Personal factors	8	0.822
Job factors	7	0.701
Unsafe acts and unsafe condition	13	0.942
Human attitude toward safety at work place	6	0.72
Training/safety awareness program	5	0.727

Table 2 Cronbach's alpha results for all variable's question

Reliability Statistics	
Cronbach's alpha	*N* of items
0.824	78

data into either excel or SPSS to review the missing or under quality items. In this study, combination of various levels from 5 various companies in Oman.

6.2 Data Analysis Method:

In order to know the factors that affect human attitude that leads to safety at workplace a self-structured questionnaire is designed on likter scale of 5. The questionnaire is divided into five sections named as personal factor, job factor, unsafe acts and unsafe condition, human attitude toward safety, and training which comprises 39 question in total (Table 1).

To know the validity of questionnaire, the reliability test is done for that by taking responses over 369 people. The CRONBACH alpha test is done in SPSS software and results of the same are as below and SPSS file result is mentioned in Table 2. As a reliability checking for this study, Cronbach alpha is measured for all variables.

The value of Cronbach's alpha is usually scaled between 0 and 1, and in this research, it is 0.82 which it is considered to be good design of questionnaire [22]. According to Sekaran and Bougie [23], if a study gets reliability over 0.8, its good and 0.7 is acceptable but less than 0.6 is second as poor.

6.3 Descriptive Statistics of Various Factors

The below Table 3 indicates the average mean values of before and after an interval (pre-test and post-test) of the administered questionnaire. A significant change in all factors has been observed by the effect of training/safety awareness program.

Table 3 Descriptive statistics of various variables

Factor	Mean–pre training	Mean–post training	Difference	P Value
Personal factor	3.253	3.631	0.378	0.000
Job factor	2.322	3.156	0.834	0.000
Unsafe act and unsafe condition	3.680	4.041	0.361	0.000
Human attitude toward safety	3.659	3.910	0.26	0.000
Training/safety awareness program	3.650	3.977	0.327	0.000

Where in total, 248 male and 121 female respondents are there from various domain like production 115 respondents, drilling 108 respondents, administration 40 respondents, maintenance 48 respondents, and operation 58 respondents having different experience at various levels in the organization.

The Table 3 gives an information about various factor pre and post-training, and it is clear from the above study that training has a significant impact on human attitude since the p value is very much significant and hence all the hypothesis is valid.

Further, the study address only the training part but it is observed during the study that few casual components show high likely hood accidents and those are:

a. Lack of training needs
b. Lack of decision-making
c. No proper standard/procedures
d. Lack of team management or supervision
e. Unclear risk assessment and hazard identification [24]

In order to have detail understanding, a new model is needed because the various accidents causation mentioned in this paper (5 why, Swiss cheese, fish bone, and Bow-tie model,) are not doubt used to identify the accident causation but each have their own limitation and also it needs expertise. So, in order to have an easy approach, attitude barrier model is proposed which work with behavior-based safety theory (Attitude Barrier Model).

7 Improved/Attitude Barrier Model

The improved method suggested is to focus more on human behavior/human attitude analysis in addition to system failures. Since major accidents that occur are largely linked to human failure and attitudes which are the root causes, so it is proposed to introduce attitude factor while analyzing the human failure.

The Table 4 is the risk matrix in identifying the hazard easily. The elements constitute the age of the person, behavioral factor, number of years of experience in the same field, work environmental condition (ergonomics), and fatigue factor.

Table 4 Risk range

Risk elements	Risk range		
	Low risk (1–80)	Medium risk (81–255)	High risk (256–625)
Age of the employee			
Experience of the employee in the task			
Work environmental condition			
Fatigue			

The scale for the above parameters assigned is 1–5. The factors vary from young age group of personnel to old age (>50 years) group of personnel. The attitudes of group of personnel in the same work environment to be determined and derive the factor. This will vary for different work environment. A person of age group between 20–30 years performing a drilling activity possess high risk than that of a person belongs to age group >30. It is up to any organization to select the scale based on the task/hazard that they have.

8 Conclusion

In identifying the root causes of an accident, there are many tools and each have their own limitations as most of them fail to give underlying causes of it which needs expert opinion while doing the analysis. As most of the tools does not consider the failure of human factor in accident causation and proposed model, Attitude Barrier Model (ABM) is to be considered as it gives analysis of accident in contribution to human factor and the matrix provides human factors detailing which can help place injuries prevention as largely the causes of accident in hydrocarbon industry which is due to failure of barriers which can be classified in details for better study.

9 Recommendations

Human factors play a significant part in accidents because human attitude (of an individual) is linked to it and improvement can be done through proper practices like on job training and off job training (considering the pandemic situation online collaboration mechanism is used). In this research, the study was limited to impact and effect of training of human factors and human attitude, but it can be extended to employee performance, time management, and organizational performance too which can even be done through qualitative study as well. In evaluation of causation of accidents, it is recommended that the industry may adopt (ABM) Attitude Barrier Model in additional to other accident causation methods/theories.

References

1. Asad, M. M. (2019, March). Oil and gas disasters and industrial hazards associated with drilling operation : An extensive literature review. In *2019 2nd International Conference on Computing, Mathematics and Engineering Technologies (ICoMET)* (pp. 1–6). https://doi.org/10.1109/ICOMET.2019.8673516
2. Mergs, M. A. (1999). Behavioural Safety And Major Accident Hazards : Magic Bullet Or Shot In The Dark ?
3. Necci, A., Tarantola, S., Vamanu, B., Krausmann, E., & Ponte, L. (2019). Lessons learned from offshore oil and gas incidents in the Arctic and other ice-prone seas Lessons learned from offshore oil and gas incidents in the Arctic and other ice-prone seas. *Ocean Engineering, 185*(August), 12–26. https://doi.org/10.1016/j.oceaneng.2019.05.021
4. Tabibzadeh, M., & Meshkati, N. (2014). Learning from the BP Deepwater Horizon accident: Risk analysis of human and organizational factors in negative pressure test. *Environment Systems and Decisions, 34*(2), 194–207. https://doi.org/10.1007/s10669-014-9497-2
5. NORSOK. (2010). *Risk and emergency preparedness assessment* (3rd ed.). Standards Norway.
6. Zubair, M., Zhijian, Z., Heo, G., Ahmed, I., & Aamir, M. (2013). A computer based living probabilistic safety assessment (LPSA) method for nuclear power plants. *Nuclear Engineering and Design, 265*, 765–771.
7. Hopkins, A. (2012). *Disastrous decisions: the human and organisational causes of the Gulf of Mexico blowout*. CCH Australia.
8. Woolfson, C. (2013). Preventable disasters in the offshore oil industry: From piper alpha to deepwater horizon. *New Solutions, 22*(4), 497–524. https://doi.org/10.2190/NS.22.4.h
9. Huynh, T. T., & Bui, V. T. (2014). Application of quantitative risk assessment on offshore oil & gas industry. *Science and Technology Development Journal, 17*(3), 62–68. https://doi.org/10.32508/stdj.v17i3.1476
10. Note, G. (2020). *Core concepts*, pp. 1–17.
11. IOGP. (2016, June). *Safety Performance Indicators—2018 Data*, pp. 4–143.
12. IOGP. (2017, June). *Safety Performance Indicators—2018 Data*, pp. 4–143.
13. IOGP. (2018, June). *Safety Performance Indicators—2018 Data*, pp. 4–143.
14. Al-Mughairi, A. M. (2018). The evaluation of training and development of employees: The case of a national oil and gas industry. Brunel University London.
15. Van Wart, M. (2017). *Leadership in public organizations: An introduction*. Taylor & Francis.
16. Akugri, A. (2017). *Examining the impact of training and development on employee's performance _a case study some selected national health insurance scheme offices within the Kumasi Metropolis*. University of Education.
17. Laing, I. F. (2009). The impact of training and development on worker performance and productivity in public sector organizations: A case study of Ghana Ports and Harbours Authority.
18. Mzimela, T., & Chikandiwa, C. (2017). Employee training and development practices in the tourism and leisure sector in KwaZulu-Natal, South Africa, *6*(4), 1–17.
19. Elnaga, A., & Imran, A. (2013). The effect of training on employee performance, *5*(4), 137–147
20. Ford, K., Kraiger, K., Merritt, S., Kozlowski, S. W. J., & Salas, E. (2010). The multidimensionality of learning outcomes revisited, 135–165.
21. Huselid, M. A., & Becker, B. E. (2011). Bridging micro and macro domains: Workforce differentiation and strategic human resource management. Sage Publications Sage.
22. Kothari CS. Research Methodology Methods and Techniques
23. Sekaran, U., & Bougie, R. (2013). *Research methods for business: A skill building approach* (6th ed., p. 436). Wiley.

24. Iqbal, M. I., Isaac, O., Al Rajawy, I., Khuthbuddin, S., & Ameen, A. (2021). Hazard identification and risk assessment with controls (Hirac) in oil industry—A proposed approach. *Materials Today: Proceedings*. https://doi.org/10.1016/j.matpr.2020.11.800
25. Ismail, M., Alrajawy, I., Isaac, O., & Ameen, A. (2021). Study the impact of safety awareness program (SAP) as moderating variable for reduction of accidents in oil and gas industry—A proposed framework. *International Journal of Management and Human Science (IJMHS)*, *5*(1). https://ejournal.lucp.net/index.php/ijmhs/article/view/1331

A Succinct Study on the Effect of Lockdown on Air Quality of Agra (Taj City) (India)

Bhawna Yadav Lamba, Madhuben Sharma, and Sapna Jain

1 Introduction

Globally, air pollution is a menace for environment, human health and climate changes. It is responsible for increased risk of mortality and morbidity. The pollution is not only restricted to air we breathe, but it is also in the water we drink and the soil in which we grow plants. All these kinds of pollutions are primarily due to human activities. As we are progressing for establishments of numerous industries, we are deteriorating our environment and hence compromising our health and life. It will not be an exaggeration to say that the cause of the premature death on the planet, causing 15% of all deaths (8.3 millions) [8] is urbanization and industrialization, that are attaining unparalleled and distressing proportions worldwide.

COVID-19, caused by severe acute respiratory syndrome Coronavirus 2 (SARSCoV-2), was declared pandemic by WHO [10]. The journey of the pandemic (from one case to 6.2 million cases in the world) and its devastating effects are not hidden from us, and we are fighting against a deadly virus with our best efforts and strategies [2].

In order to control the spread of the virus, we should avoid close contact with each other, and it is achieved by social distancing. Keeping social distancing in mind, the authorities of the affected countries have implemented lock down. In India, also, nationwide lockdown has been implemented, amid COVID-19. During the lockdown, almost all the mass transportation and industrial activities have been put on hold. The repercussions include an effective control on virus spread (considering the high population density of India), an unsaid downfall in economy and a drastic decrease

B. Y. Lamba · M. Sharma · S. Jain (✉)
School of Engineering, University of Petroleum and Energy Studies, Energy Acres, Bidholi, Dehradun, Uttarakhand 248007, India
e-mail: sapnaj22@gmail.com

B. Y. Lamba
e-mail: byadav@ddn.upes.ac.in

N. A. Siddiqui et al. (eds.), *Advances in Construction Safety*,
https://doi.org/10.1007/978-981-19-4001-9_27

in air pollution. Considering the current situation, we have analyzed the air quality in the Taj city, Agra, well known for the UNISESCO world heritage sites, viz. Taj Mahal, Agra fort and Fatehpur Sikri.

2 Materials and Methods

2.1 Description of the Study Area and Context

Agra, ((latitude 27° 12′ 12.26″, longitude 78° 00′ 21.03″ and elevation 122.26 m), a city in the Indian state of Uttar Pradesh (UP), is famous for one of the most famous tourist places of the country [15]. The city is situated on the west bank of river, the Yamuna. It has a tropical climate with a very high temperature difference between summer and winter.

It is fourth most populated city in UP and 24th in India. A high population density, large number of industries and a huge vehicular traffic have made the city highly polluted.

Uttar Pradesh Pollution Control Board (UPPCB) has established the ambient air quality monitoring systems (AAQMS) on the Nagar Nigam building, beside Sanjay palace. The monitoring station (known as Sanjay palace) is in between some office buildings and is about 0.2 km away from the main road [3].

2.2 Methodology

In the current study, we have analyzed the air quality index (AQI) of Agra (March, April and May months of five years 2016, 2017, 2018, 2019 and 2020) and the common air pollutants, viz. NO_2, SO_2, PM_{25} and ozone.

Air quality index (AQI)
Air pollution is an important environmental problem which have adversely affected the health of urban population. Numerous epidemiological studies have established that higher concentration of air pollutants have adversely affected the human health [11]. Researcher [9] has linked air pollution with poor health of the people in the developed and developing countries. Conventionally, the evaluation of sampled air quality for different parameters is done with the help of air quality regulation. This process of comparison of individual air parameters with its standards is very tedious work. Therefore, to convey the information about the overall air quality to the policy makers and general public, air quality index (AQI) came into existence. A typically AQI is constructed on air pollutants criteria. An AQI converted weight of an individual air pollutant into a sole index using a suitable aggregation method. AQI is

Table 1 AQI of Agra city for five years for four time periods

Phase	First day	Last day	Total days	2016	2017	2018	2019	2020
Before lockdown	1st March	24th March	24	143	187	189	158	106
First phase of lockdown	25th March	14th April	21	163	156	154	151	83
Second phase of lockdown	15th April	3rd May	19	216	136	138	141	91
Third phase of lockdown	4th May	17th May	14	132	159	160	167	76

calculated as per the standard formulation [4]. AQI formulation is two-step process (i) Development of sub-indices function and (ii) Development of AQI using sub-indices.

(i) Development of sub-indices function for air pollutants. Sub-index function (I_i) is expressed as per Eq. (1) for n number of pollutants.

$$Ii = f(X_i), \quad i = 1, 2, 3, 4, \ldots\ldots n \tag{1}$$

Sub-index function (*fi*) is the relationship among pollutant concentration variable X_i and corresponding sub-index I_i.

(ii) Development of AQI using sub-indices (I)

$$I = F(I_1, I_2 \ldots\ldots I_n) \tag{2}$$

Here, F is the function of sub-indices $(I_1, I_2, I_3, I_4 \ldots\ldots I_n)$.

As a result, we conducted a comparison analysis of air quality before to lockdown period (1–25th March of every year) and air quality during lockdown period (first, second and third phases) (Table 1).

To conduct a comparison research on the concentration of four criteria pollutants, we picked the month of March as the month before lockdown and the months of April and May as the months following lockdown. The findings of the investigation will provide a clear picture of the impact of the shutdown on the air quality in Agra. The Central Pollution Control Board in New Delhi has collected continuous atmospheric quality data for the chosen city, which has been made available to the public [5].

Types of air pollutants and their sources [1]

Chiefly, the air pollutants can be classified into two categories, primary and secondary (Fig. 1).

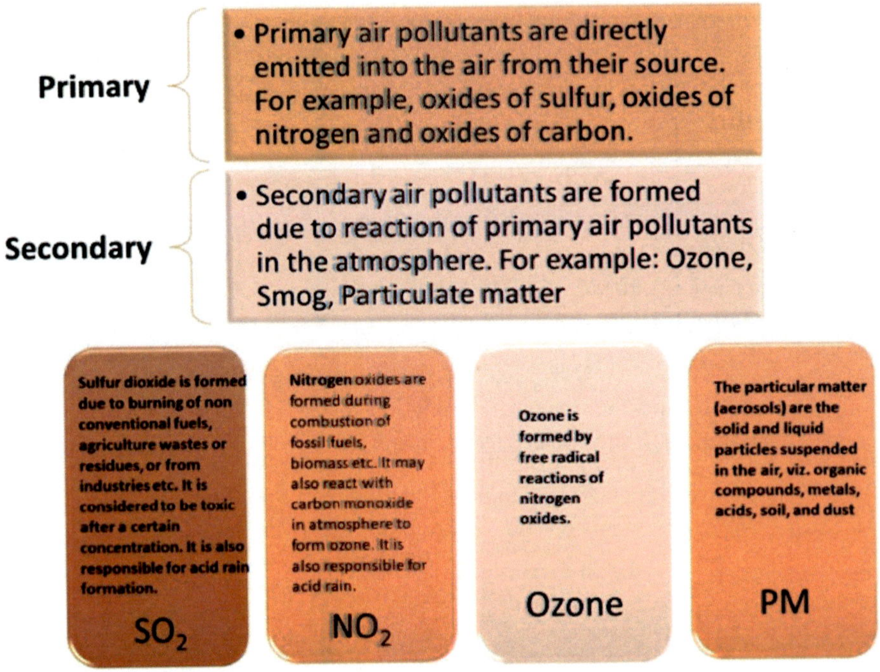

Fig. 1 Classification of air pollutants

3 Results and Discussions

The air quality index (AQI) trend in Agra, analyzed for five consecutive years (2016, 2017, 2018, 2019 and 2020), is presented in the Table 1.

The temporal analysis of the AQI revealed that air quality of year 2020 is better than the last four years (Table 1). The AQI of the city decreased during first lockdown period of year 2020 as compared to before lockdown period. A further dip in AQI was observed after third lockdown period in year 2020. However, if look upon the trend in the AQI of previous years, except 2016, there is a decrease in the AQI in the first lockdown period [25th March to 14th April]. On analysis of percentage of decrease (with respect to before lockdown period), the maximum dip was observed in year 2020 (21%). Similarly, the maximum decline of AQI was observed after third lockdown period in year 2020 (Fig. 2).

We have also analyzed the concentration of four air pollutants (SO_2, NO_2, Ozone and $PM_{2.5}$) for three months, March (before lockdown), April and May (the two months of lockdown) of five consecutive years (2016–2020). The average of these pollutants during the three months was calculated and used for analysis. The comparison has been made for five years. The data shows that the change in concentration of $PM_{2.5}$ (ca 3%) and sulfur dioxide (ca 1–7%) is not significant in COVID-19 year (2020) (Figs. 3 and 4). The main source of sulfur dioxide (SO_2) is industrial activity, e.g. thermal power generation, mineral ores and industrial activities that burn fossil

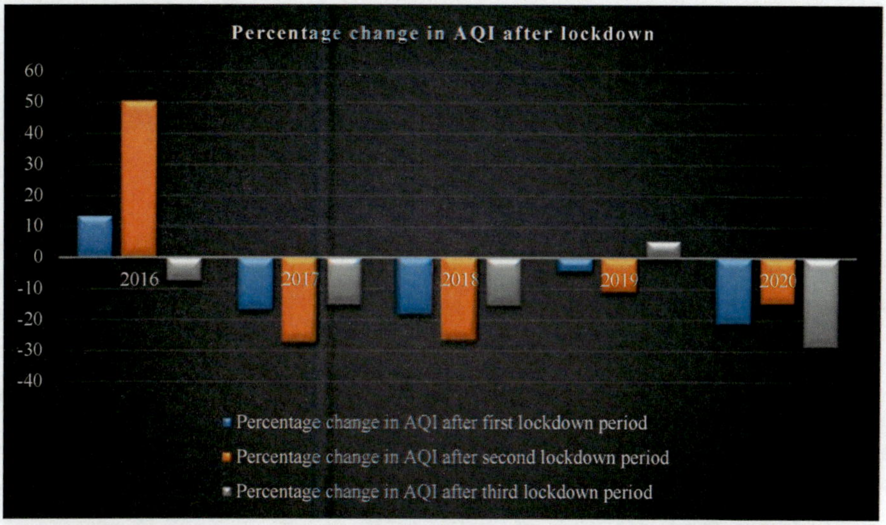

Fig. 2 Percentage change in AQI of Agra city after lockdown as compared to before lockdown period (1st–25th March)

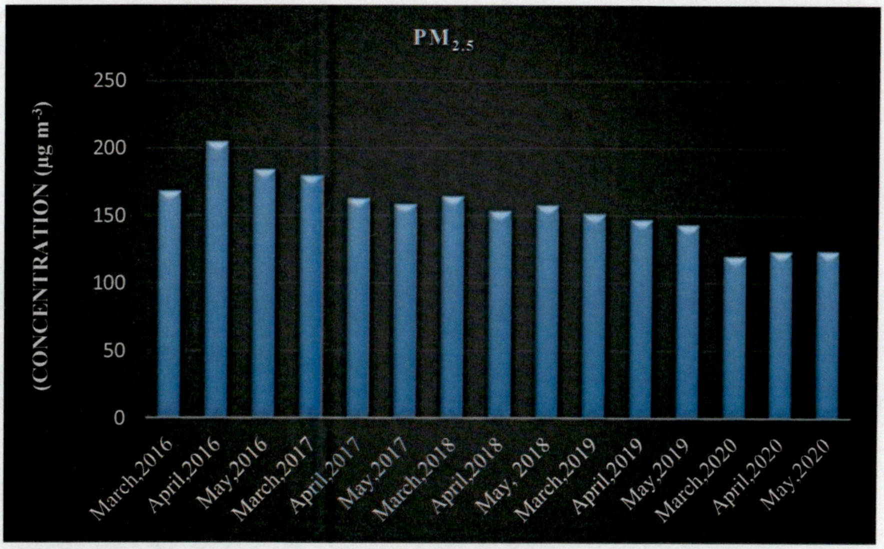

Fig. 3 Concentration of PM 2.5 (March–May, 2016–2020)

fuels, motor vehicle emissions, etc. If we exclude motor vehicle emission, the other sources of sulfur dioxide are not prevalent in the nearby area of monitoring station of Agra.

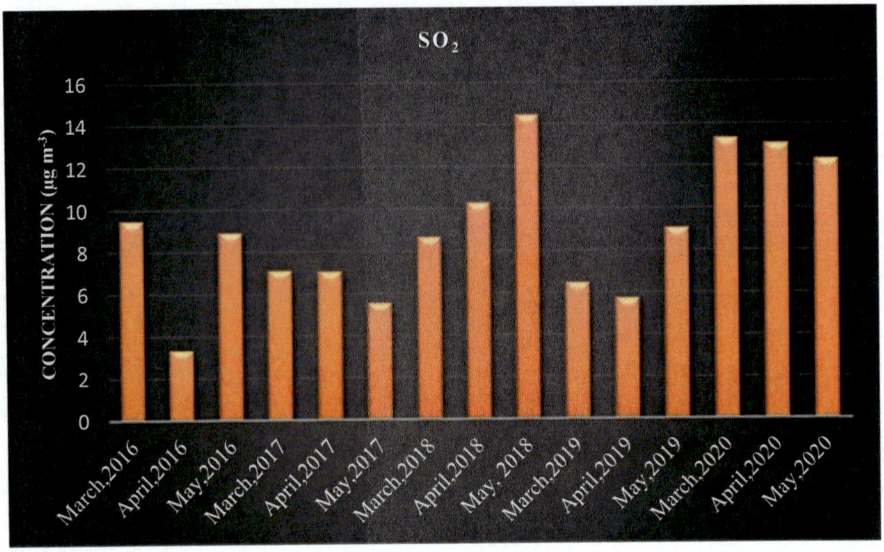

Fig. 4 Concentration of SO_2 (March–May, 2016–2020)

There is a significant increase in concentration of ozone after lockdown in the month of April and May as compared to before lockdown for year 2020 (Fig. 5) which is in well argument with the previous studies also [13, 14].

The increase in concentration of ozone can be attributed to the photocatalytic reaction involved in the generation of ozone in nature (Fig. 6). The ambient concentration

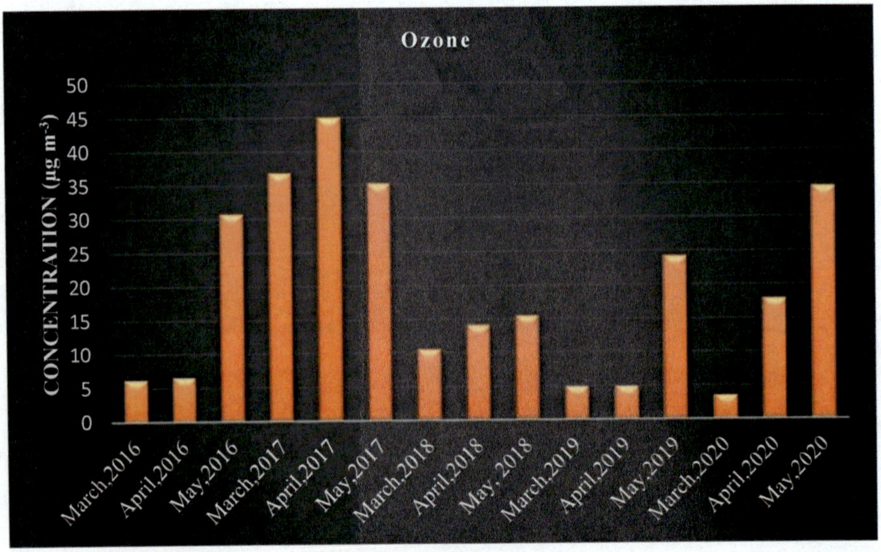

Fig. 5 Concentration of ozone (March–May, 2016–2020)

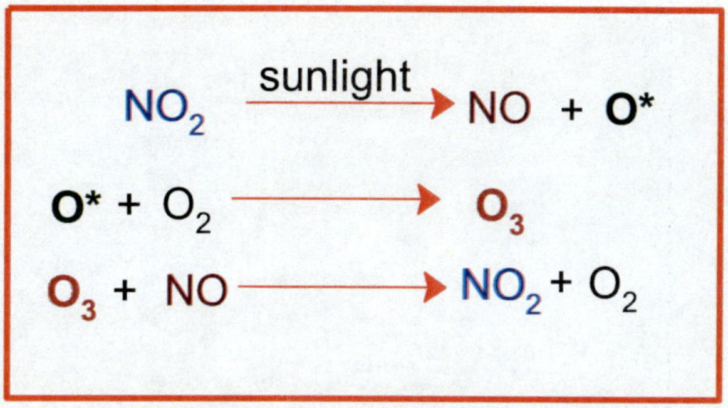

Fig. 6 Photochemical reactions of formation and decomposition of ozone

of ozone is governed by its precursor's, viz. nitrogen dioxide and volatile organic compounds, temperature, solar radiations, wind speed and other meteorological factors [12–14].

A decent decrease in concentration (ca 10–40%) of nitrogen dioxide is noticed during lockdown as compared to before lockdown month, i.e. March 2020 (Fig. 7). Previous studies have also reported that the concentration of nitrogen dioxide observed a decline during lockdown period [6] (Fig. 8).

As per a report, there is a major reduction in the average concentration of nitrogen dioxide over India during the lock down period as compared to the same time frame of the previous year (2019) [7].

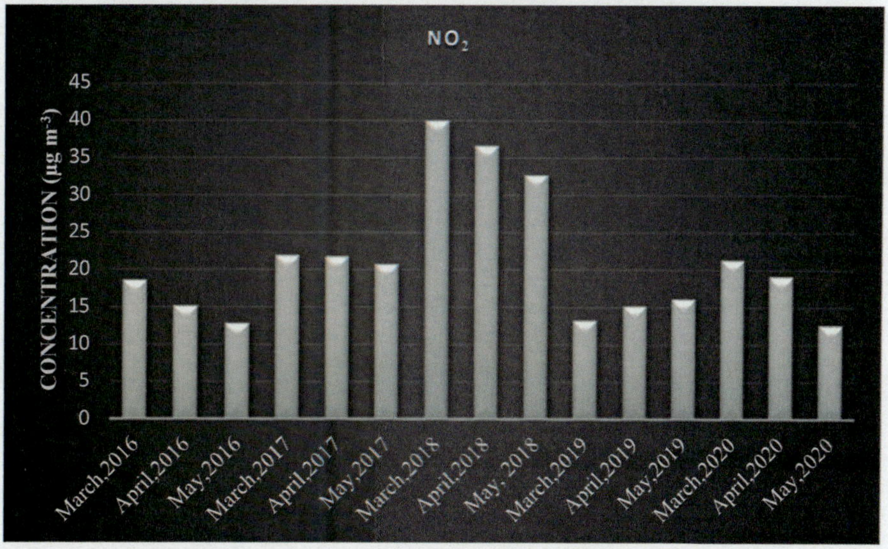

Fig. 7 Concentration of NO_2 (March–May, 2016–2020)

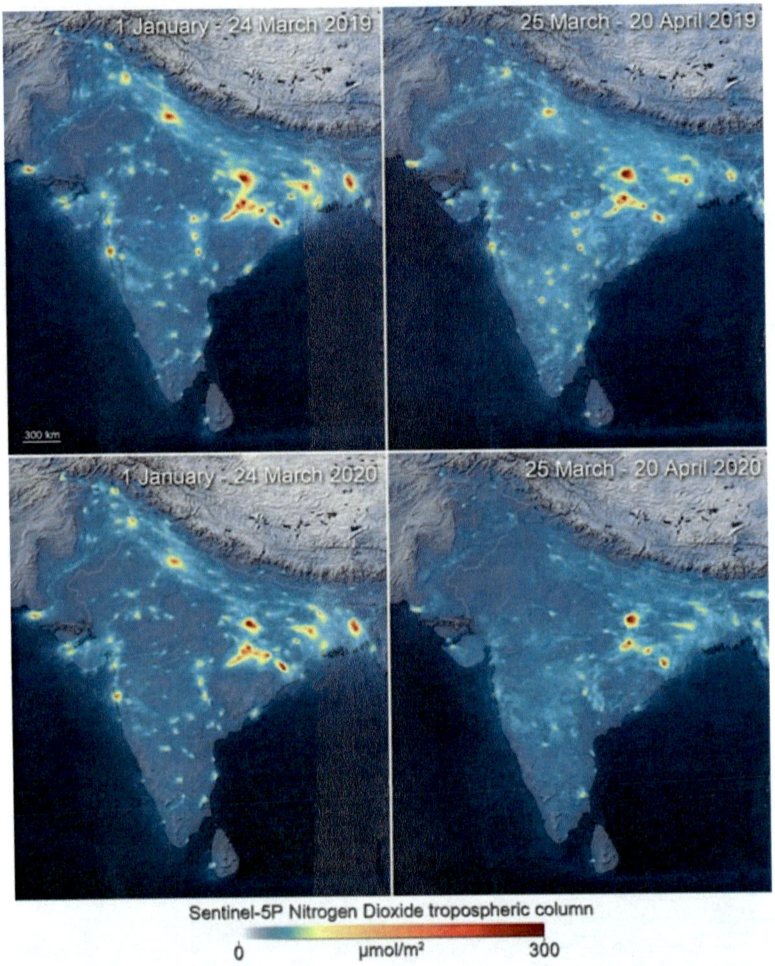

Fig. 8 Satellite maps of India [Copernicus Sentinel-5P satellite], showing averaged nitrogen dioxide concentrations January 1–March 24, 2020 (before lockdown) and March 25 (the first day of the lockdown) to April 20, 2020 with a comparison to the same time frame of year 2019 [7].

The reason of a fair decrease in concentration of nitrogen dioxide (NO_2) can be attributed to the restriction in the transport during lockdown as the nitrogen dioxide is mainly emitted from the combustion of fossil fuels.

4 Conclusion

The COVID-19 pandemic has affected all the aspects of life, adversely. However, one positive aspect is improvement in air quality during lockdown. The air quality of

Agra showed a trend of decrease in concentration of nitrogen dioxide and a fall of AQI index by 28%, during lockdown period as compared to before lockdown period due to restriction in transport, shut down of industries and construction work, refused fossil fuel burning, etc. Though, the short period of decline in concentration of air pollutants and upgrading of air quality is not enough to measure the changes on the monuments, but definitely it is an illustration by nature to understand the benefits associated with the restricted human activities. With a control on our daily activities, we can make a difference to the environment and our health. We may adapt some simple practices to control the pollution as reforming the usage of transportations (Sharing of vehicles, Use of eco-friendly vehicles for short commutations, etc.). The government interventions is highly desirable for implementing tele working/online working (as per requirement) to reduce the usage of transport and power consumptions. The current study and similar such studies are indicating that we may work together to find solutions to challenges associated with society, environment and economy to make the world better for living.

Acknowledgements We would like to thank CPCB and WHO for the air quality and COVID-19 data. We would also like to thank our organization, University of Petroleum and Energy Studies for the motivation to work on COVID-19.

References

1. Aranha, K. D. (1994). *Environmental chemistry* (pp. 213–219). New Age International Ltd.
2. Chakraborty, I., & Maity, P. (2020). COVID-19 outbreak: Migration, effects on society, global environment and prevention. *Science of the Total Environment, 728*, 138882.
3. Chitranshi, S., Sharma, S. P., & Dey, S. (2015). Satellite-based estimates of outdoor particulate pollution (PM10) for Agra City in northern India. *Air Quality, Atmosphere & Health, 8*(1), 55–65.
4. CPCB. (2014). National Air Quality Index. http://cpcb.nic.in/AQI-FINAL-BOOK.pdf
5. CPCB. (2020). https://app.cpcbccr.com/ccr/#/caaqm-dashboard-all/caaqm-landing
6. Dutheil, F., Baker, J. S., & Navel, V. (2020). COVID-19 as a factor influencing air pollution? *Environmental pollution, 263*, 114466.
7. ESA. (2020, April 24). European Space Agency.
8. GAHP. Global Alliance on Health and Population. https://www.theguardian.com/world/2019/dec/19/india-suffers-most-pollution-linked-deaths-in-world-study-finds
9. Gorai, A. K., Upadhyay, A., & Goyal, P. (2014). Design of fuzzy synthetic evaluation model for air quality assessment. *Environment Systems and Decisions, 34*(3), 456–469.
10. Gorbalenya, A. E., Baker, S. C., Baric, R. S., De Groot, R. J., Drosten, C., Gulyaeva, A. A., Lauber, C., Leontovich, A. M., Neuman, B. W.,Penzar, D., & Ziebuhr, J. (2020). The species Severe acute respiratory syndrome-related coronavirus: classifying 2019-nCoV and naming it SARS-CoV-2. *Nature Microbiology, 5*, 536–44. *Clinical Nutrition, 39*, 2324.
11. Kanchan, K., Gorai, A. K., Goyal, P., Benitez-Garcia, S. E., Kanda, I., Okazaki, Y., Wakamatsu, S., & Basaldud R. et al. (2015). A Review on air quality indexing system. *Asian Journal of Atmospheric Environment*, *9*(2).
12. Sharma, S., Chatani, S., Mahtta, R., Goel, A., & Kumar, A. (2016). Sensitivity analysis of ground level ozone in India using WRF-CMAQ models. *Atmospheric Environment, 131*, 29–40.

13. Shrestha, A. M., Shrestha, U. B., Sharma, R., Bhattarai, S., Tran, H. N. T., & Rupakheti, M. (2020). Lockdown caused by COVID-19 pandemic reduces air pollution in cities worldwide.
14. Sicard, P., De Marco, A., Agathokleous, E., Feng, Z., Xu, X., Paoletti, E., Rodriguez, J. J. D., & Calatayud, V. (2020). Amplified ozone pollution in cities during the COVID-19 lockdown. *Science of the Total Environment, 735*, 139542.
15. Taneja, A., Saini, R., & Masih, A. (2008). Indoor air quality of houses located in the urban environment of Agra, India. *Annals of the New York Academy of Sciences, 1140*(1), 228–245.

COVID-19 Impact on Indian Smart Cities: A Step Toward Build Back Better

Neha Mumtaz and Tabish Izhar

1 Introduction

United Nations (2018) stated that 55% of the global population residing in the urban locality is projected to rise resulting into 68% by 2050 [1]. According to World Health Organization (2021), a healthy city is one that is recurrently generating and refining those physico-social surroundings and escalating those public assets which facilitate individuals to equally upkeep each other in executing all the purposes of lifespan and emerging to their utmost prospectives [2]. Three hundred thirty two documents were reviewed from 2019 to 2022 using Scopus library and Mendeley desktop version. The bibliometric content analysis was done by using VOSviewer version 1.6.16., and the co-occurrence keyword threshold was taken as 50 shown in Fig. 1.

Coronavirus has elevated human torment [3], sabotaged the economy [4], turned the lives of billions of individuals around the world upside down and altogether influenced the health, monetary, ecological, and social areas [5]. This examination intends to give a thorough examination of the effect of the COVID-19 [6] episode on the ecological sector, the energy domain [7], society and the economy, and explore the worldwide preventive estimates taken to lessen the transmission of COVID-19 [8]. The data regarding the situation and trends of Corona Virus in India until December 2020 are mentioned in Fig. 2:

According to Smart Cities Mission based on Census 2011, almost 31% of Indian population constitutes to the urban population and contributes to 63% of gross

Scopus IDs: 57190794596 & 57201880766

N. Mumtaz (✉) · T. Izhar
Department of Civil Engineering, Integral University, Lucknow, Uttar Pradesh, India
e-mail: nehamumtaz@iul.ac.in

T. Izhar
e-mail: tizhar@iul.ac.in

N. A. Siddiqui et al. (eds.), *Advances in Construction Safety*,
https://doi.org/10.1007/978-981-19-4001-9_28

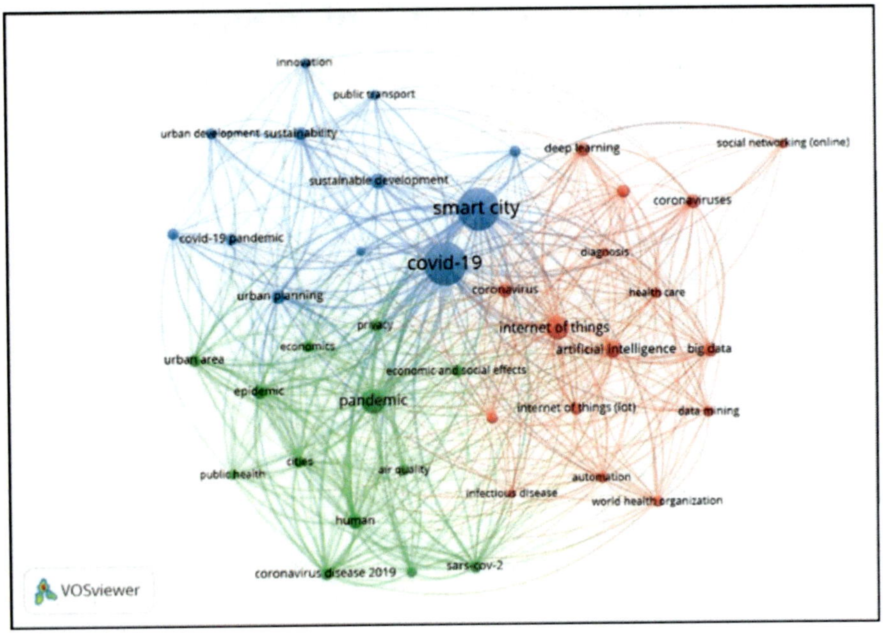

Fig. 1 Bibliometric content analysis co-occurrence map

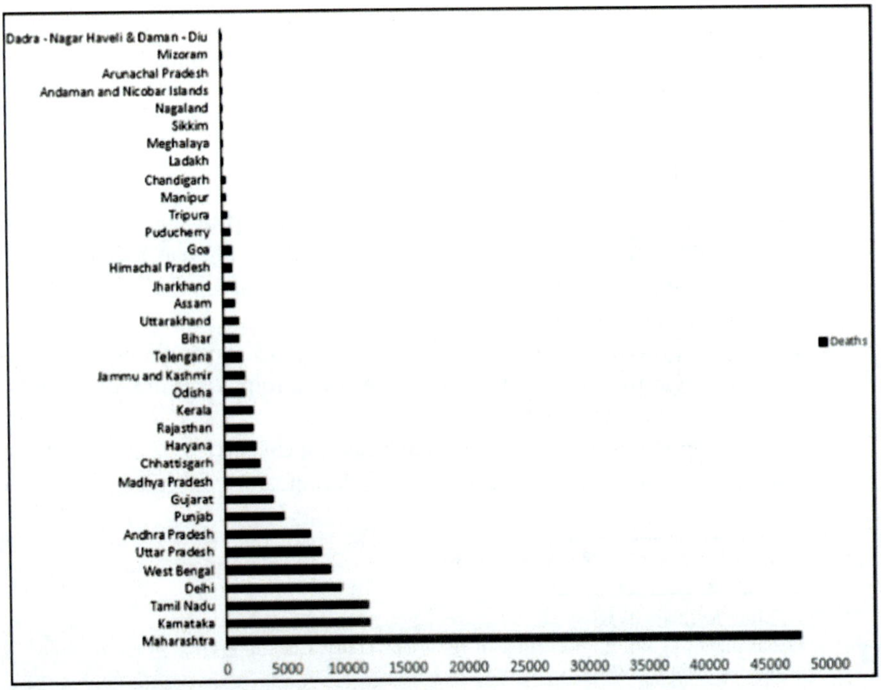

Fig. 2 Status of COVID-19—Indian scenario till December, 2020 *Source* Department of Health Research (MoHFW) Government of India, 2020 [9]

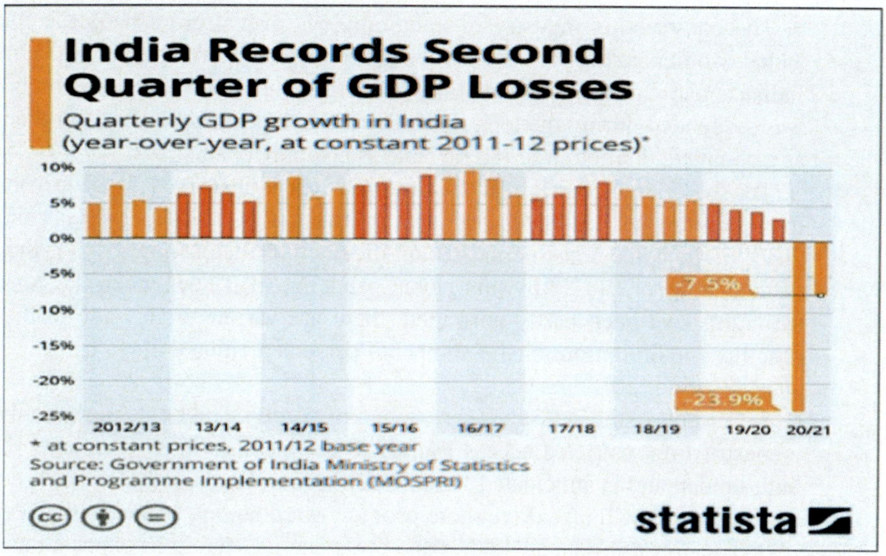

Fig. 3 GDP trend of India; Source: Ministry of Statistics and Program Implementation 2020 [12]

domestic product (GDP) [10]. Urbanization has resulted to house 40% of Indian population [11] and imparts 75% of the GDP by 2030 as mentioned in Fig. 3.

2 Construction Industry in COVID-19

Indian construction industry is the second largest employer after agriculture, and it is therefore critical to the country's economic stability. With an industry size of INR 10.5 trillion, it accounts for around 8% of the nation's GDP and employs close to 57.5 million people [12].

In addition, being a core sector, there are numerous industries that are dependent on the construction activity in the country [13]. For example, the construction equipment manufacturing industry comprises around 500 companies and is estimated to be sized at INR 375 billion by 2020 [14].

Engineering & Construction sector, which was already stumbling with numerous challenges starting from scarcity of resource, is being marred by this pandemic. COVID-19 pandemic has posed supplementary exceptional challenges for cities, including stress on their health care, education and security structure, and unreasonably affected susceptible groups [15].

As a precautionary measure during lockdown, people were instructed to stay safe at their home. In present circumstances, constructing a house can be cost inhibitive [16]. The demand for quality home construction in urban areas had witnessed a rise, prior to the pandemic outburst [17]. The accelerated urban development and 'smart cities' projects tempted construction firms and homeowners alike [18].

The construction industry in India employs over 4 crore people including 80 lakh + workers in the home construction sector. However, the pandemic brought the Indian construction industry to almost a complete halt. The home construction sector was suspended during the lockdown; there was large-scale labor migration as well as a complete disruption of the raw material supply chains [19].

The subsequent quarter of 2020 witnessed exceptional 50.3% crash in demand, and the overall impact is likely to contract the industry by 14.9% in 2020 as compared to 2019. In the home construction sector, the main challenge nowadays is the scarcity of manual labor caused by this global pandemic. Till now, when the construction activities have been partly permitted, there are various issues due to the lack of mobility and limitations on the number of persons permitted to work at any specified site—location [20].

Nevertheless, the industry is set to rebound sturdily. This epidemic has only formed a construction-restricted access leading to the increase in requirements [21]. This built-up demand is anticipated to force market escalation in 2021. The reduction in base unit costs will also drive more people toward buying homes in the years ahead [22]. The prevailing manual labor and raw material scarcity is expected to alleviate in 2021 with a complete unlock on material transportation and the inevitable of international supply chains [23].

3 COVID-19 and Indian Smart Cities

Smart City concept is an exclusive part [24] of the fourth Industrial Revolution (IR 4.0) [25]. On June 25, 2015, the Indian government announced the Smart Cities Mission. The goal was to promote equitable and resilient communities that offer essential amenities and a good standard of living for its inhabitants, as well as a safe and renewable environment and the use of 'Smart' alternatives.

According to Ministry of Urban Housing & Urban Affairs, the urban transformation missions such as Smart Cities Mission, PMAY (U) and AMRUT and have proved to be a boon in the fight against the COVID-19. The integrated command and control centers (ICCCs) developed under the Smart Cities Mission have helped cities in their fight against COVID-19 [26].

During the current COVID-19 crisis, 47 command and control centers created under the smart cities mission are serving as crisis management control rooms, helping monitor the on-ground situation while leveraging various technological innovations [27].

Additionally, various innovative strategies, outreach activities are being documented for wider dissemination and replication among stakeholders. For this purpose, 'COVID Urban Practices (CUP)-19', [28] a dedicated dashboard, has been developed and made operational to facilitate exchange of best practices; approximately, 110 initiatives were taken in 20 States/Union Territories by Government of India as mentioned in Fig. 4 as follows:

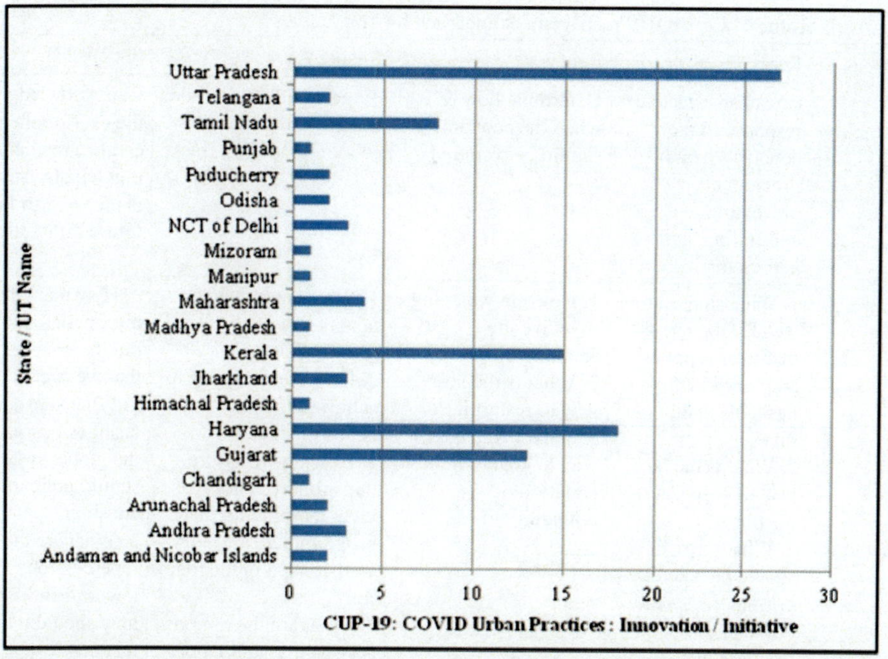

Fig. 4 Status of various COVID-19 urban practices in India [29]

The reports by Ministry of Housing and Urban Affairs (MoHUA, 2020) on COVID-19 initiatives taken by smart cities in India Agra, Bengaluru, Bhopal, Chennai, Dehradun, Jabalpur, Pimpri Chinchwad, Pune, Surat, Varanasi were studied. COVID-19 management can be organized into four stages are mentioned in Table 1:

Implementing these strategies can prove beneficial in light of Millennium Development Goals as well as strengthening to tackle post COVID-19 scenario considering the developing countries across the world.

4 Conclusion

According to Kunzmann [30], Covid-19 incidence may slow down globalization but will definitely speed up the digitalization. It acknowledged broadly that the COVID-19 epidemic requires significant reorganization of India's economic and administrative environment. There is a substantial perseverance of re-calibration of public expenses goals and priority sectors. It is indeed obvious that, in the new situation, considerably more attention will be focused on factors such as expanded medical resources and institutes, statistics-oriented public health monitoring, and smarter and safer facilities with better adaptability to future disasters.

Table 1 COVID-19 management approach for smart cities

React	Renew	Resilience	Reconsider
Create an effective response to the country's health problem by preventing, controlling, and addressing it	Determine how & when the economy should be restarted	Targeted efforts will be used to bring the economy back on track	Gain knowledge from the catastrophe to create a resilient municipal system that is prepared to face future difficulties
1. What steps can we take to ensure that vital supplies and services are available at all times? 2. What is the best way to keep track of the threat? 3. What steps should be taken to prevent the spread of the disease and its consequences?	1. How are we going to revive the economy? 2. What is the best way to restart it? 3. What is the best way to govern in the situation of ambiguity?	1. Which steps can be taken to swiftly re-establish trust? 2. How are we going to achieve moderate development? 3. How do we assure that stimulus packages yield a profit and that programmers are well-funded in the long run? 4. What are the options for dealing with labor deflection?	1. How many flaws in our operational model did the crisis expose that we need to fix? 2. What substantial changes in society and the global system should policymakers consider? 3. What are the repercussions for government's role and how should it act in response? 4. What chances does the crisis provide for speeding up reforms?

It will be particularly imperative for our cities striving to meet the Sustainable Development Goals by implementing evidence-based policy measures for legit responses to such risks. The constraints of this pandemic have highlighted the benefits of remote digital delivery of services and no-contact interactions in the socioeconomic realms, while also providing a strong argument for smart and resilient infrastructure embedded with Internet of Things (IoT) devices used in communication. These might certainly aid ongoing crisis relief operations while also enhance the effectiveness of systemic, substantial proof-based decision-making. Cities like Hong Kong, Seoul, Singapore, and Taipei, for example, could respond quickly and effectively to a pandemic, not only for their previous experience with similar virus outbreaks like severe acute respiratory syndrome (SARS) but also due to the rapid advent of smart innovation, which helped define their preparedness.

Many of the measures taken by city administrators around the world during the pandemic (unmanned aerial systems, embedded devices, machine learning monitoring, automated transportation, actual data visualizations) have been reported to continuously develop to find potential prerequisites on the go, acknowledgements to the intrinsic flexibility accessible by technology.

The pandemic has also made us realize that our routine transport and the aspects of traditional accessibility ought to be reconsidered, with smarter alternatives to minimize the risks of mass transit, decrease workplace hazards all through pandemic, and enhance distribution networks via the use of smart technologies in logistical support and transportation of goods.

After a thorough evaluation of operational efficiencies versus the risks that such smart systems may bring, this will need to be promoted by sufficient governmental support at all levels.

It may also be necessary to institutionalize excellent practices and support non-motorized modes of transportation for small distances, such as bicycling, through well-designed legislation.

Reconfiguring cities around the concepts of sustainability and adaptability is crucial, whenever these ideas are included into urban design and planning, calamities may be significantly reduced.

The pandemic's impact on urban healthcare system has been enormous, with an enormous rise in the number of cases, pressure on the healthcare staff, and severe delays in the delivery of routine non-COVID-19 healthcare services.

The epidemic has taught us that when it comes to city design, planning, and expansion, healthcare system must be prioritized, and digital technology must be embraced for sustainable and accessible healthcare provision and consumption. The pandemic also induced online working, as well as the evident underlying inequities in current labor market systems have required a rethinking of the smart city idea. Beyond the crisis, there is a chance to reconsider a 'new normal'. The technologically sound cities are required to address these coming crises and disruptions, as well as to safeguard life and livelihoods. As a result, during this epidemic, the smart cities idea will undoubtedly be a game changer in the transition from 'survival to revival' of global existence.

References

1. United Nations, Report. (2018). https://www.un.org/development/desa/en/news/population/2018-revision-of-world-urbanization-prospects.html. Accessed 23 September 2021.
2. WHO. (2021). Sign up for Save Lives: Clean Your Hands campaign. May 05, 2021. https://www.who.int/campaigns/world-hand-hygiene-day/sign-up. Accessed 19 August 2021.
3. Karaman, O., Alhudhaif, A., & Polat, K. (2021). Development of smart camera systems based on artificial intelligence network for social distance detection to fight against COVID-19. *Applied Soft Computing, 110*. https://doi.org/10.1016/j.asoc.2021.107610
4. Chauhan, A., Jakhar, S. K., & Chauhan, C. (2021). The interplay of circular economy with industry 4.0 enabled smart city drivers of healthcare waste disposal. *Journal Cleaner Production 279*. https://doi.org/10.1016/j.jclepro.2020.123854
5. Gorse, C., & Scott, L. (2021). Editorial: International sustainable ecological engineering design for society (SEEDS) conference 2018. *Frontiers in Built Environment, 7*. https://doi.org/10.3389/fbuil.2021.670559
6. Al Marzouqi, N., Bueti, C., & Menon, M. (2021). Digital transformation to build smart cities. *Studies in Computational Intelligence* (971 vol., pp. 145–169). Springer Science and Business Media Deutschland GmbH, Spectrum International Affairs—Telecom Regulatory Authority, https://doi.org/10.1007/978-3-030-72065-0_9
7. Luna Nemecio, J., & Tobón, S. (2021). Sustainable and resilient urbanization to covid-19: New hori-zons for city research. *Universidad Y Sociedad, 13*(1),

110–118. https://www.scopus.com/inward/record.uri?eid=2-s2.0-85102022137&partnerID=40&md5=d7ef62bf63c55919908352961c52fe01
8. Nikiforova, A. (2021). Smarter open government data for society 5.0: Are your open data smart enough?," *Sensors, 21*(15). https://doi.org/10.3390/s21155204
9. Department of Health Research. (2020). Department of Health Research | MoHFW | Government of India," 2020. https://dhr.gov.in/. Accessed 23 September 2021.
10. National portal of India. (2021). Smart Cities Mission: A step towards Smart India | National Portal of India. https://www.india.gov.in/spotlight/smart-cities-mission-step-towards-smart-india. Accessed 23 September 2021.
11. Gupte, J., & Mitlin, D. (2021). COVID-19: What is not being addressed. *Environment and Urbanization, 33*(1), 211–228. https://doi.org/10.1177/0956247820963961
12. Government of India. (2020). Data | Ministry of Statistics and Program Implementation | Government Of India, 2020. http://mospi.nic.in/data. Accessed 23 September 2021.
13. Goel, A., Ganesh, L. S., & Kaur, A. (2019). Deductive content analysis of research on sustainable construction in India: Current progress and future directions. *Journal of Cleaner Production, 226*, 142–158. https://doi.org/10.1016/j.jclepro.2019.03.314
14. Ismael, D., & Shealy, T. (2019). Industry perceptions of sustainable design and construction practices in Kuwait. *Journal of Green Building, 14*(4), 169–193. https://doi.org/10.3992/1943-4618.14.4.169
15. Sabolović-Krajina, D. (2021). Social impact of public libraries during the COVID-19 pandemic in the context of the smart cities concept-a comparison of singapore and Croatia. *Vjesnik Bibliotekara Hrvatske 64*(1), 250–278. https://doi.org/10.30754/vbh.64.1.853
16. Kang, M., et al. (2020). COVID-19 impact on city and region: What's next after lockdown? *International Journal of Urban Sciences, 24*(3), 297–315. https://doi.org/10.1080/12265934.2020.1803107
17. Machado, Jr. C., Melina Nassif Mantovani Ribeiro, D., & Backx Noronha Viana, A. (2021). Public health in times of crisis: An overlooked variable in city management theories? *Sustainable Cities Society, 66*. https://doi.org/10.1016/j.scs.2020.102671
18. Komarov, V. M., & Akimova, V. V. (2021). Strategies for sustainable urban mobility: Analysis of best practices. *Ekon. Polit., 2021*(1), 82–103. https://doi.org/10.18288/1994-5124-2021-1-82-103
19. Zhang, X., Hou, H., Fu, Q., & Zhang, Y. (2020). Current problems and restructuring suggestions for smart city construction: A case study on fight against COVID-19 in Several Chinese Cites. In *11th International Conference on E-Bbusiness, Management and Economics, ICEME 2020* (pp. 65–69). https://doi.org/10.1145/3414752.3414803
20. Tokazhanov, G., Tleuken, A., Guney, M., Turkyilmaz, A., & Karaca F. (2020, Oct). How is COVID-19 Experience Transforming Sustainability Requirements of Residential Buildings? A Review. *Sustainability, 12*(20), 8732. https://doi.org/10.3390/SU12208732
21. Capolongo, S., et al. (2020). COVID-19 and cities: From urban health strategies to the pandemic challenge. a decalogue of public health opportunities. *Acta Bio-Medica, 91*(2), 13–22. https://doi.org/10.23750/abm.v91i2.9515
22. Sharifi, A., Khavarian-Garmsir, A. R., & Kummitha R. K. R. (2021). Contributions of smart city solutions and technologies to resilience against the covid-19 pandemic: A literature review. *Sustainability, 13*(14). https://doi.org/10.3390/su13148018
23. Belhadi, A., Mani, V., Kamble, S. S., Khan, S. A. R., & Verma, S. (2021). Artificial intelligence-driven innovation for enhancing supply chain resilience and performance under the effect of supply chain dynamism: An empirical investigation. *Annals of Operations Research, 2021*, 1–26. https://doi.org/10.1007/S10479-021-03956-X
24. Yun, Y., & Lee, M. (2019, November) Smart City 4.0 from the Perspective of Open Innovation. *Journal Open Innovation Technology Market Complex, 5*(4), 92. https://doi.org/10.3390/JOITMC5040092
25. Correia, D. M., Teixeira, L., & Marques, J. L. (2021). Smart supply chain management: The 5W1H open and collaborative framework. In *8th IEEE International Conference on Industrial Engineering and Applications, ICIEA 2021* (pp. 401–405). https://doi.org/10.1109/ICIEA52957.2021.9436817

26. Government of India. (2021). Ministry of Housing and Urban Affairs, Goverment of India. http://mohua.gov.in/. Accessed 23 September 2021
27. Government of India. (2021). Smartcities. https://smartcities.gov.in/. Accessed 23 September 2021
28. Government of India, Covid Urban Practices -19 Innovation. https://pmay-urban.gov.in/covid-19. Accessed 23 Sep 2021
29. Government of India. (2021). State/UT Initiatives of Health. https://pmay-urban.gov.in/covid-19/state/1. Accessed 23 September 2021
30. Kunzmann, K. R. (2020). Smart cities after Covid-19: Ten narratives. *DISP*, *56*(2), 20–31. https://doi.org/10.1080/02513625.2020.1794120